STRATEGIES
FOR
BUSINESS
AND
TECHNICAL
WRITING

STRATEGIES FOR BUSINESS AND TECHNICAL WRITING

Fifth Edition

KEVIN J. HARTY

La Salle University

PEARSON

Longman

New York San Francisco Boston
London Toronto Sydney Tokyo Singapore Madrid
Mexico City Munich Paris Cape Town Hong Kong Montreal

Senior Vice President and Publisher: Joseph Opiela
Vice President and Publisher: Eben W. Ludlow
Senior Marketing Manager: Melanie Craig
Production Manager: Douglas Bell
Project Coordination, Text Design, and Electronic Page Makeup: WestWords, Inc.
Cover Design Manager: Wendy Ann Fredericks
Cover Designer: Maria Ilardi
Cover Photo: Copyright © DigitalVision/PictureQuest
Manufacturing Buyer: Roy L. Pickering, Jr.
Printer and Binder: Courier Corporation, Stoughton
Cover Printer: Coral Graphic Services, Inc.

For permission to use copyrighted material, grateful acknowledgment is made to the copyright holders on pp. 379–381, which are hereby made part of this copyright page.

Library of Congress Cataloging-in-Publication Data

Harty, Kevin J.
    Strategies for business and technical writing / Kevin J. Harty.—5th ed.
        p. cm.
    Includes bibliographical references and index.
    ISBN 0-321-24195-9 (pbk.)
    1. Commercial correspondence. 2. Business report writing. 3. Technical writing. I. Title.

HF5721.H37 2004
808'.066651—dc22
                            2004006593

Please visit our website at http://www.ablongman.com

ISBN: 0–321–24195–9

3 4 5 6 7 8 9 10—CRS—07 06 05

For Donald
and
in memory of
Jamie

Contents

PREFACE

This fifth edition of *Strategies for Business and Technical Writing* represents a thorough reworking of the previous edition. The fifth edition retains 24 of the 31 selections in the fourth edition and adds eight new selections. In revising *Strategies* for this new edition, I have been guided by the same principle that informed the selections of materials for inclusion in the four previous editions: *Strategies* should present the best advice from the best sources about the most important issues in business, technical, and other kinds of professional writing.

The changes I have made in preparing this fifth edition broaden the coverage in *Strategies* of key issues in business and technical writing and increase the usefulness of *Strategies* as both a textbook and a reference. In particular, this fifth edition extends discussions and expands coverage from previous editions on the following key topics:

- the writing process,
- assorted problems with language use and misuse, especially the dangers inherent in the use of doublespeak,
- traditional forms of correspondence such as letters and memos,
- the possibilities (and problems) offered by electronic communication,
- reports and proposal,
- graphics,
- cover letters and resumes,
- such alternatives to traditional resumes as functional and electronic resumes, and
- in an entirely new section, ethical issues in business and technical writing.

This fifth edition of *Strategies* can, like its predecessors, be used in three ways: (1) as a supplement to a business or technical writing textbook, (2) as a supplement to a composition textbook or handbook, and (3) on its own as a textbook or as a reference source.

I have designed *Strategies* to appeal to practical-minded instructors, students, and women and men already working in business and technical occupations and the professions. The contributors to this volume write with the kind of purpose and understanding that come only from a career dedicated to improving the effectiveness of business and technical communication. As a result, the selections in *Strategies* not only teach professional writing but are also fine examples of the genre.

In selecting essays for this fifth edition of *Strategies,* I have been guided by the belief that all courses in business and technical writing should firmly implant one idea in the minds of students (and of those in the world of work). *All* successful writing consists of clear and effective prose no matter in which medium it is composed or sent. However, students should recognize that business and technical writing differs in important ways from the expository prose usually taught in first-year or advanced composition courses. Because business and technical writers communicate with multiple audiences and with a variety of intentions, learning to write well for the world of work requires some broad adjustments in technique. Therefore, in addition to touching on such broad problems as style, jargon, and diction, *Strategies* includes comprehensive discussions of specific forms of writing—for example, different kinds of resumes, letters of application, letters that sell, memos, e-mail, reports, and proposals. Students learn the specific techniques used by successful men and women in business, industry, and the technical professions who communicate information as part of their jobs.

The essays in this fifth edition of *Strategies* have, like those in the four previous editions, survived the most rigorous scrutiny—that of my students in seminars and courses in business and technical writing at La Salle University, Temple University, Rhode Island College, Centenary College of Louisiana, the Federal Reserve Bank of Philadelphia, Kroll Lindquist Avey, Philadelphia Telco Credit Union, the Board of Pensions of the United Presbyterian Church (USA), First Pennsylvania Bank and Fidelity Bank (now both Wachovia Bank), Blue Cross of Greater Philadelphia, and Philadelphia Newspapers, Inc.

For permission to reprint the selections included in the fifth edition of *Strategies,* I am happy to thank the authors, agents, and editors who handled my permissions requests with efficiency and good cheer. I am grateful to Professor Stephen A. Bernhardt at the University of Delaware and Professor Susan Loudermilk at Texas A&M—Corpus Christi for their thoughtful critiques of the fourth edition of *Strategies.* Their helpful comments inform many of the changes that I have incorporated into this fifth edition. At La Salle University, I owe a continuing debt to my many good friends and colleagues, especially to Joe Brogan, Jim Butler, Gabe Fagan, Bob Fallon, John Keenan, John Kleis, Francine Lottier, Emery Mollenhauer, and Lynne Texter. I am especially grateful to the staff of the University library, in particular to Stephen Breedlove and Nancy Tresnan in Interlibrary Loan Services.

Finally, I am indebted to the editorial and production staff at Longman, especially to Eben Ludlow, with whom it has been a genuine pleasure to work for the past 25 years.

Kevin J. Harty

Introduction

Writing consumes a substantial portion of the working day for almost all college-educated workers.[1]

Ability to express ideas cogently and goals persuasively—in plain English—is the most important skill to leadership. I know of no greater obstacle to the progress of good ideas and good people than the inability to compose a plain English sentence.[2]

Whether you are planning a career in business, in industry, or in some technical field, or whether you already have such a career, you may be ill-prepared for what is perhaps the most difficult part of your job. Although you may be an excellent accountant, salesperson, manager, engineer, or scientist, all of your training may not help you when you sit down to write. If this is the case, you are not alone in being frustrated by the writing demands of your job. Nevertheless, writing is essential to most technical and professional occupations; the ability to communicate effectively both in person and on paper can help you advance in your profession. Good writing can mean the difference between winning or losing a major sale, a pleased customer, or a challenging new position with a higher salary.

The selections in this book will help you make your letters, reports, memos, and other professional documents more effective. The selections represent some of the best advice veteran teachers and practitioners of business and technical writing have to offer. None of the selections, however, offers a quick cure for writing ills. There is no such cure. Like any worthwhile skill, good professional writing requires time, practice, and, most of all, discipline.

To begin with, we need to realize that there are only two kinds of professional writing:

1. clear, effective writing that meets the combined needs of the reader and the writer, and
2. bad writing.

Bad writing is unclear and ineffective; it wastes time and money. Even worse, it ignores its readers, usually raising more questions than answers.

---

[1]Paul V. Anderson, "What Survey Research Tells Us about Writing at Work," in *Writing in Nonacademic Settings*, ed. Lee Odell and Dixie Goswarmi (New York: Guilford, 1985), p. 30.

[2]The late John D. deButts, former Chairman, AT&T. Quoted as the epigraph to the third edition of the Bell Laboratories' *Editorial Style Guide* (Whippany, NJ: 1979).

To be effective writers, each of us needs to develop a variety of strategies for the different writing situations we face. We must also respect writing as something more than words typed or written on the page or fed into a dictaphone or word processor. Writing takes thought and planning.

THREE KEY QUESTIONS

Business and technical writing situations present us with a variety of challenges. One of the ways to meet these challenges is to begin by asking ourselves these three key questions:

1. Exactly who is my audience?
2. What is the most important thing I want to tell my audience?
3. What is the best way of making sure my audience understands what I have to say?

Exactly Who Is My Audience?

Whatever we write on the job, somebody—across the hall, across the city, across the country, or even across the world—will eventually read what we have written. Most likely, our readers will receive our message in their own offices or homes. Since we won't be on hand to explain anything that is unclear or that doesn't make sense, our message will have to speak for itself.

Whenever we sit down to write, we become experts. That is in large part why we are the writers and not the readers. Our readers depend on us for clear, effective writing, so that they may share in our expertise. Although this point seems obvious, business, industrial, and technical practice shows that the single most common cause of bad writing is ignoring the reader. There are any number of ways to ignore our audience: we can talk over their heads; we can talk down to them; we can beat around the bush; we can leave out important details—and, as a result, we can undermine our credibility.

When you next sit down to write, therefore, it is important to ask yourself what you know about your audience—you must determine your audience's situation and anticipate your audience's potential reaction to your message. To paraphrase the golden rule, make sure you are prepared to write unto others the way you would have them write unto you.

What Is the Most Important Thing I Want to Tell My Audience?

Whether you write one page or one hundred pages, you should be able to condense your most important idea into about 25 words. If you can't, you aren't ready to write, since you haven't figured out what you are trying to say. You may not be able to explain every reason for a decision or a proposal in 25 words, but 25 words should be enough for a statement of the decision or proposal itself. For instance, you may have several reasons for recommending that your company replace its Brand X word processors with Brand Y word processors. You can make the recommendation in 22 words flat:

> I recommend that we immediately replace all our Brand X word processors with new Brand Y units for the reasons presented below.

In short, you can cut through all the bull and come right out with your recommendation. Then, in the body of your memo or report, you can present the relevant facts. Busy people dislike having their time wasted, and they hate surprises. So, in professional writing, the bottom line should also be the top line.

What Is the Best Way of Making Sure My Audience Understands What I Have to Say?

Different writing situations require different strategies. After answering the first two questions, you may decide that you shouldn't write at all—that you should phone instead or make a personal visit. But once you do commit yourself to writing, you need to determine how to put the particular document together. You will need to use one strategy for good news and a different one for bad news, one for giving information and another for changing someone's mind, one for accepting a job offer and another for turning the offer down. But in each case you will need to get your main idea out as soon as possible in as few words as possible, and then later defend or elaborate on that idea as clearly and as effectively as possible.

To illustrate the importance of answering these three key questions properly, let's look at a brief excerpt from a booklet formerly used by the New York State Income Tax Bureau:

> The return for the period before the change of residence must include all items of income, gain, loss or deduction accrued to the taxpayer up to the time of his change of residence. This includes any amounts not otherwise includable in the return because of an election to report income on an installment basis.
>
> Stated another way, the return for the period prior to the change of residence must be made on the accrual basis whether or not that is the taxpayer's established method of reporting. However, in the case of a taxpayer changing from nonresident to resident status, these accruals need not be made with respect to items derived from or connected with New York sources.

How did you respond to this information? If you were a taxpayer in need of quick and concise information to solve a particular problem, you would surely be disappointed and probably even annoyed. Tax publications and forms need not be complicated; since ordinary people normally read the publications and fill out these forms, they should be written for the average taxpayer, not for a trained accountant. In fact, all professional writing should be reader-directed.

Furthermore, the result of such convoluted writing is that these two paragraphs do not fulfill their purpose; they do not tell taxpayers how to adjust their returns when changing residence. Although the grammar is correct, the message is certainly not effective. Among other faults, the writer forgot to ask, "Who is my audience?" Unfamiliar words or usages such as *includable, accrual basis,* and *election* appear without explanation. If taxpayers cannot understand such language, they will complete their returns incorrectly and cheat either themselves or the state out of money. In the end, these two paragraphs—and others like them in the tax booklet—will have missed their mark, and unnecessary complications for both taxpayers and the state will be the result.

GUIDELINES FOR EFFECTIVE WRITING

To make sure you answer the three key questions correctly, keep these guidelines in mind:

- *Put your main idea up front.* Then provide whatever background or evidence your reader needs to understand your main idea.
- *Avoid using technical terms with nontechnical readers.* If you are an accountant writing to another accountant, use all the technical terminology you need. But if you are an accountant writing to a layperson, use ordinary language that the layperson will understand. If you must use technical terms, define them carefully.
- *Limit the length and complexity of your sentences.* Whenever possible, try to include only one main idea in each sentence. More mechanically, try also to limit your sentences to 25 to 35 words (or about 2½ typed lines). Generally speaking, the shorter the sentences, the fewer the difficulties you and your reader will get into. To ensure that your sentences are easily understood, move the main subject and verb to the front of the sentence and cut out any unnecessary words.
- *Limit the length and content of your paragraph.* Nothing intimidates a reader more than a half-page-long block of dense prose. Rather than using one paragraph to discuss every aspect of a topic, use a separate paragraph for each aspect of your topic. In that way, you can limit your paragraphs to five or six sentences. If you really need a longer paragraph, use indentation or enumeration so that your subpoints stand out from the pack.
- *Give your reader all necessary information in a clear and unmistakable way.* If your reader needs to let you know something by June 12, tell the reader so plainly. If you want your reader to check back with you for additional information, make sure you provide your e-mail address or telephone number, and your travel or vacation schedule if you are going to be out of the office. If you are attaching or enclosing anything, specify exactly what. If your message is a follow-up to some previous correspondence or to a recurring problem, make sure you give the reader this information too.

The selections that follow suggest many ways in which you can make your professional writing more effective. They also suggest ways you can make the process more enjoyable. If you view your professional writing as a chore, its effectiveness will be limited. But the more you get into the habit of asking the three key questions before you write your memos, letters, reports, and other documents, the easier, the more effective—and the more enjoyable—writing will become for you.

Part 1

Process as Well as Product

The old adage tells us to think before we speak. We should also think before we write. Actually, we should think before we write, while we write, and after we write. Thinking is what the writing process is all about. A process approach to writing is simply a writing strategy that asks writers to make sure they carefully plan each written message, rather than simply letting words fall where they may in the hope they will somehow organize themselves effectively and make sense to their intended audience.

People in business and industry are, however, trained to think in terms of finished products, bottom lines, and deadlines, so at first they may react with skepticism—if not with antagonism—to a process that requires them to slow down and think, especially when it is 4:30 and they need to get a memo to the boss by 5:00. Sometimes, sheer momentum and adrenaline will actually help, and a writer will produce an acceptable, even an effective, document. More times than not, though, the document will fail to achieve its purpose. The writer who balked at spending an extra ten minutes thinking a document through ends up spending an additional hour writing a follow-up document to clarify what should have been clear in the first document or, even worse, to mend fences because the original document created new problems.

In the introduction, I suggested an approach to the writing process that involved having writers ask themselves three key questions:

1. Exactly who is the audience?
2. What is the most important thing they want to tell this audience?
3. What is the best way of making sure this audience understands what they have to say?

These three key questions are simply one of any number of approaches to the writing process. The essays that follow in this section of *Strategies* offer other approaches and additional insights into a process approach to writing.

In the first essay, John Keenan offers a five-part approach to the writing process that asks writers to consider purpose, audience, format, evidence, and organization. Michael Adelstein follows Keenan with another five-part approach to writing that lays down a schedule for any writing task: 15% worrying, 10% planning, 25% writing, 45% revising, and 5% proofreading. The number of steps involved is, of course, unimportant. What is important is that there be a conscious effort to understand that, in business and technical writing, the process is as important as the goal. Keenan's approach is more open-ended, allowing writers considerable freedom in the amount of time they spend on the different steps in the writing process. While allowing for some variation among writing situations, Adelstein suggests a more regimented approach that will appeal to other writers.

Peter Elbow takes a different tack than Keenan or Adelstein by simply dividing writing time in half. Using what Elbow calls "the direct writing process," writers spend the first half of their available time "fast writing" in an attempt to get their ideas on the page before worrying about organization, correctness, or precision. Elbow suggests that these latter issues are best left to the second step in his approach to the writing process, revision. Revision is the key to successful writing, and Linda Flower and John Ackerman suggest ways in which the process of revision can help writers produce more effective documents.

In an often reprinted and adapted essay, John S. Harris next examines the writing process from a different perspective. The fault in poor quality documents may lie as much with the managers who assign them as with those who produce them. To counter problems in the management of writing, Harris offers a project worksheet designed to facilitate efficient writing management.

Sheryl Lindsell-Roberts then concludes this section of *Strategies* with a handy checklist business and technical writers can use to double- (or triple-) check that their documents will meet the needs of their intended readers.

John Keenan

Using PAFEO Planning

A former technical writer for SmithKline, John Keenan is now Professor Emeritus of English at La Salle University in Philadelphia and coauthor with Kevin J. Harty of Writing for Business and Industry: Process and Product *(Macmillan, 1987).*

Surveys tell us that "lack of clarity" is the most frequently cited weakness when executives evaluate their employees' writing. But since lack of clarity results from various bad habits, it is often used as a convenient catchall criticism. Pressed to explain what they meant by lack of clarity, readers would probably come up with comments such as these:

"I had to read the damn thing three times before I got what he was driving at. At least I *think* I got his point; I'm not sure."

"I couldn't follow her line of thought."

"I'm just too busy to wade through a lot of self-serving explanation and justification. If he has something to say, why didn't he just say it without all that buildup?"

Most unclear writing results from unclear thinking. But it is also true that habits of clear writing help one to think clearly. When a writer gets thoughts in order and defines the purpose and goal, what he wants to say may no longer be exactly the same as it was originally. Sometimes the writer may even decide it is better *not* to write. Whatever the case, the work done *before* the first draft in the prewriting process and *after* the draft in the revising process is often the difference between successful communication and time-wasting confusion. . . .

THINK PAFEO

Some years ago I made up a nonsense word to help my students remember the important steps in constructing a piece of writing. It seemed to stick in their minds. In fact, I keep running into former students who can't remember my name, but remember PAFEO because it proved to be so useful. Here are the magic ingredients:

P stands for purpose
A stands for audience
F stands for format
E stands for evidence
O stands for organization.

Put together, they spell PAFEO, and it means the world to me!

PURPOSE

Put the question to yourself, "Why am I writing this?" ("Because I have to," is not a sufficient answer.) This writing you're going to do must aim at accomplishing something; you must be seeking a particular response from your reader. Suppose, for example, you're being a good citizen and writing to your congressman on a matter of interest to you. Your letter may encompass several purposes. You are writing to inform him of your thinking and perhaps to persuade him to vote for or against a certain bill. In explaining your position, you may find it useful to narrate a personal experience or to describe a situation vividly enough to engage his emotions. Your letter may therefore include the four main kinds of writing— exposition, argument, narration, and description—but one of them is likely to be primary. Your other purposes would be subordinated to your effort to persuade the congressman to vote your way. So your precise purpose might be stated: "I am writing to Congressman Green to persuade him to vote for increased social security benefits provided by House Bill 5883."

Write it out in a sentence just that way. Pin it down. And make it as precise as you can. You will save yourself a good deal of grief in the writing process by getting as clear a focus as you can on your purpose. You'll know better what to include and what to omit. The many choices that combine to form the writing process will be made easier because you have taken the time and the thought to determine the exact reason you have for writing this communication.

When you're writing something longer than a letter, like a report, you can make things easier for yourself and your reader by making a clear statement of the central idea you're trying to develop. This is your thesis. To be most helpful, it must be a complete sentence, not just the subject of a sentence. "Disappointing sales" may be the subject you are writing about, but it is not a thesis. Stating it that way won't help you organize the information in your report. But if you really say something specific about "disappointing sales," you will have a thesis that will provide a framework for development. For example:

> Disappointing sales may be attributed to insufficient advertising, poor selection of merchandise, and inadequate staffing during peak shopping hours.

The frame for the rest of the report now exists. By the time you have finished explaining how each cause contributes to the result (disappointing sales), the report is finished. Not only finished, but clear; it follows an orderly pattern. The arguments, details, illustrations, statistics, or quotations are relevant because the thesis has given you a way of keeping your purpose clearly in focus.

Of course the ideal method is to get your purpose and your thesis down before you write a first draft, but I can tell you from experience that it doesn't always work that way. Sometimes the only way to capture a thesis statement is to sneak up on it by means of a meandering first draft. When you've finished that chore, read the whole thing over and try to ambush the thesis by saying, "OK, now just what am I trying to say?" Then *say it* as quickly and concisely as you can—out loud, so you can hear how it sounds. Then try to write it down in the most precise and concise way you can. Even if you don't capture the exact thesis statement you're looking for, there's a good chance that you will have clarified your purpose sufficiently to make your next revision better.

AUDIENCE

Most of the writing the average person does is aimed at a specific reader: the congressman, the boss, a business subordinate, a customer, a supplier. Having a limited audience can be an advantage. It enables the writer to analyze the reader and shape the writing so that it effectively achieves the purpose with that particular reader or group of readers.

Here are some questions that will help you communicate with the reader:

1. How much background do I need to give this reader, considering his or her position, attitude toward the subject, and experience with this subject?
2. What does the reader need to know, and how can I best give him or her this information?
3. How is my credibility with this reader? Must I build it gradually as I proceed, or can I assume that he or she will accept certain judgments based on my interpretations?
4. Is the reader likely to agree or disagree with my position? What tone would be most appropriate in view of this agreement or disagreement?

The last question, I think, is very important. It implies that the writer will try to see the reader's point of view, will bend every effort to look at the subject the way the reader will probably look at it. That isn't easy to do. Doing it takes both imagination and some understanding of psychology. But it is worth the considerable effort it involves; it is a gateway to true communication.

Let me briefly suggest a point here that is easily overlooked in the search for better techniques of communication. It is simply this: *better writing depends on better reading*. Technique isn't everything. One cannot learn to be imaginative and understanding in ten easy lessons, but reading imaginative literature— fiction, poetry, drama—is one way of developing the ability to put yourself in someone else's place so you can see how the problem looks from that person's angle. The writer who limits reading to his or her own technical field is building walls around the imagination. The unique value of imaginative literature is the escape that it allows from the prison of the self into the experience and emotions of persons of different backgrounds. That ability to understand why a reader is likely to think and act in a certain way is constantly useful to the writer in the struggle to communicate.

FORMAT

Having thought about your purpose and your audience, consider carefully the appropriate format for this particular communication you are writing. Of course the range of choice may be limited by company procedures, but even the usual business formats allow some room for the writer to use ingenuity and intelligence.

The key that unlocks this ingenuity is making the format as well as the words work toward achieving your purpose. You can call attention to important points by the way you arrange your material on the page.

Writers sometimes neglect this opportunity, turning the draft over to the typist without taking the trouble to plan the finished page. Remember the reader, his or her desk piled high with things to be read, and try to make your ideas as clear as possible on the page as the reader glances at it.

External signals such as headings, underlining, and numerals are ways of giving the reader a quick preview. Use them when they are appropriate to your material.

Don't be afraid of using white space as a way of drawing attention to key ideas.

To see how experts use format for clarity and emphasis, take a look at some of the sales letters all of us receive daily. Note how these persuasive messages draw the reader's attention to each advantage of the product. You may not want to go that far, but the principle of using the format to save the reader's time is worth your careful consideration.

Overburdened executives are always looking for shortcuts through the sea of paper. Many admit that they cannot possibly read every report or memo. They scan, they skip, they look for summaries. A good format identifies the main points quickly and gives an idea of the organizational structure and content. If the format looks logical and interesting, readers may be lured into spending more time on your ideas. You've taken the trouble to separate the important from the unimportant, thus saving them time. The clear format holds promise of a clear analysis, and what boss can be too busy for that?

EVIDENCE

Unless you're an authority on your subject, your opinions carry only as much weight as the evidence you can marshal to support them. The more evidence you can collect before writing, the easier the writing task will be. Evidence consists of the facts and information you gather in three ways: (1) through careful observation, (2) through intelligent fieldwork (talking to the appropriate persons), and (3) through . . . research.

When you follow the evidence where it leads and form a hypothesis, you are using inductive reasoning, the scientific method. If your evidence is adequate, representative, and related to the issue, your conclusion must still be considered a probability, not a certainty, since you can never possess or weigh *all* the evidence. At some point you make the "inductive leap" and conclude that the weight of evidence points to a theory as a probability. The probability is likely to be strong if you are aware of the rules of evidence.

The Rules of Evidence

Rule 1. Look at the Evidence and Follow Where It Leads.

The trick here is not to let your own bias seduce you into selecting only the evidence you agree with. If you aren't careful, you unconsciously start forcing the evidence to fit the design that seems to be emerging. When fact A and fact B both point toward the same conclusion, there is always the temptation to *make* fact C fit. Biographer Marchette Chute's warning is worth heeding: ". . . you will never succeed in getting at the truth if you think you know, ahead of time, what the truth ought to be."

A reliable generalization ought to be based on a number of verifiable, relevant facts—the more the better. Logicians tell us that evidence supporting a generalization must be (1) known or available, (2) sufficient, (3) relevant, and (4) representative. Let's see how these apply to the generalization, "Cigarette smoking is hazardous to your health."

The evidence accumulated in animal studies and in comparative studies of smokers and nonsmokers has been published in medical journals: it is therefore known and available. Is it sufficient? The government thinks so; the tobacco industry does not. Which is more likely to have a bias that might make it difficult to follow where the evidence leads?

Tobacco industry spokesmen argue that animal studies are not necessarily relevant to human beings and point to individuals of advanced age who have been smoking since childhood as living refutations of the supposed evidence. But the number of cases studied is now in the thousands, and studies have been done with representative samplings. The individuals who have smoked without apparent damage to their health are real enough, but can they be said to be numerous enough to be representative of the usual effects of tobacco on humans? "Proof" beyond doubt lies only in mathematics, but the evidence in this example is the kind on which a sound generalization can be based.

Rule 2. Look for the Simplest Explanation that Accounts for All the Evidence.

When the lights go out, the sudden darkness might be taken as evidence of a power failure. But a quick investigation turns up other evidence that must be accounted for: the streetlights are still on; the refrigerator is still functioning. So a simpler explanation may exist, and a check of the circuit breakers or fuse box would be appropriate.

Rule 3. Look at All Likely Alternatives.

Likely alternatives in the example just discussed would include such things as burned-out bulbs, loose plugs, and defective outlets.

Rule 4. Beware of Absolute Statements.

In the complexity of the real world, it is seldom possible to marshal sufficient evidence to permit an absolute generalization. Be wary of writing general statements using words like *all, never* or *always*. Sometimes these words are implied rather than stated, as in this example:

> Jogging is good exercise for both men and women.
>
> [Because it is unqualified, the statement means *all* men and women. Since jogging is not good for people with certain health problems, this statement would be better if it said *most* or *many* men and women.]

Still, caution is always necessary. Induction has its limitations, and a hypothesis is best considered a probability subject to change on the basis of new evidence.

The other kind of reasoning we do is called deduction. Instead of starting with particulars and arriving at a generalization, deduction starts with a general premise or set of premises and works toward the conclusion necessarily implied by them. If the premise is true, it follows that the conclusion must be true. This logical relationship is called an inference. A fallacy is an erroneous or unjustified inference. When you are reasoning deductively, you can get into trouble if your premise is faulty or if the route from premise to conclusion contains a fallacy. Keep the following two basic principles in mind, and watch out for the common fallacies that can act as booby traps along the path from premise to conclusion.

Two Principles for Sound Deduction

1. The ideas must be true; that is, they must be based on facts that are known, sufficient, relevant, and representative.
2. The two ideas must have a strong logical connection.

If your inductive reasoning has been sound enough to take care of the first principle, let me warn you that the second is not so simple as it sounds. We have all been exposed to so much propaganda and advertising that distort this principle that we may have trouble recognizing the weak or illogical connection. Nothing is easier than to tumble into one of the commoner logic fallacies. As a matter of fact, no barroom argument would be complete without one of these bits of twisted logic.

Twisted Logic

Begging the question. Trying to prove a point by repeating it in different words.

> Women are the weaker sex because they are not as strong as men.
> Our company is more successful because we outsell the competition.

Non sequitur. The conclusion does not follow logically.

> I had my best sales year; the company's stock should be a big gainer this year.
> I was shortchanged at the supermarket yesterday. You can't trust these young cashiers at all.

Post hoc. Because an event happened first, it is presumed to be the cause of the second event.

> Elect the Democrats and you get war; elect the Republicans and you get a depression.

Oversimplification. Treating truth as an either-or proposition, without any degrees in between.

> Better dead than Red.
> America—love it or leave it.
> If we do not establish our sales leadership in New York, we might as well close up our East Coast outlets.

False analogy. Because of one or two similarities, two different things are assumed to be *entirely* similar.

> You can lead a horse to water but you can't make him drink; there is no sense in forcing kids to go to school.
> Giving the Canal back to Panama would be like giving the Great Lakes back to the Indians.

Seen in raw form, as in the examples, these fallacies seem easy to avoid. Yet all of us fall into them with dismaying regularity. A writer must be on guard against them whenever he or she attempts to draw conclusions from data.

ORGANIZATION

Experienced writers often use index cards when they collect and organize information. Cards can be easily arranged and rearranged. By arranging them in piles, you can create an organizational plan. Here's how you play the game:

1. You can write only one point on each card—one fact, one observation, one opinion, one statistic, one whatever.
2. Arrange the cards into piles, putting all closely related points together. All evidence related to marketing goes in one pile, all evidence related to product development goes in another pile, and so on.
3. Now you can move the piles around, putting them in sequence. What kind of sequence? Consider one of the following commonly used principles:

   Chronological. From past to present to future.
   Background, present status, prospects
   Spatial. By location.
   New England territory, Middle Atlantic, Southern
   Logical. Depends on the topic.
   Classification and division according to a consistent principle (divisions of a company classified according to function)
   Cause and effect (useful in troubleshooting manuals)
   Problem-analysis-solution. Description of problem, why it exists, what to do about it.
   Order of importance. From least important to most important or from most important to least important.

   The choice of sequence will depend largely on the logic of the subject matter and the needs of your audience.
4. Go through each pile and arrange the cards in an understandable sequence. Which points need to precede others in order to present a clear picture?
5. That's it! You now have an outline that is both orderly and flexible. You can add, subtract, or rearrange whenever necessary.

SUMMARY

Think PAFEO. Use this word to remind you to clarify your purpose and analyze your audience before you write. When you have a clear sense of purpose, create a thesis that will act as a frame for your ideas.

Choose a format for your communication that will help the reader identify the main points quickly.

Collect your evidence before you write by observing, interviewing, and doing library research. Use the index card system to help you organize your evidence. Keep in mind the rules of evidence before you draw conclusions. Be alert to the common logic fallacies that can easily undermine the deductive process.

Try the file card system as a way of organizing your material. Write one point on each card and then place closely related points together in a pile. Place the piles in sequence according to a principle of organization you select. Some useful organizing principles include: chronological, spatial, logical, problem-analysis-solution, and order of importance. The sequence you choose will depend on the logic of the subject matter and the needs of your audience.

Michael E. Adelstein

The Writing Process

Michael E. Adelstein was formerly Professor of English at the University of Kentucky and coauthor with W. Keats Sparrow of Business Communications *(Harcourt Brace Jovanovich, 1983).*

THE WRITING PROCESS

Their prevalent belief in the myths of genius, inspiration, and correctness prevents many people from writing effectively. If these individuals could realize that they can learn to write, that they cannot wait to be inspired, and that they must focus on other aspects besides correctness, then they can overcome many obstacles confronting them. But they should be aware that writing, like many other forms of work, is a process. To accomplish it well, people should plan on completing each one of its five stages. The time allocations may vary with the deadline, subject, and purpose, but the work schedule should generally follow this pattern:

1. Worrying—15%
2. Planning—10%
3. Writing—25%
4. Revising—45%
5. Proofreading—5%

Note that only 25% of the time should be spent in writing; the rest—75%—should be spent in getting ready for the task and perfecting the initial effort. Observe also that more time is spent in revising than in any other stage, including writing.

Stage One—Worrying (15%)

Worrying is a more appropriate term for the first operation than *thinking* because it suggests more precisely what you must do. When you receive a writing assignment, you must avoid blocking it out of your mind until you're ready to write;

instead, allow it to simmer while you try to cook up some ideas. As you stew about the subject while brushing your teeth, taking a shower, getting dressed, walking to class, or preparing for bed, jot down any pertinent thoughts. You need only note them on scrap paper or the back of an envelope. But if you fail to write them down, you'll forget them.

Another way to relieve your worrying is to read about the subject or discuss it with friends. The chances are that others have wrestled with similar problems and written their ideas. Why not benefit from their experience? But if you think that there is nothing in print about the subject or a related one—either because your assignment is restricted, localized, or topical—then you should at least discuss it with friends.

Let's say, for example, that as a member of the student government on your campus, you've been asked to investigate the possibility of opening a student-operated bookstore to combat high textbook prices. During free hours, you might drop into the library to read about the retail book business and, if possible, about college bookstores. If your student government is affiliated with the National Student Association, you might write to it for information about policies and practices on other campuses. In addition, you might start talking about the problem to the manager of your college bookstore, owners of other bookstores in your community, student transfers from other schools, managers of college bookstores in nearby communities, and some of your professors. Because you might be concerned about the public relations repercussions of a student bookstore, you should propose the venture to college officials. At this point, you will have heard many different ideas. To discuss some of them, bring up the subject of a student bookstore with friends. As you talk, you will be clarifying and organizing your thoughts.

The same technique can be applied to less involved subjects, such as a description of an ideal summer job, or an explanation of a technical matter like input-output analysis. By worrying about the subject as you proceed through your daily routine, by striving to get ideas from books and people or both, and by clarifying your thoughts in conversation with friends, you will be ready for the next step. Remember: you can proceed only if you have some ideas. You must have something to write about.

Stage Two—Planning (10%)

Planning is another term for organizing or for the task that students dread so—*outlining*. You should already have had some experience with this process. Like many students, unfortunately, you may have learned only how to outline a paper after writing it. This practice is a waste of time. If you want to operate efficiently, then it follows that you will have to specialize in each of the writing stages in turn. When you force yourself simultaneously to conceive of ideas, to arrange them in logical order, and to express them in words, you cannot perform all of these complex activities as effectively as if you had concentrated on each one separately. By dividing the writing process into five stages, and by focusing on each one individually, the odds are that the result will be much better than you had thought possible. Of course you will hear about people who never plan their writing but who do well anyway, just as you may know someone who skis or plays golf well without having had a lesson. There are always some naturals who can flaunt the rules. But perhaps these people could have significantly improved their writing, golf, or skiing if they had received some formal instruction and proceeded in the prescribed manner.

An efficient person plans his work. Many executives take a few minutes before leaving the office or retiring at night to jot down problems to attend to the following day. Many vacationers list tasks to do before leaving home: stop the newspaper, turn down the refrigerator, inform the mailman, cut off the hot water, check the car, mail the mortgage payment, notify the police, and the like. Of course, you can go on vacation without planning—just as you can write without planning—but the chances are that the more carefully you organize, the better the results will be.

In writing, planning consists mainly in examining all your ideas, eliminating the irrelevant ones, and arranging the others in a clear, logical order. Whether you write a formal outline or merely jot down your ideas on scrap paper is up to you. The outline is for your benefit; follow the procedure that helps you the most. But realize that the harder you work on perfecting your outline, the easier the writing will be. As you write, therefore, you can concentrate on formulating sentences instead of also being concerned with thinking of ideas and trying to organize them. For the present, realize the importance of planning. . . .

Stage Three—Writing (25%)

When you know what you want to say and have planned how to say it, then you are ready to write. This third step in the writing process is self-explanatory. You need only dash away your thoughts, using pen, pencil, or typewriter, whichever you prefer. Let your mind flow along the outline. Don't pause to check spelling, worry about punctuation, or search for exact words. If new thoughts pop into your mind, as they often do, check them with your outline, and work them into your paper if they are relevant. Otherwise discard them. Keep going until either the end or fatigue halts you. And don't worry if you write slowly: many people do.

Stage Four—Revising (45%)

Few authors are so talented that they can express themselves clearly and effectively in a first draft. Most know that they must revise their papers extensively. So they roll up their sleeves and begin slashing away, cutting out excess verbiage, tearing into fuzzy sentences, stabbing at structural weaknesses, and knifing into obfuscation. Revision is painful: removing pet phrases and savory sentences is like getting rid of cherished possessions. Just as we dislike to discard old magazines, books, shoes, and clothes, so we hate to get rid of phrases and sentences. But no matter how distasteful the process may be, professional writers know that revising is crucial; it is usually the difference between a mediocre paper and an excellent one.

Few students revise their work well; they fail to realize the importance of this process, lack the necessary zeal, and are unaware of how to proceed. Once you understand why you seem to be naturally adverse to revision, you may overcome your resistance.

Because writing requires concentrated thought, it is about the most enervating work that humans perform. Consequently, upon completing a first draft, we are so relieved at having produced something that we tidy it up slightly, copy it quickly, and get rid of it immediately. After all, why bother with it anymore? But there's the rub. The experienced writer knows that he has to sweat through it again and again, looking for trouble, willing to rework cumbersome passages, and striving always to find a better word, a more felicitous phrase, and a smoother sentence. John Galbraith, for example, regularly writes five drafts, the last one reserved for inserting "that note of spontaneity" that ironically makes his work appear natural and effortless. The record for revision is probably held by Ernest Hemingway, who wrote the last page of *A Farewell to Arms* thirty-nine times before he was satisfied!

To revise effectively, you must not only change your attitude, but also your perspective. Usually we reread a paper from our own viewpoint, feeling proud of our accomplishment. We admire our cleverness, enjoy our graceful flights, and glow at our fanciful turns. How easy it is to delude ourselves! The experienced writer reads his draft objectively, looking at it from the standpoint of his reader. To accomplish this, he gets away from the paper for a while, usually leaving it until the following morning. You may not be able to budget your time this ideally, but you can put the paper aside while you visit a friend, grab a bite to eat, or phone someone. Unless you divorce yourself from the paper, you will probably be under its spell; that is, you will reread it with your mind rather than with your eyes. You will see only what you think is on the page instead of what is actually there. And you will not be able to transport yourself from your role of writer to that of critic.

Only by attacking your paper from the viewpoint of another person can you revise it effectively. You must be anxious to find fault and you must be honest with yourself. If you cannot or will not realize the weaknesses in your writing, then you cannot correct them. This textbook should open your eyes to many things that can go wrong, but above all you must *want* to find them. If you blind yourself to the bad, then your work will never be good. You must convince yourself that good writing depends on good rewriting. This point cannot be repeated too frequently or emphasized too much. Tolstoy revised *War and Peace*—one of the world's longest and greatest novels—five times. The brilliant French stylist Flaubert struggled for hours, even days, trying to perfect a single sentence. But question your professors about their own writing. Many—like me—revise papers three, four, or five times before submitting them for publication. You may not have this opportunity, your deadlines in school and business may not permit this luxury; but you can develop a respect for revision and you can devote more time and effort to this crucial stage in the writing process. . . .

Stage Five—Proofreading (5%)

Worrying, planning, writing, and revising are not sufficient. The final task, although not as taxing as the others, is just as vital. Proofreading is like a quick check in the mirror before leaving for a date. A sloppy appearance can spoil a favorable acceptance of you or your paper. If, through your fault or your typist's, words are missing or repeated, letters are transposed, or sentences juxtaposed, a reader may be perturbed enough to ignore or resist your ideas or information. Poor proofreading of an application letter can cost you a job. Such penalties may seem unfair, but we all react in this fashion. What would you think of a doctor with dirty hands? His carelessness would not only disgust you but would also raise questions about his professional competence. Similarly, carelessness in your writing antagonizes readers and raises questions in their minds about your competence. The merit of a person or paper should not depend on appearances, but frequently it does. Being aware of this possibility, you should keep yourself and your writing well groomed.

Poor proofreading results from failure to realize its importance, from inadequate time, and from improper effort. Unless you are convinced that scrutinizing your final copy is important, you cannot proofread effectively. If you realize the significance of this fifth step in the writing process, then you will not only have the incentive to work hard at proofreading but also will allow time for it in planning.

Like most things, proofreading takes time. We usually run out of time for things that we don't care about. But there's always time for what we enjoy or what is important to us. So with proofreading; lack of time is seldom a legitimate excuse for doing it badly.

But lack of technique is. To proofread well, you must forget the ideas in a paper and focus on words. Slow down your reading speed, stare hard at the black print, and search for trouble rather than trying to finish quickly. . . . Some professional proofreaders start with the last sentence and read backward to the first. Whatever technique you adopt, work painstakingly, so that a few careless errors will not spoil your efforts. If you realize the importance of proofreading, view it as one of the stages in the writing process, and labor at it conscientiously, your paper will reflect your care and concern. . . .

Each Stage Is Important

These then are the five stages in the writing process: worrying, planning, writing, revising, and proofreading. Each is vital. You need to become proficient at each of them, although, at some later point in your life, an editor or efficient secretary may relieve you of some proofreading chores. But in your career as a student and in your early business years, you will be on your own to guide your paper through the five stages. Since it is your paper, no one else will be as interested in it as you. It's your offspring—to nourish, to cherish, to coddle, to bring up, and finally, to turn over to someone else. You will be proud of it only if you have done your best throughout its growth and development.

Peter Elbow

The Direct Writing Process for Getting Words on Paper

Peter Elbow has taught at the University of Massachusetts at Amherst. He is also the author of Writing without Teachers *(Oxford, 1973), an innovative introduction to the writing process.*

The direct writing process is most useful if you don't have much time or if you have plenty to say about your topic. It's a kind of let's-get-this-thing-over-with writing process. I think of it for tasks like memos, reports, somewhat difficult letters, or essays where I don't want to engage in much new thinking. It's also a good approach if you are inexperienced or nervous about writing because it is simple. . . .

Unfortunately, its most common use will be for those situations that aren't supposed to happen but do: when you have to write something you *don't* yet understand, but you also don't have much time. The direct writing process may not always lead to a satisfactory piece of writing when you are in this fix, but it's the best approach I know.

The process is very simple. Just divide your available time in half. The first half is for fast writing without worrying about organization, language, correctness, or precision. The second half is for revising.

Start off by thinking carefully about the audience (if there is one) and the purpose for this piece of writing. Doing so may help you figure out exactly what you need to say. But if it doesn't, then let yourself put them out of mind. You may find that you get the most benefit from ignoring your audience and purpose at this early stage of the writing process. . . .

In any event spend the first half of your time making yourself write down everything you can think of that might belong or pertain to your writing task: incidents that come to mind for your story, images for your poem, ideas and facts for your essay or report. Write fast. Don't waste any time or energy on how to organize it, what to start with, paragraphing, wording, spelling, grammar, or any other matters of presentation. Just get things down helter-skelter. If you can't find the right word just leave a blank. If you can't say it the way you want to say it, say it the wrong way. (If it makes you feel better, put a wavy line under those wrong bits to remind you to fix them.)[1]

I'm not saying you must never pause in this writing. No need to make this a frantic process. Sometimes it is very fruitful to pause and return in your mind to some productive feeling or idea that you've lost. But don't stop to worry or criticize or correct what you've already written.

While doing this helter-skelter writing, don't allow too much digression. Follow your pencil where it leads, but when you suddenly realize, "Hey, this has nothing to do with what I want to write about," just stop, drop the whole thing, skip a line or two, and get yourself back onto some aspect of the topic or theme.

Similarly, don't allow too much repetition. As you write quickly, you may sometimes find yourself coming back to something you've already treated. Perhaps you are saying it better or in a better context the second or third time. But once you realize you've done it before, stop and go on to something else.

When you are trying to put down everything quickly, it often happens that a new or tangentially related thought comes to mind while you are just in the middle of some train of thought. Sometimes two or three new thoughts crowd in on you. This can be confusing: you don't want to interrupt what you are on, but you fear you'll forget the intruding thoughts if you don't write them down. I've found it helpful to note them without spending much time on them. I stop right at the moment they arrive—wherever I am in my writing—and jot down a couple of words or phrases to remind me of them, and then I continue on with what I am writing. Sometimes I jot the reminder on a separate piece of paper. When I write at the typewriter I often just put the reminder in caps inside double parentheses ((LIKE THIS)) in the middle of my sentence. Or I simply start a new line
LIKE THIS
and then start another new line to continue my old train of thought. But sometimes the intruding idea seems so important or fragile that I really want to go to work on it right away so I don't lose it. If so, I drop what I'm engaged in and start working on the new item. I know I can later recapture the original thought because I've already written part of it. The important point here is that what you produce during this first half of the writing cycle can be very fragmented and incoherent without any damage at all.

There is a small detail about the physical process of writing down words that I have found important. Gradually I have learned not to stop and cross out something I've just written when I change my mind. I just leave it there and

---

[1] An excerpt from a letter giving me feedback on an earlier draft: "I tried the direct writing process. Though it sounds simple enough, I . . . see now that in the past I've often interrupted the flow of writing by spending disproportionate time on spelling, punctuation, etc. I can spend hours on an opening paragraph stroking the words to death; then, if there's a deadline, have to rush through the remainder." (Joanne Turpin, 7/24/78.)

write my new word or phrase on a new line. So my page is likely to have lots of passages that look like

Many of my pages
Still I don't mean that you should stop and rewrite every passage till you are happy with it.
This kind of appearance.

What is involved here is developing an increased tolerance for letting mistakes show. If you find yourself crumpling up your sheet of paper and throwing it away and starting with a new one every time you change your mind, you are really saying, "I must destroy all evidence of mistakes." Not quite so extreme is the person who scribbles over every mistake so avidly that not even the tail of the *y* is visible. Stopping to cross out mistakes doesn't just waste psychic energy, it distracts you from full concentration on what you are trying to say.

What's more, I've found that leaving mistakes uncrossed out somehow makes it easier for me to revise. When I cross out all my mistakes I end up with a *draft*. And a draft is hard to revise because it is a complete whole. But when I leave my first choices there littering my page along with some second and third choices, I don't have a draft, I just have a succession of ingredients. Often it is easier to whip that succession of ingredients into something usable than, as it were, to *undo* that completed draft and turn it into a better draft. It turns out I can just trundle through that pile of ingredients, slash out some words and sections, rearrange some bits, and end up with something quite usable. And quite often I discover in retrospect that my original "mistaken phrase" is really better than what I replaced it with: more lively or closer to what I end up saying.

* * *

If you only have half an hour to write a memo, you have now forced yourself in fifteen minutes to cram down every hunch, insight, and train of thought that you think might belong in it. If you have only this evening to write a substantial report or paper, it is now 10:30 P.M., you have used up two or two and a half hours putting down as much as you can, and you only have two or more hours to give to this thing. You must stop your raw writing now, even if you feel frustrated at not having written enough or figured out yet exactly what you mean to say. If you started out with no real understanding of your topic, you certainly won't feel satisfied with what is probably a complete mess at this point. You'll just have to accept the fact that of course you will do a poor job compared to what you could have done if you'd started yesterday. But what's more to the point now is to recognize that you'll do an even crummier job if you steal any of your revising time for more raw writing. Besides, you will have an opportunity during the revising process to figure out what you want to say—what all these ingredients add up to—and to add a few missing pieces. It's important to note that when I talk about revising . . . I mean something much more substantial than just tidying up your sentences.

So if your total time is half gone, stop now no matter how frustrated you are and change to the revising process. That means changing gears into an entirely different consciousness. You must transform yourself from a fast-and-loose-thinking person who is open to every whim and feeling into a ruthless, tough-minded, rigorously logical editor. Since you are working under time pressure, you will probably use quick revising or cut-and-paste revising. . . .

* * *

Direct writing and quick revising are probably good processes to start with if you have an especially hard time writing. They help you to prove yourself that you *can* get things written quickly and acceptably. The results may not be the very best you can do, but they work, they get you by. Once you've proved you can get the job done you will be more willing to use other processes for getting words down on paper and for revising—processes that make greater demands on your time and energy and emotions. And if writing is usually a great struggle, you have probably been thrown off balance many times by getting into too much chaos. The direct writing process is a way to allow a limited amount of chaos to occur in a very controlled fashion.

It's easiest to explain the direct writing process in terms of pragmatic writing: you are in a hurry, you know most of what you want to say, you aren't trying for much creativity or brilliance. But I also want to stress that the direct writing process can work well for very important pieces of writing and ones where you haven't yet worked out your thinking at all. But one condition is crucial: you must be confident that you'll have no trouble finding lots to say once you start writing. . . .

As I wrote many parts of this book [i.e., *Writing with Power*], for example, I didn't have my thinking clear or worked out by any means, I couldn't have made an outline at gunpoint, and I cared deeply about the results. But I knew that there was lots of *stuff* there swirling around in my head ready to go down on paper. I used the direct process. I just wrote down everything that came to mind and went on to revise.

But if you want to use the direct writing process for important pieces of writing, you need plenty of time. You probably won't be able to get them the way you want them with just quick revising. You'll need thorough revising or revising with feedback. . . . For important writing I invariably spend more time revising than I do getting my thoughts down on paper the first time.

MAIN STEPS IN THE DIRECT WRITING PROCESS

- If you have a deadline, divide your total available time: half for raw writing, half for revising.
- Bring to mind your audience and purpose in writing but then go on to ignore them if that helps your raw writing.

- Write down as quickly as you can everything you can think of that pertains to your topic or theme.
- Don't let yourself repeat or digress or get lost, but don't worry about the order of what you write, the wording, or about crossing out what you decide is wrong.
- Make sure you stop when your time is half gone and change to revising, even if you are not done.
- The direct writing process is most helpful when you don't have difficulty coming up with material or when you are working under a tight deadline.

Linda Flower and John Ackerman

Evaluating and Testing as You Revise

Both Linda Flower, a member of the faculty at Carnegie Mellon University, and John Ackerman, a member of the faculty at the University of Utah, are nationally recognized authorities on the teaching of writing.

IMAGINING A READER'S RESPONSE

Even if you have a model or text to work from, as a writer you will work hard to construct a text: generating and discarding ideas, trying to figure out your point, sketching out alternative organizations, and then trying to signal that point and structure to your reader. But the story does not end there because your various readers have to work equally hard to construct a meaning based on your text. That is, readers want to construct *in their own minds* a coherent text, with a hierarchical organization (like an issue tree) based on key points. And they need to see a purpose for reading. Readers want to know *Why read this? What is the point?* and *How is all of this connected?*—and they want to find answers as quickly as possible.

Readers begin to predict the structure of a text and its meaning as soon as they face a page. They will look for cues that the text might offer about point, purpose, and structure, but if they do not find them, they will go ahead and construct their own version of the text *and assume that is what you intended*. It may be helpful to imagine your readers as needing to *write* your text for themselves. Thus, their own goals and interests will strongly influence what they look for in a text and the meaning they make out of it. Your goal, then, as a writer is to do the best job you can to make sure a reader's process through your text and his or her understanding matches or comes close to what you intended.

Once your ideas are down in a draft of some form, **revision** lets you anticipate how readers will respond and adjust the text to get the response you hope for. **Local revision** involves editing, correcting spelling and grammar, and making local improvements in wording or sentence style. **Global revision** involves looking at the big picture, but that does not mean throwing the draft out and starting again, and it may not even take a lot of time. Global revision, however, does mean looking at the text as a whole—thinking globally about the major rhetorical decisions and plans you made—now that the draft is complete. Global revisions alter the focus, organization, argument, or detail to improve the overall text.

Revision begins with a tricky reading process, a close reading of your own text. You know from your own experience in school that readers can read for different purposes: to skim a text to prepare for class discussion and then to read more carefully for an examination. In business, because writing usually augments a transaction between people, readers look for information but they also look for how well a piece of writing is adapted to their needs. The key to your success as a reviser is your ability to read your own writing critically for different features and purposes. At the simplest level, revision is *re-vision*—the process of stepping back from one's text and seeing it anew. . . .

STRATEGY 1: LOOK FOR WRITER-BASED PROSE

Why is it that even experienced writers typically choose to draft and revise rather than write a final text in one pass? Suppose you put time into planning your paper; you followed a good model; you thought about your reader as you chose what to say. Why should your first draft not do the job? One reason is because first drafts often contain large sections of **writer-based prose.** Writer-based prose appears when writers are essentially talking to themselves, talking through a problem, exploring their own knowledge, or trying to get their ideas out. In fact, producing writer-based prose is often a smart problem-solving strategy. Instead of getting blocked or spending hours staring at a blank page trying to write a perfect text the first time through, writers can literally walk through their own memory, talking out on paper what they know. Other concerns, including what the reader needs to hear, are put on temporary hold. Writer-based prose is, in fact, a very effective strategy for searching your memory and for dealing with difficult topics or lots of information. Writer-based prose may also be the best a writer can do in a new situation. A writer facing an unfamiliar task may produce writer-based prose, uncertain of what readers expect.

Whether it is a strategy or consequence, the downside of this strategy is that the text it produces is typically focused on the writer's thinking—not the readers' questions or needs—and it often comes out organized as a narrative or a river of connections. A number of studies of writers new to organizations demonstrate the reasoning behind writer-based prose and its ill-effect. New employees write consciously and unconsciously to report their discovery process or to survey all they know. And their readers impatiently wade through the river looking for the specific ideas and information they need.

Here [see Figure 1] is the first draft of a progress report written by four students in an organizational psychology course who were doing a consulting project with a local organization, the Oskaloosa Brewing Company. As a reader, put yourself first in the position of Professor Charns. He is reading the report to answer three questions: As analysts, what assumptions and decisions did these students make in setting up their study? Why did they make them? And where

---

Draft 1

Group Progress Report

(1)   Work began on our project with the initial group decision to evaluate the Oskaloosa Brewing Company. Oskaloosa Brewing Company is a regionally located brewery manufacturing several different types of beer, notably River City and Brough Cream Ale. This beer is marketed under various names in Pennsylvania and other neighboring states. As a group, we decided to analyze this organization because two of our group members had had frequent customer contact with the sales department. Also, we were aware that Oskaloosa Brewing had been losing money for the past five years, and we felt we might be able to find some obvious problems in its organizational structure.

(2)   Our first meeting, held February 17th, was with the head of the sales department, Jim Tucker. Generally, he gave us an outline of the organization, from president to worker, and discussed the various departments that we might ultimately decide to analyze. The two that seemed the most promising and more applicable to the project were the sales and production departments. After a few group meetings and discussions with the personnel manager, Susan Harris, and our advisor, Professor Charns, we felt it best suited our needs and Oskaloosa Brewing's needs to evaluate their bottling department.

(3)   During the next week we had a discussion with the superintendent of production, Henry Holt, and made plans for interviewing the supervisors and line workers. Also, we had a tour of the bottling department that gave us a first-hand look at the production process. Before beginning our interviewing, our group met several times to formulate appropriate questions to use in interviewing, for both the supervisors and the workers. We also had a meeting with Professor Charns to discuss this matter.

(4)   The next step was the actual interviewing process. During the weeks of March 14–18 and March 21–25, our group met several times at Oskaloosa Brewing and interviewed ten supervisors and twelve workers. Finally, during this past week, we have had several group meetings to discuss our findings and the potential problem areas within the bottling department. Also, we have spent time organizing the writing of our progress report.

1 of 2

FIGURE I   The Oskaloosa Brewing Progress Memo, Draft I

(5) The bottling and packaging division is located in a separate building, adjacent to the brewery, where the beer is actually manufactured. From the brewery the beer is piped into one of five lines (four bottling lines and one canning line) in the bottling house, where the bottles are filled, crowned, pasteurized, labeled, packaged in cases, and either shipped out or stored in the warehouse. The head of this operation, and others, is production manager Phil Smith. Next in line under him in direct control of the bottling house is the superintendent of bottling and packaging, Henry Holt. In addition, there are a total of ten supervisors who report directly to Henry Holt and who oversee the daily operations and coordinate and direct the twenty to thirty union workers who operate the lines.

(6) During production, each supervisor fills out a data sheet to explain what was actually produced during each hour. This form also includes the exact time when a breakdown occurred, what it was caused by, and when production was resumed. Some supervisors' positions are production-staff-oriented. One takes care of supplying the raw material (bottles, caps, labels, and boxes) for production. Another is responsible for the union workers' assignments each day.

These workers are not all permanently assigned to a production-line position. Workers called "floaters" are used, filling in for a sick worker or helping out after a breakdown.

(7) The union employees are generally older than 35, some in their late fifties. Most have been with the company many years and are accustomed to having more workers per a slower moving line. . . .

2 of 2

FIGURE I   *(Continued)*

are they in the project now? Then take on the role of the client, the company vice president who follows their progress and wants to know: O.K. What is the problem (i.e., how did they define it)? And what did they conclude? Would this draft answer these questions for either of its intended readers?

Put yourself in the shoes of the professor. What would you look for in a progress report? According to Charns, he used this report to evaluate the group's progress: Were they on schedule; were they on task; did he need to intervene? However, he didn't need a blow-by-blow story to do that. As an evaluator he wanted to see whether they knew how to analyze an organization: Were they making good decisions (that is, decisions they could justify in this report); had they made any discoveries about this company? His needs as a reader, then, reflected his dual role as a teacher (supervisor) and an evaluator.

When we showed this draft to a manager (with comparable experience and responsibility to the Oskaloosa VP), we got a very different response. Here was a reader looking quickly for the information she wanted and building an image of the writers' business savvy based on their text. Here is part of her response as she read and thought aloud (the student text she reads is underlined):

> Work began on our project with the initial group decision to evaluate . . . *Ok. Our project What project is this? I must have a dozen "projects" I keep tabs on. And who is this group?* This beer is marketed . . . *blah, blah, I'm tempted to skim. This must be a student project. But why am I reading about the fact somebody bought a lot of beer for their frat? Maybe the next paragraph.*
>
> Our first meeting . . . *Ok, they saw Jim and Susan,* . . . *looked at bottling,* . . . *wrote their paper. And now they are telling me where my packaging division is located! This is like a shaggy plant tour story. They are just wasting my time. And I suppose I should say that I am also forming an image of them as rather naive, sort of bumbling around the plant, interrupting my staff with questions. I mean, what are they after? Do they have any idea of what they are doing?*

Now put yourself in the shoes of the professor. What would you look for in a progress report? How would you evaluate this group as decision makers? Have they learned anything about analyzing organizations? How would you evaluate their progress at this point? Have they made any discoveries, or are they just going through the steps?

Fortunately, this is not the draft the writers turned in to their professor or the company manager. The revised draft [as shown in Figure 2] was written after a short conference with a writing instructor who instead of offering advice, asked the writers to predict what each of their readers would be looking for. It took ten minutes to step back from their draft and to rethink it from the perspective of their professor and the Oskaloosa manager. They found that they needed global revision—a revision that kept the substance of their writer-based first draft, but transformed it into reader-based text. As you compare these two drafts, notice the narrative and survey organization in the first draft and the "I did it" focus that are often a tip-off to writer-based prose, and how they improved the second draft.

How would you characterize the differences between drafts one and two? One clear difference is the use of a conventional memo/report format to focus the reader's attention. But beyond the visual display of information, the writers moved away from narrative organization, an "I" focus, and a survey form or "textbook" pattern of organization. Watch for these three patterns as you revise.

Narrative Organization

The first four paragraphs of the first draft are organized as a narrative, starting with the phrase "Work began. . . ." We are given a story of the writers' discovery process. Notice how all of the facts are presented in terms of when they were discovered, not in terms of their implications or logical connections. The writers want to tell us what happened and when; the reader, on the other hand, wants to ask "why?" and "so what?"

A narrative organization is tempting to write because it is a prefabricated order and easy to generate. All of us walk around with stories in our head, and chronology is a common rhetorical move. Instead of creating a **hierarchical** organization among ideas or worrying about a reader, the writer can simply

Draft 2

MEMORANDUM

TO: Professor Martin Charns
FROM: Nancy Lowenberg, Todd Scott, Rosemary Nisson,
            Larry Vollen
DATE: March 31, 1987
RE: Progress Report: The Oskaloosa Brewing Company

Why Oskaloosa Brewing?

Oskaloosa Brewing Company is a regionally located brewery manufacturing several different types of beer, notably River City and Brough Cream Ale. As a group, we decided to analyze this organization because two of our group members have frequent contact with the sales department. Also, we were aware that Oskaloosa Brewing had been losing money for the past five years and we felt we might be able to find some obvious problems in its organizational structure.

Initial Steps: Where to Concentrate?

After several interviews with top management and a group discussion, we felt it best suited our needs, and Oskaloosa Brewing's needs, to evaluate the production department. Our first meeting, held February 17, was with the head of the sales department, Jim Tucker. He gave us an outline of the organization and described the two major departments, sales and production. He indicated that there were more obvious problems in the production department, a belief also suggested by Susan Harris, the personnel manager.

Next Step

The next step involved a familiarization with the plant and its employees. First, we toured the plant to gain an understanding of the brewing and bottling processes. Next, during the weeks of March 14–18 and March 21–25, we interviewed ten supervisors and twelve workers. Finally, during the past week we had group meetings to exchange information and discuss potential problems.

The Production Process

Knowledge of the actual production process is imperative in understanding the effects of various problems on efficient production. Therefore, we have included a brief summary of this process.

The bottling and packaging division is located in a separate building, adjacent to the brewery, where the beer is actually manufactured. From the brewery the beer is piped into one of five lines (four bottling lines and one canning line) in the bottling house, where the bottles are filled, crowned, pasteurized, labeled, packaged in cases, and either shipped out or stored in the warehouse.

Problems

Through extensive interviews with supervisors and union employees, we have recognized four apparent problems within the bottling house operations. The first is that the employees' goals do not match those of the company. . . . This is especially apparent in the union employees, whose loyalty lies with the union instead of the company. This attitude is well-founded, as the union ensures them of job security and benefits. . . .

FIGURE 2   The Oskaloosa Brewing Progress Memo, Draft 2

remember his or her own discovery process and write a story. Remember that in a hierarchical structure, such as an issue tree, the ideas at the top of the structure work as the organizing concepts that include other ideas. The alternative is often a string of ideas simply linked by association or by the order in which the writer thought about them. Papers that start out, "In studying the reasons for the current decline in our return customers, ..." are often a dead giveaway. They tell us we are going to watch the writer's mind at work and follow him or her through the process of thinking out conclusions. Following one's own associations makes the text easier to write. But another reason new employees are tempted to write narrative reports is that they were often rewarded for narratives at some point in their career *as students*. They fail to realize that in business the reader is someone who expects to *use* this text (not check off whether or not they did the assignment).

A narrative pattern, of course, has the virtue of any form of drama—it keeps you in suspense by withholding closure. But this drama is an effective strategy only if the audience is willing to wait that long for the point. Most professional and academic readers are impatient, and they tend to interpret such narrative, step-by-step structures either as wandering and confused (Is there a point?) or as a form of hedging. Narrative structures may be read as veiled attempts to hide what really happened or the writers' actual position. Although a progress report naturally involves narrative, how has Draft 2 been able to *use* the narrative to answer readers' questions?

The "I" Focus

The second feature of Draft 1 is that it is a discovery story starring the writers. Its drama, such as it is, is squarely focused on the writer: "I did/I thought/I felt. . . ." Of the fourteen sentences in the first three paragraphs, ten are grammatically focused on the writer's thoughts and actions rather than on the issues. For example: "Work began . . . ," "We decided . . . ," Also, "we were aware . . . and we felt. . . ." Generally speaking, the reader is more interested in issues and ideas than in the fact that the writer thought them.

In pointing out the "I" focus in Draft 1, we are not saying that writers cannot refer to themselves or begin a sentence with "I," as many learned in school. Sometimes a specific reference to oneself is exactly the information a reader needs, and a reader may respond to the honesty and directness. Use "I" or "we" to make a claim or when it is an important piece of information, not just as a convenient way to start a sentence. In Draft 2, the students are clearly present as people doing the research, but the focus is on the information the reader wants to hear.

Survey Form or Textbook Organization

In the fifth paragraph of Draft 1 the writers begin to organize their material in a new way. Instead of a narrative, we are given a survey of what the writers observed. Here, the raw facts of the bottling process dictated the organization of the paragraph. Yet the client-reader already knows this, and the professor probably does not care. In the language of computer science we could say the writers are performing a "memory dump": printing out information in the exact form in which they stored it in memory. Notice how in the revised version the writers try to use their observations to understand production problems.

The problem with a survey or "textbook" pattern is that it ignores the reader's need for a different organization of the information. Suppose, for example, you are writing to model airplane builders about wind resistance. The information you need comes out of a physics text, but that text is organized around the field of physics; it starts with subatomic particles and works up from there. To meet the needs of your reader, you have to adapt that knowledge, not lift it intact from the text. Sometimes writers can simply survey their knowledge, but generally the writer's main task is to use knowledge rather than reprint it.

To sum up, in Draft 2 of the Oskaloosa report, the writers made a real attempt to write for their readers. Among other things, the report is now organized around major questions readers might have, it uses headings to display the overall organization of the report, and it makes better use of topic sentences that tell the reader what each paragraph contains and why to read it. Most important, it focuses more on the crucial information the reader wants to obtain.

Obviously this version could still be improved. But it shows the writers attempting to transform writer-based prose into reader-based prose and change their narrative and survey pattern into a more issue-centered top-to-bottom organization.

STRATEGY 2: TEST YOUR TEXT FOR A READER-BASED STRUCTURE

Reading your text for writer-based prose lets you spot places where you were still exploring ideas or talking to yourself—places that probably call for some sort of global or structural revision to make this text a reader-based document. But that does not tell you how to revise. **Reader-based prose** foregrounds and makes explicit the information a reader needs or expects to find. Reader-based prose tries to anticipate and support an active reader, one who probably will use your writing for some specific end. What do you want your reader to see, think, or do? How will your reader respond? One important approach to global revision is to look at your text as a conversation with the reader in which you set up some initial agreements and expectations and then fulfill your promises. We offer three ways to test for a reader-based structure: *testing your drafts against your initial plans, using clues that reveal this plan to a reader, and keeping the promises you made in your writing.*

Does Your Text Reflect Your Plans?

Texts have a way of running off by themselves. The more text you produce, the more convinced you are that your prose is complete, readable, even entertaining. This is natural given the commitment it takes to write anything, but because texts often drift away from your intentions, you need a way to keep your text honest or to realize that you have come up with a better approach. So instead of reading your text "as written" and just going with the flow, start by setting up an image of your purpose and plans in your mind's eye, then test your text against that image. Can you find any evidence of your plans in the text?

To get a good image of your plan, return to . . . your planning notes to review your plans consciously. In your mind or on paper, restate the plans *to do* and *say* that you produced prior to your current draft, and find the exact places in your text where your writing satisfies (or departs from) your plans to reach a reader. You could set the mental exercise up as a checklist [Figure 3].

Holding your draft up against the backdrop of your own plans can help you notice how well the two fit together. Did you have important goals, or good points in your notes that have just not appeared in your text yet? Did you simply forget ideas? Or did you find—as many writers do—that the act of writing was itself an inventive, generative process? Your plans may have changed as a result of writing. If that happened, what should you do? At this point, inexperienced writers often abandon their old plans and follow wherever the text seems to be taking them. However, experienced writers make another move. They go back into planning and look for possible ways to consolidate their new ideas with other old plans. They try to build a new plan that makes use of the good parts

---

My plans for writing, to do and say were...

> The important Goals and Purposes I gave myself...
> My Key Points...
> How I wanted my Reader to respond (and what other responses I anticipated)...
> My choice of Text Conventions...
> How I planned to make all of this work together...

As a checklist, I could evaluate my writing this way:

Plans *to do* and *say*	Draft	Text Reference
Goals...	✓	~~~~~~~~~~~~~~~~~~~~~~~~~~
Key points...	✓	~~~~~~~~~~~~~~~~~~~~~~~~~~~
Intended Response...	✓	~~~~~~~~~~~~~~~~~~~~
Text Conventions, Models...	✓	~~~~~~~~~~~~~~~~~~~~~~~~~
Making it all work together...	????	*oops, gotta work on this!!!*

FIGURE 3 Checking Your Text Against Your Plans

of both ideas. They use this round of planning to guide their global revisions consciously.

This strategy is obviously one you will use more than once—a kind of in-process evaluation that lets you keep checking in with your goals for writing and checking your text against the big picture of what you want it to do.

Cues That Reveal Your Plans to Your Reader

Maybe you are satisfied that you have indeed <u>defined</u> a real and shared <u>problem</u> for your reader, and <u>compared</u> some <u>alternative</u> ways to respond to it, <u>supported</u> them with <u>examples</u>, while <u>proposing</u> your favored course of <u>action</u>. But will the reader recognize all of those rhetorical moves? It is possible that you "talked about" this information, but did not make your good rhetorical plan to define, compare, support and propose fully apparent or explicit.

In most pieces of writing there are two conversations going on. One is the information that the reader needs or expects to find: the recommendations that you want to make, the results of your study, the idea you are proposing, the specifics of a solution. As we shall see, this conversation is held together, usually, by a strong chain of topics and an appropriate rhetorical pattern. But the second conversation is explicitly between the writer and the reader, announcing what the reader will find, in what order, and reminding the reader of where he or she is in the text.

This second conversation is often called **metadiscourse.** *Meta-* is a Greek prefix that means "along with" or "among," and the basic strategy is to include explicit statements and cues to the reader that announce and reinforce your intentions along with your content. There are two main ways writers give such cues. One is to talk directly to the reader, inserting metacomments that preview what will come, remind, predict, or summarize. This lets the writer step back, make the plan of the text more visible, and direct traffic, by telling the reader, "In the next section I will argue that. . . ." The other kind of cues work more like traffic signals—they are the conventional words and phrases that signal transitions, or logic, or the structure of ideas. Readers often expect to find these metacomments and signalling cues in some standard places. Some common places to insert cues to the reader include:

Title, title page
Table of contents
Abstract
Introduction or first paragraphs
Headings
The beginning and end of paragraphs (i.e., topic sentences)
Entire paragraphs in between long sections

So review your text first to see if you have used enough cues to make your plan clear, and second to see if you have included cues in places readers expect to see some guidance on what to look for and how to read this document [Figure 4].

Cues that signal your plan and guide the reader:

<u>Cues that lead the reader forward</u>

To show addition:		*To show time:*	
Again	Moreover,	At length	And then
And	Nor,	Immediately thereafter,	Later,
And then,	Too,	Soon,	Previously,
Besides	Next,	After a few hours,	Formerly,
Equally important,	First, second, etc.	Afterwards,	First, second, etc.
Finally	Lastly,	Finally,	Next, etc.
Further,	What's more,	Then	
Furthermore,			

<u>Cues that make the reader stop and compare</u>

But	Notwithstanding,	Although
Yet,	On the other hand,	Although this is true,
And yet,	On the contrary,	While this is true,
However,	After all,	Conversely,
Still	For all that,	Simultaneously,
Nevertheless,	In contrast,	Meanwhile
Nonetheless,	At the same time,	In the meantime,

<u>Cues that develop and summarize</u>

To give examples:	*To emphasize:*	*To repeat:*	*To signal in a relationship:*
For instance,	Obviously,	In brief,	Finally
For example,	In fact,	In short,	Because
To demonstrate,	As a matter of fact,	As I have said,	Yet
To illustrate,	Indeed,	As I have noted,	For instance,
As an illustration,	In any case,	In other words,	
	In any event,	That is,	

To introduce conclusions:	*To summarize:*
Hence,	In brief,
Therefore,	On the whole,
Accordingly,	Summing up,
Consequently,	To conclude,
Thus,	In conclusion,
As a result,	

FIGURE 4   Giving Cues to the Reader

Metacomment cues that announce and reinforce your intentions

To ask a question about your topic or the argument unfolding:

What series of events led to event? . . .
To answer that question . . .

To preview what will come:

In the next section, we will see how this formula applies . . .
The third paragraph will reveal how . . .

To summarize what has been said thus far:

In the preceding pages, I've described . . .
Thus far, I have argued . . .

To comment on your writing and thinking as it unfolds:

I haven't mentioned yet that . . .
I'm talking about . . .
My main point is . . .

FIGURE 4   *(Continued)*

There is an endless variety of sentences and phrases that can be invented and inserted to announce and reinforce your main ideas and the progression that you want your reader to follow. To show you the power of metadiscourse and how it plays out in a text, here is an excerpt from a shareowner's letter with the metadiscourse highlighted [as shown in Figure 5]. We offer the original paragraph numbers with the cues underlined and the rhetorical purpose of the cues as we read them. . . .

Did You Keep Your Promises?

Your text started out with the best of intentions—a strong rhetorical plan and cues that keep your reader on track. The next test is to see if you followed through on the promises that you made. Read your text as if you were outlining its key points and promises and then look back to see if you have delivered the necessary detail. For instance:

- Your problem/purpose statement promises four main points and an extended example: Does each paragraph keep that promise? By referring back to your announced plan and delivering four main (i.e., well-developed) points in the same order that you promised?
- Your topic sentence in the seventh paragraph promises the two key instances that support a legal precedent: Does the paragraph deliver them in the order and detail necessary?

Chairman's Letter

¶   Rhetorical Purpose/Cues

1		Often the greatest opportunities..........................
4	ties to history	Our strong 1984 results speak for themselves ......
5		Pacific Telesis Group earned $829 million............
6	forces question	What happened? How did the corporation that many observers predicted would be the biggest loser in the AT&T breakup turn out to be one of the biggest winners?
7	links date with solution	To answer that question, you have to go back to 1980, when we developed..........................
8	signals logic	Our employees mobilized to make it work. And work it did. More specifically:..............
8c	signals order & emphasis	Third, and very important, we've built a relationship with the California Public Utilities Commission ....
	previews, emphasizes	a subject I'll return to later on in this letter.............

The Future of Your Investment

34	previews	A corporation, particularly one as new as ours, must operate with a clearly articulated and widely understood vision of its future....................
37	summarizes	In the preceding pages I've described to you our strategies for deploying technology, marketing technology, and diversifying into new lines ....
38	summarizes, emphasizes	I've talked about our determination ....
	signals logic, emphasis uses authority	But there is another very significant factor I haven't mentioned. I'm talking about the people who work for the Pacific Telesis Group

FIGURE 5   Metadiscourse in the Chairman's Letter

John S. Harris

The Project Worksheet for Efficient Writing Management

John S. Harris is Professor Emeritus of English at Brigham Young University. The author of several books on technical writing, he was the founding president of the Association of Teachers of Technical Writing.

When employees produce poor quality documents, the fault may lie not with their incompetence or lack of devotion to the job, but with inadequate and unclear assignments from their project managers or publications managers. The Project Worksheet is designed to help managers give initial writing assignments, effectively manage documents during production, and evaluate finished products. The following pages describe the Worksheet and suggest ways publications managers can use it effectively.

THE PROJECT WORKSHEET

The Project Worksheet consists of a series of questions that help writers and managers consider the factors involved in planning a document. I developed the Worksheet years ago, and faculty in the technical writing program at Brigham Young University have successfully used it to teach technical writing for some years. After graduation, many of our students continue to use it in industry to plan documents and make document assignments.

What the Writer Must Consider

The manager and writer must determine the document's intended readers, its purpose, scope, form, length, graphic aids, and sources of information.[1] They must know such things before the writing begins. Failure to consider them results in inefficient document production and poor-quality documents.

Experienced writers in an organization usually answer these planning questions intuitively or subconsciously. Or they get the answers by direct conference, by study of past documents, or through the grapevine. They then efficiently plan and write good documents.

But less-experienced writers learn the answers to the questions only by the inefficient and expensive trial-and-error method of submitting draft documents and then repeatedly amending them as management requires.

Managers—whether of engineering projects, scientific studies, or computer software documentation—can reduce the trial-and-error writing cycle by providing complete answers for such planning questions during a writing assignment conference. But managers cannot provide the answers to the planning questions without first knowing the questions and their consequences. The Project Worksheet in Figure 1 guides both writers and managers in considering the important factors in planning documents. It can, of course, be adapted for special situations. Managers may want to reformat it to allow more space for answers, and some may wish to put it into a word processor as a template. Let us first examine the Worksheet questions and then consider procedures for using the Worksheet for document management.

Housekeeping Headings

Headings such as assignment date, due date, writer, and assignment authority are self-explanatory. Still, listing them together permits the use of the Worksheet as a tracking document for projects. The *document subject* is the file subject. The *document title* may be the same, or it may be something less prosaic—especially if the document is going out of house. Thus a document subject could be *Dental Hygiene,* and the document title could be *Your Teeth: An Owner's Manual.*

Primary Reader

No factor is more important for the writer than full understanding of the nature and needs of the document's primary reader. Experienced writers know this, of course. But even they do not always know everything that management knows and believes about the readers of the projected document. Too frequently both managers and writers assume they understand who the readers will be, but they may have two quite different audiences in mind. Using the Worksheet, managers and writers can reach an understanding of this and other questions.

---

[1]Although this . . . [essay] explains how managers can use the Project Worksheet for managing documents, the books listed under Selected Readings explain how writers can use the Worksheet and other methods to design technical documents for various audiences and purposes.

PROJECT WORKSHEET

Subject _____ Title _____
Assigned by _____ Approved _____ Assignment date _____
Writer _____ Agreed to _____ Due date _____

Primary Reader:

    Technical level (education, experience, etc.):
    Position (title or organizational relationship):
    Attitude toward subject:
    Other factors:

Secondary Readers (others who may read the document):

    Technical level:
    Position:
    Attitude toward subject:
    Other factors:

Reader's Purpose:

    What should the reader know after reading?
    What should the reader be able to do after reading?
    What attitude should the reader have after reading?
    How will the reader access the material?

Writer's Purpose:

    Intellectual purpose:
    Career or monetary purpose:

Logistics:

    Sources (lab reports, library research, etc.):
    Physical size limits (if any):
    Form or medium prescribed or desirable:
    Available aids (graphics, etc.):
    Means of production:
    Outline (Preliminary):

Distribution and Disposition:

FIGURE I   Project Worksheet

Reader's Technical Level

By carefully considering the technical level of the reader, the manager and writer can decide the proper level of explanation needed. How much information does the reader already have? Are definitions of *photo-grammetry* or *perihelion* or *debentures* needed? Considering such things can help avoid losing the reader in a maze of technical jargon, or alternatively avoid alienating a reader with overly simplistic explanations. In the trade-off between those two extremes, erring on the side of simplicity is nearly always better.

Reader's Organizational Relationship

The writer must also consider whether the document will go to the executive tower or to the shop floor. Since writers often prepare documents for someone else's signature, the manager—and the writer—must consider the organizational relationship of the reader *to the document's signer*. Thus, if the document will be signed by the boss, the writer must—for the task—carefully assume the persona of the boss. Is the signer addressing superiors, subordinates, or peers? Such situations raise delicate questions, and the manager must recognize the entailed problems and pass on an understanding of those problems to the writer.

Reader's Attitude

Similarly, the attitude of the reader must be considered. Is the reader hostile to the new manufacturing procedure because it will result in downsizing her department? Or is the reader skeptical about the efficacy of new cutting-edge technology because it has not yet been proved? Or is the reader torpid and apathetic and needing a stiff jolt to see the promise—or the threat—of the subject of the document?

Other Factors

Often even more subtle factors about the reader must be considered. Some of these may overlap the preceding questions, but they may also affect the kind of document needed. Is the reader a white, conservative Republican woman of sixty, raised on an Iowa farm, or a young Afro-American male from the South Bronx, or a highly educated but condescending academic? Is the reader an Ivy-League MBA obsessed with this year's bottom line, a high school dropout production-line worker who is fearful that increased automation will eliminate jobs, a displaced homemaker with an acute case of computer anxiety, or an immigrant with limited understanding of English? Such demographic factors may require delicate adjustments in the document.

Secondary Reader

Often a secondary reader must also be considered. The company annual report may be read by both stockholders and the CEO. An advertising brochure intended for customers may also be read by the competition. An environmental impact statement may be read by both the Douglas Fir Plywood Association and the Sierra Club. A report intended for Level 2 management may have to get past a Level 3 gatekeeper. Such situations require that the writer juggle the factors of two or more audiences at once, and this requires some skill. If the readers differ in attitude, the writer must watch out for red-flag statements that may cause a secondary reader to charge. If the differences are in technical level, the secondary reader's need for more elementary explanations or more technical information can often be taken care of in footnotes, glossaries, sidebars, or appendices.

The Reader's Purpose

Organizational documents have purposes as varied as securing approval for a new manufacturing process, justifying spending public funds for a reclamation project, instructing a software purchase on how to use a spreadsheet, or easing the worries of environmentalists about the effect of a new highway near a prime trout-fishing stream. Four basic questions about the reader's purposes need consideration:

- What should the reader know after the reading?
- What should the reader be able to do after reading?
- What attitude should the reader have after reading?
- How will the reader access the information?

What Should the Reader Know after Reading?

What factual information does the reader need to obtain from the document? Does the reader need to know the percentage of radial keratotomy patients who can pass a 20/40 eye examination without glasses after the operation? The percentage of impurities remaining in palladium after the macrocycle solvent extraction process? The change in numbers of predator kills of sheep after reintroduction of wolves to the grazing range? Often the specific information the reader is to gain from the document can be presented in lists, outlines, tables, or graphs.

What Should the Reader Be Able to Do after Reading?

What the reader can *do* after reading may be quite different from what the reader *knows*. Knowing the names for the stages of mitosis is different from being able to recognize telophase through a microscope. Knowing the stoichiometric ratio for combustion of a gasoline/air mixture is different from being able to adjust a fuel-injection system to achieve the ideal mixture.

What a reader can do after reading may depend on technical skills. It may also depend on the reader's capability to make executive decisions based on the information contained in the document. Thus the reader's needs probably extend beyond gaining general information to applying specific knowledge. A reader may use the information on color changes in a chemical solution to perform titration in the assay laboratory. Or a reader may use the information on the size of natural gas reserves of the overthrust belt to decide the feasibility of building a 24-inch pipeline from Wyoming to California.

What Attitude Should the Reader Have after Reading?

Earlier the Worksheet asked about the reader's attitude before reading. Here the Worksheet asks about the attitude the writer and manager want the reader to have *after* reading the document. Do they want the reader to believe and feel that the proposed SDI weapons system is an effective deterrent to foreign aggression? Or do they want the reader to believe that it is a horrendously expensive and impractical pork-barrel boondoggle? Though such attitudes in technical documents may be based on hard data, they are nonetheless attitudes—sometimes emotionally charged attitudes—and the writer must realize that shaping those attitudes is sometimes part of the job, even a moral responsibility. The technical writer uses different tools for shaping attitudes than the politician or advertiser does, and may have a higher regard for truth, but like the politician or advertiser, the technical writer may be an attitude shaper.

And one other attitude deserves consideration: the attitude that the reader should have toward the preparer of the document. The writer wants the reader to think that the document was prepared by a credible and meticulous professional who cares about the subject, the needs of the reader, and the needs of his or her employer. Though this point may seem obvious, writers do not always automatically consider it.

How Will the Reader Access the Information?

We read novels, murder mysteries, and perhaps newspaper editorials beginning-to-end, but almost everything else, we read in some other fashion. We do not read dictionaries, phone books, repair manuals, computer documentation, or encyclopedias beginning-to-end. We often read textbooks and journal articles in a nonlinear fashion too. Sometimes the text is well designed to help us get the critical information quickly. Or, exasperatingly, the text may bury critical information in the middle of a full-page paragraph. Whatever way the text is designed, readers scan, or skip around trying to find what they want without reading every blessed word.

Realizing what information the reader wants, the writer or manager can design documents so that the information is easy to find. Information accessibility relies on such devices as

- Headings
- Outlines
- Indexes
- Graphics
- Underlining
- Sidebars
- Glosses
- Varied typefaces
- Cross-references
- Bulleted lists (like this).

USA Today uses many of these devices. The *New York Times* uses fewer. *USA Today* lacks the substance of the *Times,* but it is easier to get the news from it quickly. A similar comparison could be made between *Popular Science* and *Scientific American.* The differences in format reflect how the writers and editors anticipate the publications will be read.

The writer and manager must carefully consider whether the reader will read beginning-to-end or access pieces of data individually. The basic consideration should be the importance of the information being conveyed. The more important the information, the more accessible it should be. The document manager must help the writer anticipate *all* the ways that *all the readers* will want to access the information and require the writer to provide the machinery that will allow that access. Most writers of technical and scientific documents should pay much more attention to accessibility. Few readers of technical documents read them beginning-to-end.

Writer's Purpose

The writer's and reader's purposes for a document differ just as the buyer's and seller's purposes differ. Both have their agendas. The writer—or the manager—may want to have a project approved. The target reader may want to know if the project is feasible.

The writer probably has potential salary, prestige, and promotion purposes for writing, but may also be writing because the task is challenging or intellectually interesting. The writer and manager should decide in advance whether the project is routine and can be done rather perfunctorily, or whether it will affect the safety of the user, the prosperity of the company, the security of the nation, the preservation of the environment—or the continued employment of the writer. These questions help the writer and manager determine the priorities of the task.

Logistics

The logistics are the nuts and bolts things—such as where the information comes from, the size of the projected document, and what form the document will take. These too must be considered in advance.

Sources of Information

Will the document be based on observation, personal opinion, government reports, a literature search, engineers' notes, customer interviews, public opinion polls, laboratory tests, compilations of vendors' brochures, or what? Frequently, lead time may be required, so early consideration of sources is wise to allow tests to be run, surveys to be conducted, or publications to be ordered.

Size of Document Expected

A two-page memo will not do if a twenty-page report is expected, and a twenty-page report will not do if a two-page memo is expected. Again, the question should be resolved between the writer and the manager before the writing begins.

Form or Medium Prescribed or Desirable

The form or medium of the eventual document should also be considered in advance. Should the document follow the company style sheet? Would a film, a wall chart, a wallet card, or a video be a more useful medium? *Such choices should be made according to subject, purpose, reader (user), and cost.*

Graphic Aids Available

Would photographs or art work be useful? Are facilities available to prepare them? Is there time? Are funds available? Again, these must be considered early.

Means of Document Production

The means of production of the document also need to be considered early. Will the document be laid out on a Macintosh with PageMaker software? Will it be photocopied or multilithed? How will the means of production affect the legibility and credibility of the document, and how much will it cost?

Outline

Even at the early planning stages, the writer and manager should consider the topics to be treated, their order, and their proportion of the expected document. Such an outline can, of course, be revised during preparation.

Distribution and Disposition

How many copies will be needed? Where do they go? What happens to them? The manager and writer should consider these questions—and their many consequences—in advance. Will the document be read by the competition? Will it come back to haunt everybody ten years later? Will it be subpoenaed in court? The wise consider such things.

PROJECT PLANNING CONFERENCE

The manager and the writer must come to an agreement about a document's audience, purpose, and scope, and the logistics of its development, production, and distribution. Such negotiation is most efficient in a planning conference. Usually the manager calls the conference, but the negotiation can take a variety of forms, depending on the situation and the experience of the writers.

With an inexperienced writer, the manager may say, "Willoughby, I want you to write a proposal for my signature recommending adoption of the hyperbaric procedure. It should be addressed to Allardyce, the plant manger. He is a chemical engineer with twenty years' experience on the job. But it will also be read by Sung. She is the company comptroller and has ultimate approval power on expenditures. She is pretty conservative, but her attention to cash-flow probably saved the company during the last recession."

With a more experienced writer, the manager can more inductively and democratically ask, "Kim, we need a brochure about our prefabricated forms for concrete. What ways do you see to make a brochure that will meet the needs of all of its readers?" Obviously, here the manager respects the writer's knowledge and expects useful suggestions.

Whatever the method, the manager and writer should agree upon the answers, write them down—Yes, and sign off—each keeping a copy. Such a signed-off Worksheet is then a kind of contract agreed to by both parties, and both can feel more secure with it. The manager is now more secure knowing that the assignment is clearly understood, and the writer is more secure knowing what is expected.

POSTCONFERENCE MANAGEMENT

Obviously other factors may arise during the writing. The manager may receive new data, or the writer may see new ways to deal with the problems. Depending on their impact on the document, such things may require additional conferences and a renegotiating of the specifications.

As needed, the manager may discreetly ask about the progress of the document, or may even ask to see sections in draft. However, if the initial assignment has been done carefully, a responsible and skilled writer should be able to produce a sound document with little intermediate prodding.

The writer properly attaches the Project Worksheet to the draft form of the document when submitting it. Then the manager checks to see if it fulfills the assignment. If it does, the manager approves it for production.

Quite often, however, problems appear. Of these, failures to match the assignment are the most obvious, and the manager can ask for a rewrite to make the document match the specifications. The most common problem seems to be a failure to consider adequately the technical levels of all the readers.

Or it may become clear that the specifications need changing. If for good reasons they do need changing, the manager can change them and ask that the new specifications be followed, but the manager who has agreed to the original specifications is less likely to make such changes capriciously or arbitrarily. If lessening the power to make arbitrary changes seems to take the fun out of being boss, then the manager should recognize that the resultant empowerment of the writer is not only enlightened and trendy, but also pragmatically effective.

OTHER APPLICATIONS

If the writer's manager does not make such detailed assignments, the writer obviously can still use the Project Worksheet for planning. Or the writer can use the Worksheet to initiate a conference and negotiate with the manager. We have found that managers are often impressed when writers request such conferences and demonstrate through their questions—the Worksheet questions—that they have a clear understanding of the situation.

The Worksheet is also effective in negotiating contracts in freelance writing assignments, or as an assignment device between editor and writer. In the university, a professor can make writing assignments following the Worksheet questions. Or a sharp student can use the questions to clarify a vaguely given assignment. And many problems of master's theses and doctoral dissertations would be avoided if candidates and supervising professors negotiated the specifications for theses and dissertations through the Worksheet.

In much technical writing, the most important work is the planning done before the first words of the document are written. If managers will use the Project Worksheet approach to document management, their writers will work more efficiently and produce better documents.

SELECTED READINGS

Anderson, P. V. (ed.), *Teaching Technical Writing: Teaching Audience Analysis and Adaptation*, ATTW Anthology No. 1, Association of Teachers of Technical Writing, St. Paul, Minnesota, 1980.

Caernarven-Smith, P., *Audience Analysis and Response*, Firman Technical Publications, Pembroke, Massachusetts, 1983.

Mathes, J. C., and D. Stevenson, *Designing Technical Reports: Writing for Audiences in Organizations* (2nd Edition), Macmillan, New York, 1991.

Pearsall, T., *Audience Analysis for Technical Writing*, Glencoe Press, Beverly Hills, California, 1969.

Souther, J. W., and M. L. White, *Technical Report Writing* (2nd Edition), John Wiley & Sons, New York, 1977.

Spilka, R., Orality and Literacy in the Workplace: Process- and Text-Based Strategies for Multiple-Audience Adaptation. *Journal of Business and Technical Communication*, 4:1, pp. 44–67, 1990.

Sheryl Lindsell-Roberts

Ten Ways to Make Your Technical Documents Shout "Read Me!"

A professional writer, Sheryl Lindsell-Roberts, is the author of 18 books on a variety of topics.

The computer does not substitute for judgment any more than a pencil substitutes for literacy.

—Robert McNamara, American politician and statesman

Whenever you pick up a technical document that someone else wrote, notice what you like and don't like about it. Let each document be your teacher. Following are guidelines to make you the poster child of technical documents. . . .

"SEE" YOUR TARGET SO YOU KNOW WHERE TO AIM

Reading requires a minimum of two people—a reader and a writer. (And you already know which one you are.) . . . If you don't know your readers or can't envision them, you can't write to them. Take a moment to close your eyes and ask yourself these questions:

- What do my readers look like?
- What do they do for fun?
- What kinds of cars do they drive?

When you envision your readers, put smiles on their faces. They're not a pack of growling faultfinders out to get you. Unless you tell your readers that the Internal Revenue Service wants to see them immediately, they're on your side and they want to read what you write. After all, you're sharing information of value.

CREATE STRUCTURE WITH BONES

No human being stands straight without a skeleton—the bones that provide structure. Writing also needs structure. Technical writing's structure comes in the form of an outline. The outline may be the traditional one you learned in school (with roman numerals, uppercase and lowercase letters, and all those indentations), or it may be an annotated table of contents, such as the one you see at the beginning of . . . [a] book. Whichever method works for you, you must provide a framework from which to begin writing. . . .

ADD MEAT TO THE BONES

After you have a solid outline, writing the draft is like adding meat to the bones or filling in the blanks. . . . Start with one section—the one that's easiest for you to write. Your readers will never know where you started. Then proceed to the second easiest, and so forth. Create a comfortable environment, get all your stuff together, set attainable goals, and stay focused.

Don't procrastinate by saying that you have to rearrange your sock drawer or take your cat in for a whisker-ectomy. If you can't get started immediately, try freewriting. *Freewriting* is a warm-up exercise for your mind, just as stretching is a warm-up exercise for joggers. Sit at your computer and start writing . . . anything. Eventually, you start to come around to your topic. Some of my best ideas often come from freewriting.

MAKE THE BONES VISUALLY APPEALING

You must give your document visual appeal or no one will read it . . . :

- Create headings and subheadings so that readers can find information at a glance. These serve as mileposts and help readers skim a document for valuable information just as they do a newspaper.
- Use bulleted and numbered lists, when appropriate.
- Limit paragraphs to no more than 8 lines and sentences to no more than 25 words.
- Include tables, figures, charts, and photos, when appropriate.
- Use color judiciously.
- Include navigational aids in your electronic documents. Web-savvy readers expect to find these elements.

HONE THE TONE

Certain types of documents often dictate the tone of the writing. Stereotypically speaking, for example, a directory is boooooring. An academic paper is pompous. Advertising copy is hype. Technical documents are mind boggling. Prove the expectations wrong for technical documents: Write documents so they're clear, concise, and geared for the level of your reader. When your writing isn't clear, it's because your thoughts aren't clear . . . :

- Say what you need to say in as few words as possible.
- Create a positive tone with positive words. (That may be as simple as stating what to do, rather than what not to do.)
- Use the active voice to bring life to your documents.
- Choose words that are appropriate for your readers.
- Use humor and jargon sparingly (if at all).

Even if you determine that your readers don't have the intelligence to pass an inkblot test, you must still use a tone that shows respect.

PROOFREAD 'TIL YOUR EYES POP OUT

. . . It's important to proofread until your eyes pop out so you don't wind up with errors such as the ones that follow. The writers of these remarks were either numskulls or poor proofreaders, but in either case, they lost their credibility. Proofread carefully so that you don't lose your credibility. The magic is in the minutiae.

- Charles Darwin was a naturalist who wrote the Organ of the Spices.
- It is a well-known fact that a deceased body warps the mind.
- For snakebite: Bleed the wound and rape the victim in a blanket for shock.
- To collect sulphur, hold a deacon over a flame in a test tube.
- Three kinds of blood vessels are arteries, veins, and caterpillars.
- The thermometer is an instrument for raising temperance.
- Heredity means that if your grandfather didn't have children, then your father probably didn't have any. So you probably won't have any.
- Algebra was the wife of Euclid.
- To be a nurse, you must be absolutely sterile.
- Artificial insemination is when the farmer does it to the cow, not the bull.

GIVE THE DOCUMENT THE LITMUS TEST

Always get a second opinion. When you write a short document, ask a co-worker or some other reader to give it a sanity check. Is the message clear? Is the information well organized? For a large document, go through a peer review after the first draft is complete. This is the time to get everyone's input on the general content, formatting, and any other issues. If you can, assemble all the reviewers in the same room. If not, gather as many as you can.

Two heads may be better than one, but two egos are worse. Typically, everyone who reviews a document feels compelled to comment. This compulsion is just human nature and doesn't necessarily reflect the quality of your work. Never take comments from others personally, or you may wind up selling pencils on the street corner.

FOR E-DOCS ONLY

Readers can't flip through electronic documents as they can through paper documents. Following are ways to make your electronic documents shout "Read me!"

Make the Home Page Revealing

Prepare the home page so that when readers view it they know what the document is about and where to find information. (This may be as simple as creating hyperlinks or buttons.) In technical terms, this is called "before the scroll." That's a term derived from the newspaper language "above the fold," which is what you see when the newspaper is folded and sitting on the rack.

Create Summaries

In addition to hyperlinks and buttons, consider writing brief summaries—a hyperlink followed by a line or two explaining what the readers may expect to find. This is similar to a descriptive abstract that gives the highlights of an article in a sentence or two. . . .

Create Links That Are Easy to Follow

When you write an electronic document, minimize the number of links the readers must follow. You don't want readers paging through too many links to get to the "pot of gold," and you don't want them missing valuable information along the way.

When you label links clearly and specifically, you tell readers what they'll find so that they can quickly decide whether to follow the link. Notice how a key piece of information turns the weak link that follows into an informative link.

Weak link: Conclusions
Informative link: Conclusions: Omica has a 95 percent success rate

Part 2

Problems with Language

In Lewis Carroll's *Through the Looking Glass,* Humpty Dumpty and Alice have the following exchange:

> "When *I* use a word," Humpty Dumpty said, in a rather scornful tone, "it means just what I choose it to mean—nothing more or less."
>
> "The question is," said Alice, "whether you *can* make words mean so many different things."

After reading some examples of business and technical writing, it is easy to suspect that Humpty Dumpty has been promoted to project manager—or even company president.

It is not always clear why business and technical writers get themselves into the problems with language that they do. At times, they mistakenly assume their readers will know what they are talking about, or that their goal is to impress rather than inform their readers. At other times, they simply imitate their superiors or model their documents after those in the files, as if the filing cabinet were a repository for sacred writings handed down from on high.

To solve problems with language, business and technical writers must remember their readers and then be guided by two principles:

- they should write everything as clearly, straightforwardly, and simply as possible; and
- they should write in a manner or style with which they would be comfortable speaking.

These two principles will help writers avoid problems with jargon, gobbledy-gook, legalese, and sexist and biased language. They will also serve as a guide in selecting the appropriate style for different writing situations.

Jargon is simply technical language unique to a profession or occupation. It only becomes a problem when writers use it in writing to an audience without the necessary background or training to understand it. A systems engineer writing to another systems engineer could rightly assume that the following passage would make sense to his or her reader:

> If you are using a CONFIG.SYS file to modify version 2.0 of the Personal Computer DOS, you may want to create a CONFIG.SYS file to use with the program instead of the AUTOEXEC.BAT file.

Unfortunately, this passage comes from an introductory computer manual.

Jargon in the extreme becomes gobbledygook, mindless gibberish akin to double-talk and characterized by pretentiousness. For example, the host of a popular television game show tells contestants who are losing that they are in "a deficit situation"; ordinary No. 2 pencils become, on the requisition forms of some bureaucrats, "writing implements, standard issue"; politicians wary of angering voters call tax increases "revenue enhancements" and "misspeak themselves" rather than lie when they are caught making false claims or leveling inaccurate charges against their opponents; "turn out the lights when you leave" becomes "illumination is required to be extinguished upon vacation of the premises"; the threat that "anyone caught smoking on the premises will be dealt with accordingly" may at first cause someone to think twice before lighting up, but, in the final analysis, we may well wonder just what the threat is really supposed to mean.

Gobbledygook also includes words and phrases that writers mistakenly think sound the way business and technical writing should sound. They may be in vogue or out of date, but either way, they have unfortunately come to be viewed by some writers as examples of good professional writing practice. People no longer meet for discussions—they "interface"—and every issue about which they interface has "parameters," "viable" and otherwise. Some of these same people see nothing wrong with beginning their letters as follows:

> Pursuant to yours of the twelfth, enclosed herewith please find our check in the amount of $24.95.

Subsequently in the same letter, they "are pleased to advise that," and they go on to refer to data "hereunder discussed" or "above referenced."

Legalese is an overreliance on legal terminology—or legal *sounding* terminology—when Plain English will serve the reader and the writer as well.

Sexist or biased language is linguistic discrimination usually, although not exclusively, against women and groups that are perceived to be minorities. Just as all managers are not men, so all secretaries are not women. Common sense, sensitivity to audience and language, and fairness can help writers avoid sexism and bias.

   The five essays in this section of *Strategies* address further the issues raised in these introductory comments. In a now classic piece, Stuart Chase sends up gobbledygook, the linguistic overkill that is all too common in the writing of government officials, lawyers, and, sadly, even some academics. William Zinsser follows with an essay that suggests that it is possible to write for the world of work and still sound like a human being. William Lutz next discusses the dangers of doublespeak. Alan Siegel follows with a brief history of the Plain English movement in his account of the work his company has done in helping clients reduce legalese in the documents they produce. Guidelines developed at the University of Wisconsin offer ways to avoid language that stereotypes or denigrates both men and women.

Stuart Chase

Gobbledygook

Stuart Chase worked for many years as a consultant to various government agencies; his other books include The Tyranny of Words *(1938) and* Democracy Under Pressure *(1945).*

Said Franklin Roosevelt, in one of his early presidential speeches: "I see one-third of a nation ill-housed, ill-clad, ill-nourished." Translated into standard bureaucratic prose his statement would read:

> It is evident that a substantial number of persons within the Continental boundaries of the United States have inadequate financial resources with which to purchase the products of agricultural communities and industrial establishments. It would appear that for a considerable segment of the population, possibly as much as 33.3333* of the total, there are inadequate housing facilities, and an equally significant proportion is deprived of the proper types of clothing and nutriment.
> *Not carried beyond four places.

This rousing satire on gobbledygook—or talk among the bureaucrats—is adapted from a report[1] prepared by the Federal Security Agency in an attempt to break out of the verbal squirrel cage. "Gobbledygook" was coined by an exasperated Congressman, Maury Maverick of Texas, and means using two, or three, or ten words in the place of one, or using a five-syllable word where a single syllable would suffice. Maverick was censuring the forbidding prose of executive departments in Washington, but the term has now spread to windy and pretentious language in general.

"Gobbledygook" itself is a good example of the way a language grows. There was no word for the event before Maverick's invention; one had to say: "You know, that terrible, involved, polysyllabic language those government people use down in Washington." Now one word takes the place of a dozen.

---

[1]This and succeeding quotations from FSA report by special permission of the author, Milton Hall.

A British member of Parliament, A. P. Herbert, also exasperated with bureaucratic jargon, translated Nelson's immortal phrase, "England expects every man to do his duty":

> England anticipates that, as regards the current emergency, personnel will face up to the issues, and exercise appropriately the functions allocated to their respective occupational groups.

A New Zealand official made the following report after surveying a plot of ground for an athletic field:[2]

> It is obvious from the difference in elevation with relation to the short depth of the property that the contour is such as to preclude any reasonable developmental potential for active recreation.

Seems the plot was too steep.

An office manager sent this memo to his chief:

> Verbal contact with Mr. Blank regarding the attached notification of promotion has elicited the attached representation intimating that he prefers to decline the assignment.

Seems Mr. Blank didn't want the job.

> A doctor testified at an English trial that one of the parties was suffering from "circum-morbital haematoma."

Seems the party had a black eye.

> In August 1952 the U.S. Department of Agriculture put out a pamphlet entitled: "Cultural and Pathogenic Variability in Single-Condial and Hyphaltip Isolates of Hemlin-Thosporium Turcicum Pass."

Seems it was about corn leaf disease.

On reaching the top of the Finsteraarhorn in 1845, M. Dollfus-Ausset, when he got his breath, exclaimed:

> The soul communes in the infinite with those icy peaks which seem to have their roots in the bowels of eternity.

Seems he enjoyed the view.

A government department announced:

> Voucherable expenditures necessary to provide adequate dental treatment required as adjunct to medical treatment being rendered a pay patient in in-patient status may be incurred as required at the expense of the Public Health Service.

Seems you can charge your dentist bill to the Public Health Service. Or can you?

---

[2]This item and the next two are from the piece on gobbledygook by W. E. Farbstein, *New York Times*, March 29, 1953.

LEGAL TALK

Gobbledygook not only flourishes in government bureaus but grows wild and lush in the law, the universities, and sometimes among the literati. Mr. Micawber was a master of gobbledygook, which he hoped would improve his fortunes. It is almost always found in offices too big for face-to-face talk. Gobbledygook can be defined as squandering words, packing a message with excess baggage and so introducing semantic "noise." Or it can be scrambling words in a message so that meaning does not come through. The directions on cans, bottles, and packages for putting the contents to use are often a good illustration. Gobbledygook must not be confused with double talk, however, for the intentions of the sender are usually honest.

I offer you a round fruit and say, "Have an orange." Not so an expert in legal phraseology, as parodied by editors of *Labor:*

> I hereby give and convey to you, all and singular, my estate and interests, right, title, claim and advantages of and in said orange, together with all rind, juice, pulp and pits, and all rights and advantages therein . . . anything hereinbefore or hereinafter or in any other deed or deeds, instrument or instruments of whatever nature or kind whatsoever, to the contrary, in any wise, notwithstanding.

The state of Ohio, after five years of work, has redrafted its legal code in modern English, eliminating 4,500 sections and doubtless a blizzard of "where-ases" and "hereinafters." Legal terms of necessity must be closely tied to their referents, but the early solons tried to do this the hard way, by adding synonyms. They hoped to trap the physical event in a net of words, but instead they created a mumbo-jumbo beyond the power of the layman, and even many a lawyer, to translate. Legal talk is studded with tautologies, such as "cease and desist," "give and convey," "irrelevant, incompetent, and immaterial." Furthermore, legal jargon is a dead language; it is not spoken and it is not growing. An official of one of the big insurance companies calls their branch of it "bafflegab." Here is a sample from his collection.[3]

> One-half to his mother, if living, if not to his father, and one-half to his mother-in-law, if living, if not to his mother, if living, if not to his father. Thereafter payment is to be made in a single sum to his brothers. On the one-half payable to his mother, if living, if not to his father, he does not bring in his mother-in-law as the next payee to receive, although on the one-half to his mother-in-law, he does bring in the mother or father.

You apply for an insurance policy, pass the tests, and instead of a straightforward "here is your policy," you receive something like this:

> This policy is issued in consideration of the application therefor, copy of which application is attached hereto and made part hereof, and of the payment for said insurance on the life of the above-named insured.

---

[3]Interview with Clifford B. Reeves by Sylvia F. Porter, *New York Evening Post*, March 14, 1952.

ACADEMIC TALK

The pedagogues may be less repetitious than the lawyers, but many use even longer words. It is a symbol of their calling to prefer Greek and Latin derivatives to Anglo-Saxon. Thus instead of saying: "I like short clear words," many a professor would think it more seemly to say: "I prefer an abbreviated phraseology, distinguished for its lucidity." Your professor is sometimes right, the longer word may carry the meaning better—but not because it is long. Allen Upward in his book *The New Word* warmly advocates Anglo-Saxon English as against what he calls "Mediterranean" English, with its polysyllables built up like a skyscraper.

Professional pedagogy, still alternating between the Middle Ages and modern science, can produce what Henshaw Ward once called the most repellent prose known to man. It takes an iron will to read as much as a page of it. Here is a sample of what is known in some quarters as "pedageese":

> Realization has grown that the curriculum or the experiences of learners change and improve only as those who are most directly involved examine their goals, improve their understandings and increase their skill in performing the tasks necessary to reach newly defined goals. This places the focus upon teacher, lay citizen and learner as partners in curricular improvement and as the individuals who must change, if there is to be curriculum change.

I think there is an idea concealed here somewhere. I think it means: "If we are going to change the curriculum, teacher, parent, and student must all help." The reader is invited to get out his semantic decoder and check on my translation. Observe there is no technical language in this gem of pedageese, beyond possibly the word "curriculum." It is just a simple idea heavily ververbalized.

In another kind of academic talk the author may display his learning to conceal a lack of ideas. A bright instructor, for instance, in need of prestige may select a common sense proposition for the subject of a learned monograph—say, "Modern cities are hard to live in"—and adorn it with imposing polysyllables: "Urban existence in the perpendicular declivities of megalopolis . . ." et cetera. He coins some new terms to transfix the reader—"mega-decibel" or "stratocosmopolis"—and works them vigorously. He is careful to add a page or two of differential equations to show the "scatter." And then he publishes, with 147 footnotes and a bibliography to knock your eye out. If the authorities are dozing, it can be worth an associate professorship.

While we are on the campus, however, we must not forget that the technical language of the natural sciences and some terms in the social sciences, forbidding as they may sound to the layman, are quite necessary. Without them, specialists could not communicate what they find. Trouble arises when experts expect the uninitiated to understand the words; when they tell the jury, for instance, that the defendant is suffering from "circumorbital haematoma."

Here are two authentic quotations. Which was written by a distinguished modern author, and which by a patient in a mental hospital? You will find the answer at the end of the [selection].

1. Have just been to supper. Did not knowing what the woodchuck sent me here. How when the blue blue blue on the said anyone can do it that tries. Such is the presidential candidate.
2. No history of a family to close with those and close. Never shall he be alone to be alone to be alone to be alone to be alone to lend a hand and leave it left and wasted.

REDUCING THE GOBBLE

As government and business offices grow larger, the need for doing something about gobbledygook increases. Fortunately the biggest office in the world is working hard to reduce it. The Federal Security Agency in Washington,[4] with nearly 100 million clients on its books, began analyzing its communication lines some years ago, with gratifying results. Surveys find trouble in three main areas: correspondence with clients about their social security problems, office memos, official reports.

Clarity and brevity, as well as common humanity, are urgently needed in this vast establishment which deals with disability, old age, and unemployment. The surveys found instead many cases of long-windedness, foggy meanings, clichés, and singsong phrases, and gross neglect of the reader's point of view. Rather than talking to a real person, the writer was talking to himself. "We often write like a man walking on stilts."

Here is a typical case of long-windedness:

> *Gobbledygook as found:* "We are wondering if sufficient time has passed so that you are in a position to indicate whether favorable action may now be taken on our recommendation for the reclassification of Mrs. Blank, junior clerk-stenographer, CAF 2, to assistant clerk-stenographer, CAF 3?"
>
> *Suggested improvement:* "Have you yet been able to act on our recommendation to reclassify Mrs. Blank?"

Another case:

> Although the Central Efficiency Rating Committee recognizes that there are many desirable changes that could be made in the present efficiency rating system in order to make it more realistic and more workable than it now is, this committee is of the opinion that no further change should be made in the present system during the current year. Because of conditions prevailing throughout the country and the resultant turnover in personnel, and difficulty in administering the Federal programs, further mechanical improvement in the present rating system would require staff retraining and other administrative expense which would seem best withheld until the official termination of hostilities, and until restoration of regular operations.

The FSA invites us to squeeze the gobbledygook out of this statement. Here is my attempt:

---

[4]Now the Department of Health and Human Services.

> The Central Efficiency Rating Committee recognizes that desirable changes could be made in the present system. We believe, however, that no change should be attempted until the war is over.

This cuts the statement from 111 to 30 words, about one-quarter of the original, but perhaps the reader can do still better. What of importance have I left out?

Sometimes in a book which I am reading for information—not for literary pleasure—I run a pencil through the surplus words. Often I can cut a section to half its length with an improvement in clarity. Magazines like *The Reader's Digest* have reduced this process to an art. Are long-windedness and obscurity a cultural lag from the days when writing was reserved for priests and cloistered scholars? The more words and the deeper the mystery, the greater their prestige and the firmer the hold on their jobs. And the better the candidates's chance today to have his doctoral thesis accepted.

The FSA surveys found that a great deal of writing was obscure although not necessarily prolix. Here is a letter sent to more than 100,000 inquirers, a classic example of murky prose. To clarify it, one needs to *add* words, not cut them:

> In order to be fully insured, an individual must have earned $50 or more in covered employment for as many quarters of coverage as half the calendar quarters elapsing between 1936 and the quarter in which he reaches age 65 or dies, whichever first occurs.

Probably no one without the technical jargon of the office could translate this: nevertheless, it was sent out to drive clients mad for seven years. One poor fellow wrote back: "I am no longer in covered employment. I have an outside job now."

Many words and phrases in officialese seem to come out automatically, as if from lower centers of the brain. In this standardized prose people never *get* jobs, they "secure employment"; *before* and *after* become "prior to" and "subsequent to"; one does not *do,* one "performs"; nobody *knows* a thing, he is "fully cognizant"; one never *says,* he "indicates." A great favorite at present is "implement."

Some charming boners occur in this talking-in-one's-sleep. For instance:

> The problem of extending coverage to all employees, regardless of size, is not as simple as surface appearances indicate.
> Though the proportions of all males and females in ages 16–45 are essentially the same . . .
> Dairy cattle, usually and commonly embraced in dairying . . .

In its manual to employees, the FSA suggests the following:

Instead of	**Use**
give consideration to	consider
make inquiry regarding	inquire
is of the opinion	believes
comes into conflict with	conflicts
information which is of a confidential nature	confidential information

Professional or office gobbledygook often arises from using the passive rather than the active voice. Instead of looking you in the eye, as it were, and writing "This act requires . . . ," the office worker looks out of the window and writes: "It is required by this statute that . . ." When the bureau chief says, "We expect Congress to cut your budget," the message is only too clear; but usually he says, "It is expected that the departmental budget estimates will be reduced by Congress."

> GOBBLED: "All letters prepared for the signature of the Administrator will be single spaced."
> UNGOBBLED: "Single space all letters for the Administrator." (Thus cutting 13 words to 7.)

Only People Can Read

The FSA surveys pick up the point . . . that human communication involves a listener as well as a speaker. Only people can read, though a lot of writing seems to be addressed to beings in outer space. To whom are you talking? The sender of the officialese message often forgets the chap on the other end of the line.

A woman with two small children wrote the FSA asking what she should do about payments, as her husband had lost his memory. "If he never gets able to work," she said, "and stays in an institution would I be able to draw any benefits? . . . I don't know how I am going to live and raise my children since he is disable to work. Please give me some information. . . ."

To this human appeal, she received a shattering blast of gobbledygook, beginning, "State unemployment compensation laws do not provide any benefits for sick or disabled individuals . . . in order to qualify an individual must have a certain number of quarters of coverage . . ." et cetera, et cetera. Certainly if the writer had been thinking about the poor woman he would not have dragged in unessential material about old-age insurance. If he had pictured a mother without means to care for her children, he would have told her where she might get help—from the local office which handles aid to dependent children, for instance.

Gobbledygook of this kind would largely evaporate if we thought of our messages as two way—in the above case, if we pictured ourselves talking on the doorstep of a shabby house to a woman with two children tugging at her skirts, who in her distress does not know which way to turn.

Results of the Survey

The FSA survey showed that office documents could be cut 20 to 50 percent, with an improvement in clarity and a great saving to taxpayers in paper and payrolls.

A handbook was prepared and distributed to key officials.[5] They read it, thought about it, and presently began calling section meetings to discuss gob-

---

[5]By Milton Hall.

bledygook. More booklets were ordered, and the local output of documents began to improve. A Correspondence Review Section was established as a kind of laboratory to test murky messages. A supervisor could send up samples for analysis and suggestions. The handbook is now used for training new members; and many employees keep it on their desks along with the dictionary. Outside the Bureau some 25,000 copies have been sold (at 20 cents each) to individuals, governments, business firms, all over the world. It is now used officially in the Veterans Administration and in the Department of Agriculture.

The handbook makes clear the enormous amount of gobbledygook which automatically spreads in any large office, together with ways and means to keep it under control. I would guess that at least half of all the words circulating around the bureaus of the world are "irrelevant, incompetent, and immaterial"—to use a favorite legalism; or are just plain "unnecessary"—to ungobble it.

My favorite story of removing the gobble from gobbledygook concerns the Bureau of Standards at Washington. I have told it before but perhaps the reader will forgive the repetition. A New York plumber wrote the Bureau that he had found hydrochloric acid fine for cleaning drains, and was it harmless? Washington replied: "The efficacy of hydrochloric acid is indisputable, but the chlorine residue is incompatible with metallic permanence."

The plumber wrote back that he was mighty glad the Bureau agreed with him. The Bureau replied with a note of alarm: "We cannot assume responsibility for the production of toxic and noxious residues with hydrochloric acid, and suggest that you use an alternate procedure." The plumber was happy to learn that the Bureau still agreed with him.

Whereupon Washington exploded: "Don't use hydrochloric acid; it eats hell out of the pipes!"[6]

---

[6] *Note:* The second quotation on page 60 comes from Gertrude Stein's *Lucy Church Amiably.*

William Zinsser

Writing in Your Job

William Zinsser is a noted author, columnist, editor, teacher of writing, and critic.

The memo, the business letter, the administrative report, the financial analysis, the marketing proposal, the note to the boss, the fax, the Post-it—all the pieces of paper that circulate through your office every day are forms of writing. Take them seriously. Countless careers rise or fall on the ability or inability of employees to state a set of facts, summarize a meeting or present an idea coherently.

Most people work for institutions: businesses, banks, insurance firms, law firms, government agencies, school systems, nonprofit organizations and other entities. Many of them are managers whose writing goes out to the public: the president addressing the stockholders, the banker explaining a change in procedure, the school principal writing a newsletter to parents. Whoever they are, they are often so afraid of writing that their sentences lack all humanity—and so do their institutions. It's hard to imagine that these are real places where real men and women come to work every morning.

But just because people work for an institution they don't have to write like one. Institutions can be warmed up. Administrators can be turned into human beings. Information can be imparted clearly and without pomposity. It's a question of remembering that readers identify with people, not with abstractions like "profitability," or with Latinate nouns like "utilization" and "implementation," or with inert constructions in which nobody can be visualized doing something ("pre-feasibility studies are in the paperwork stage").

Nobody has made the point better than George Orwell in his translation into modern bureaucratic fuzz of this famous verse from Ecclesiastes:

> I returned and saw under the sun that the race is not to the swift, nor the battle to the strong, neither yet bread to the wise, nor yet riches to men of understanding, nor yet favor to men of skill; but time and chance happeneth to them all.

Orwell's version goes:

> Objective consideration of contemporary phenomena compels the conclusion that success or failure in competitive activities exhibits no tendency to be commensurate with innate capacity, but that a considerable element of the unpredictable must invariably be taken into account.

First notice how the two passages look. The one at the top invites us to read it. The words are short and have air around them; they convey the rhythms of human speech. The second one is clotted with long words. It tells us instantly that a ponderous mind is at work. We don't want to go anywhere with a mind that expresses itself in such suffocating language. We don't even start to read.

Also notice what the two passages say. Gone from the second one are the short words and vivid images from everyday life—the race and the battle, the bread and the riches—and in their place have waddled the long and flabby nouns of generalized meaning. Gone is any sense of what one person did ("I returned") or what he realized ("saw") about one of life's central mysteries: the capriciousness of fate.

Let me illustrate how this disease infects the writing that most people do in their jobs. I'll use school principals as my first example, not because they are the worst offenders (they aren't) but because I happen to have such an example. My points are intended for all the men and women who work in all the organizations where language has lost its humanity and nobody knows what the people in charge are talking about.

My encounter with the principals began when I got a call from Ernest B. Fleishman, superintendent of schools in Greenwich, Connecticut. "We'd like you to come and 'dejargonize' us," he said. "We don't think we can teach students to write unless all of us at the top of the school system clean up our own writing." He said he would send me some typical materials that had originated within the system. His idea was for me to analyze the writing and then conduct a workshop.

What appealed to me was the willingness of Dr. Fleishman and his colleagues to make themselves vulnerable. Vulnerability has a strength of its own. We decided on a date, and soon a fat envelope arrived. It contained various internal memos and mimeographed newsletters that had been mailed to parents from the 16 elementary, junior, and senior high schools.

The newsletters had a cheery and informal look. Obviously the system was making an effort to communicate warmly with its families. But even at first glance certain chilly phrases caught my eye—"prioritized evaluative procedures," "modified departmentalized schedule"—and one principal promised that his school would provide "enhanced positive learning environments." Just as obviously the system wasn't communicating as warmly as it thought it was.

I studied the principals' material and divided it into good and bad examples. On the appointed morning in Greenwich I found 40 principals and curriculum coordinators assembled and eager to learn. I told them I could only applaud them for submitting to a process that so threatened their identity. In the national clamor over why Johnny can't write, Dr. Fleishman was the first adult in my experience who admitted that youth has no monopoly on verbal sludge and that the problem must also be attacked at the top.

I told the principals that we want to think of the men and women who run our children's schools as people not unlike ourselves. We are suspicious of pretentiousness, of all the fad words that the social scientists have coined to avoid making themselves clear to ordinary mortals. I urged them to be natural. How we write and how we talk is how we define ourselves.

I asked them to listen to how they were defining themselves to the community. I had made copies of certain bad examples, changing the names of the schools and the principals. I explained that I would read some of the examples aloud. Later we would see if they could turn what they had written into plain English. This was my first example:

> Dear Parent:
> We have established a special phone communication system to provide additional opportunities for parent input. During this year we will give added emphasis to the goal of communication and utilize a variety of means to accomplish this goal. Your inputs, from the unique position as a parent, will help us to plan and implement an educational plan that meets the needs of your child. An open dialogue, feedback and sharing of information between parents and teachers will enable us to work with your child in the most effective manner.
>
> Dr. George B. Jones
> Principal

That's the kind of communication I don't want to receive, unique though my parent inputs might be. I'd like to be told that the school is going to make it easier for me to telephone the teachers and that they hope I'll call often to discuss how my children are getting along. Instead the parent gets junk: "special phone communication system," "added emphasis to the goal of communication," "plan and implement an educational plan." As for "open dialogue, feedback and sharing of information," they are three ways of saying the same thing.

Dr. Jones is clearly a man who means well, and his plan is one we all want: a chance to pick up the phone and tell the principal what a great kid Johnny is despite that unfortunate incident in the playground last Tuesday. But Dr. Jones doesn't sound like a person I want to call. In fact, he doesn't sound like a person. His message could have been tapped out by a computer. He is squandering a rich resource: himself.

Another example that I chose was a "Principal's Greeting" sent to parents at the start of the school year. It consisted of two paragraphs that were very different:

> Fundamentally, Foster is a good school. Pupils who require help in certain subjects or study skills areas are receiving special attention. In the school year ahead we seek to provide enhanced positive learning environments. Children, and staff, must work in an atmosphere that is conducive to learning. Wide varieties of instructional materials are needed. Careful attention to individual abilities and learning styles is required. Cooperation between school and home is extremely important to the learning process. All of us should be aware of desired educational objectives for every child.

Keep informed about what is planned for our children this year and let us know about your own questions and about any special needs your child may have. I have met many of you in the first few weeks. Please continue to stop in to introduce yourself or to talk about Foster. I look forward to a very productive year for all of us.

Dr. Ray B. Dawson
Principal

In the second paragraph I'm being greeted by a person; in the first I'm hearing from an educator. I like the real Dr. Dawson of Paragraph 2. He talks in warm and comfortable phrases: "Keep informed," "let us know," "I have met," "Please continue," "I look forward."

By contrast, Educator Dawson of Paragraph 1 never uses "I" or even suggests a sense of "I." He falls back on the jargon of his profession, where he feels safe, not stopping to notice that he really isn't telling the parent anything. What are "study skills areas" and how do they differ from "subjects"? What are "enhanced positive learning environments" and how do they differ from "an atmosphere that is conducive to learning"? What are "wide varieties of instructional materials": pencils, textbooks, filmstrips? What exactly are "learning styles"? What "educational objectives" are "desired," and who desires them?

The second paragraph, in short, is warm and personal; the other is pedantic and vague. This was a pattern I found repeatedly. Whenever the principals wrote to notify the parents of some human detail, they wrote with humanity:

It seems that traffic is beginning to pile up again in front of the school. If you can possibly do so, please come to the rear of the school for your child at the end of the day.

I would appreciate it if you would speak with your children about their behavior in the cafeteria. Many of you would be totally dismayed if you could observe the manners of your children while they are eating. Check occasionally to see if they owe money for lunch. Sometimes children are very slow in repaying.

But when the educators wrote to explain how they proposed to do their educating, they vanished without a trace:

In this document you will find the program goals and objectives that have been identified and prioritized. Evaluative procedures for the objectives were also established based on acceptable criteria.

Prior to the implementation of the above practice, students were given very little exposure to multiple choice questions. It is felt that the use of practice questions correlated to the unit that a student is presently studying has had an extremely positive effect as the test scores confirm.

After I had read various good and bad examples, the principals began to hear the difference between their true selves and their educator selves. The problem was how to close the gap. I recited my four articles of faith: clarity, simplicity, brevity and humanity. I explained about using active verbs and avoiding windy "concept nouns." I told them not to use the private vocabulary of education as a crutch; almost any subject can be made accessible in good English.

These were all basic tenets, but the principals wrote them down as if they had never heard them before—and maybe they hadn't, or at least not for many years. Perhaps that's why bureaucratic prose becomes so turgid, whatever the bureaucracy. Once an administrator rises to a certain level, nobody ever points out to him again the beauty of a simple declarative sentence or shows him how his writing has become swollen with pompous generalizations.

Finally our workshop got down to work. I distributed my copies and asked the principals to rewrite the more knotty sentences. It was a grim moment. They had met the enemy for the first time. They scribbled on their pads and scratched out what they had scribbled. Some didn't write anything. Some crumpled their paper. They began to look like writers. An awful silence hung over the room, broken only by the crossing out of sentences and the crumpling of paper. They began to sound like writers.

As the day went on, they slowly relaxed. They began to write in the first person and to use active verbs. For a while they still couldn't loosen their grip on long words and vague nouns ("parent communication response"). But gradually their sentences became human. When I asked them to tackle "Evaluative procedures for the objectives were also established based on acceptable criteria," one of them wrote: "At the end of the year we will evaluate our progress." Another wrote: "We will see how well we have succeeded."

That's the kind of plain talk a parent wants. It's also what stockholders want from their corporation, what customers want from their bank, what the widow wants from the agency that's handling her social security. There is a deep yearning for human contact and a resentment of bombast. Recently I got a "Dear Customer" letter from the company that supplies my computer needs. It began: "Effective March 30 we will be migrating our end user order entry and supplies referral processing to a new telemarketing center." I finally figured out that they had a new 800 number and that the end user was me. Any institution that won't take the trouble in its writing to be both clear and personal will lose friends, customers, and money. Let me put it another way for business executives: a shortfall will be experienced in anticipated profitability.

Here's an example of how organizations throw away their humanity with pretentious language. It's a "customer bulletin" distributed by a major corporation. The sole purpose of a customer bulletin is to give helpful information to a customer. This one begins: "Companies are increasingly turning to capacity planning techniques to determine when future processing loads will exceed processing capabilities." That sentence is no favor to the customer; it's congealed with Orwellian nouns like "capacity" and "capabilities" that convey no procedures that a customer can picture. What *are* capacity planning techniques? Whose capacity is being planned? By whom? The second sentence says: "Capacity planning adds objectivity to the decision-making process." More dead nouns. The third sentence says: "Management is given enhanced decision participation in key areas of information system resources."

The customer has to stop after every sentence and translate it. The bulletin might as well be in French. He starts with the first sentence—the one about capacity planning techniques. Translated, that means "It helps to know when you're giving your computer more than it can handle." The second sentence—"Capacity planning adds objectivity to the decision-making process"—means you should know the facts before you decide. The third sentence—the one about enhanced decision participation—means "The more you know about your system, the better it will work." It could also mean several other things.

But the customer isn't going to keep translating. Soon he's going to start looking for another company. He thinks, "If these guys are so smart, why can't they tell me what they do? Maybe they're *not* so smart." The bulletin goes on to say that "for future cost avoidance, productivity has been enhanced." That seems to mean that the product will be free: all costs have been avoided. Next the bulletin assures the customer that "the system is delivered with functionality." That means it works. I would hope so.

Finally, at the end, we get a glimmer of humanity. The writer of the bulletin asks a satisfied customer why he chose this system. The man says he chose it because of the company's reputation for service. He says: "A computer is like a sophisticated pencil. You don't care how it works, but if it breaks you want someone there to fix it." Notice how refreshing that sentence is after all the garbage that preceded it: in its language (comfortable words), in its details that we can visualize (the pencil), and above all in its humanity. The writer has taken the coldness out of a technical process by relating it to an experience we're all familiar with—waiting for the repairman when something breaks. I'm reminded of a sign I saw in the New York subway which proves that even a huge municipal bureaucracy can talk to its constituents humanely: "If you ride the subway regularly you may have seen signs directing you to trains you've never heard of before. These are only new names for very familiar trains."

Still, plain talk will not be easily achieved in corporate America. Too much vanity is on the line. Managers at every level are prisoners of the notion that a simple style reflects a simple mind. Actually a simple style is the result of hard work and hard thinking; a muddled style reflects a muddled thinker or a person too arrogant or too dumb or too lazy to organize his thoughts. Remember that what you write is often the only chance you'll get to present yourself to someone whose business or money or goodwill you want. If what you write is ornate or pompous or fuzzy, that's how you'll be perceived. The reader has no other choice.

I learned about corporate America by venturing out into it, after Greenwich, to conduct workshops for some major corporations, which also asked to be dejargonized. "We don't even understand our own memos anymore," they told me. I worked with the men and women who write the vast amounts of material these companies generate for internal and external consumption. The internal material consists of house organs and newsletters whose purpose is to tell employees what's happening at their "facility" and to give them a sense of belonging. The external material includes the glossy magazines and annual reports that go to stockholders, the speeches delivered by high executives, the releases sent to the press, and the consumer manuals that explain how the product works. I found almost all of it lacking in human juices and much of it impenetrable.

Typical of the sentences in the newsletters was this one:

Announced concurrently with the above enhancements were changes to the System Support Program, a program product which operates in conjunction with the NCP. Among the additional functional enhancements are dynamic reconfiguration and inter-systems communications.

There's no joy for the writer in such work, and certainly none for the reader. It's language out of *Star Trek*, and if I were an employee I wouldn't be cheered—or informed—by these efforts to raise my morale. I would soon stop reading them. I told the corporate writers that they had to find the people behind the fine achievements being described. "Go to the engineer who conceived the new system," I said, "or to the designer who designed it, or to the technician who assembled it, and get them to tell you in their own words how the idea came to them, or how they put it together, or how it will be used by real people out in the real world." The way to warm up any institution is to locate the missing "I." Remember: "I" is the most interesting element in any story.

The writers explained that they often did interview the engineer but couldn't get him to talk English. They showed me some typical quotes. The engineers spoke in an arcane language studded with acronyms ("Sub-system support is available only with VSAG or TNA"). I said that the writers had to keep going back to the engineer until he finally made himself intelligible. They said the engineer didn't *want* to be made intelligible: if he spoke too simply he would look like a jerk to his peers. I said that their responsibility was to the facts and to the reader, not to the vanity of the engineer. I urged them to believe in themselves as writers and not to relinquish control. They replied that this was easier said than done in hierarchical corporations, where approval of written reports is required at a succession of higher levels. I sensed an undercurrent of fear: Do things the company way and don't risk your job trying to make the company human.

High executives were equally victimized by wanting to sound important. One corporation had a monthly newsletter to enable "management" to share its concerns with middle managers and lower employees. Prominent in every issue was a message of exhortation from the division vice-president, whom I'll call Thomas Bell. Judging by his monthly message, he was a pompous ass, saying nothing and saying it in inflated verbiage.

When I mentioned this, the writers said that Thomas Bell was actually a diffident man and a good executive. They pointed out that he doesn't write the message himself; it's written for him. I said that Mr. Bell was being done a disservice—that the writers should go to him every month (with a tape recorder, if necessary) and stay there until he talked about his concerns in the same language he would use when he got home and talked to Mrs. Bell.

What I realized was that most executives in America don't write what appears over their signature or what they say in their speeches. They have surrendered the qualities that make them unique. If they and their institutions seem cold, it's because they acquiesce in the process of being pumped up and dried out. Preoccupied with their high technology, they forget that some of the most powerful tools they possess—for good and for bad—are words.

If you work for an institution, whatever your job, whatever your level, be yourself when you write. You will stand out as a real person among the robots, and your example might even persuade Thomas Bell to write his own stuff.

William Lutz

The World of Doublespeak

Chair of the Committee on Public Doublespeak of the National Council of Teachers of English, William Lutz was a member of the faculty of English at Rutgers University.

There are no potholes in the streets of Tucson, Arizona, just "pavement deficiencies." The Reagan Administration didn't propose any new taxes, just "revenue enhancement" through new "user's fees." Those aren't bums on the street, just "non-goal oriented members of society." There are no more poor people, just "fiscal underachievers." There was no robbery of an automatic teller machine, just an "unauthorized withdrawal." The patient didn't die because of medical malpractice, it was just a "diagnostic misadventure of a high magnitude." The U.S. Army doesn't kill the enemy anymore, it just "services the target." And the doublespeak goes on.

Doublespeak is language that pretends to communicate but really doesn't. It is language that makes the bad seem good, the negative appear positive, the unpleasant appear attractive or at least tolerable. Doublespeak is language that avoids or shifts responsibility, language that is at variance with its real or purported meaning. It is language that conceals or prevents thought; rather than extending thought, doublespeak limits it.

Doublespeak is not a matter of subjects and verbs agreeing; it is a matter of words and facts agreeing. Basic to doublespeak is incongruity, the incongruity between what is said or left unsaid, and what really is. It is the incongruity between the word and the referent, between seem and be, between the essential function of language—communication—and what doublespeak does—mislead, distort, deceive, inflate, circumvent, obfuscate.

HOW TO SPOT DOUBLESPEAK

How can you spot doublespeak? Most of the time you will recognize doublespeak when you see or hear it. But, if you have any doubts, you can identify doublespeak just by answering these questions: Who is saying what to whom, under what conditions and circumstances, with what intent, and with what results? Answering these questions will usually help you identify as doublespeak language that appears to be legitimate or that at first glance doesn't even appear to be doublespeak.

First Kind of Doublespeak

There are at least four kinds of doublespeak. The first is the euphemism, an inoffensive or positive word or phrase used to avoid a harsh, unpleasant, or distasteful reality. But a euphemism can also be a tactful word or phrase which avoids directly mentioning a painful reality, or it can be an expression used out of concern for the feelings of someone else, or to avoid directly discussing a topic subject to a social or cultural taboo.

When you use a euphemism because of your sensitivity for someone's feelings or out of concern for a recognized social or cultural taboo, it is not doublespeak. For example, you express your condolences that someone has "passed away" because you do not want to say to a grieving person, "I'm sorry your father is dead." When you use the euphemism "passed away," no one is misled. Moreover, the euphemism functions here not just to protect the feelings of another person, but to communicate also your concern for that person's feelings during a period of mourning. When you excuse yourself to go to the "rest room," or you mention that someone is "sleeping with" or "involved with" someone else, you do not mislead anyone about your meaning, but you do respect the social taboos about discussing bodily functions and sex in direct terms. You also indicate your sensitivity to the feelings of your audience, which is usually considered a mark of courtesy and good manners.

However, when a euphemism is used to mislead or deceive, it becomes doublespeak. For example, in 1984 the U.S. State Department announced that it would no longer use the word "killing" in its annual report on the status of human rights in countries around the world. Instead, it would use the phrase "unlawful or arbitrary deprivation of life," which the department claimed was more accurate. Its real purpose for using this phrase was simply to avoid discussing the embarrassing situation of government-sanctioned killings in countries that are supported by the United States and have been certified by the United States as respecting the human rights of their citizens. This use of a euphemism constitutes doublespeak, since it is designed to mislead, to cover up the unpleasant. Its real intent is at variance with its apparent intent. It is language designed to alter our perception of reality.

The Pentagon, too, avoids discussing unpleasant realities when it refers to bombs and artillery shells that fall on civilian targets as "incontinent ordnance." And in 1977 the Pentagon tried to slip funding for the neutron bomb unnoticed into an appropriations bill by calling it a "radiation enhancement device."

Second Kind of Doublespeak

A second kind of doublespeak is jargon, the specialized language of a trade, profession, or similar group, such as that used by doctors, lawyers, engineers, educators, or car mechanics. Jargon can serve an important and useful function. Within a group, jargon functions as a kind of verbal shorthand that allows members of the group to communicate with each other clearly, efficiently, and quickly. Indeed, it is a mark of membership in the group to be able to use and understand the group's jargon.

But jargon, like the euphemism, can also be doublespeak. It can be—and often is—pretentious, obscure, and esoteric terminology used to give an air of profundity, authority, and prestige to speakers and their subject matter. Jargon as doublespeak often makes the simple appear complex, the ordinary profound, the obvious insightful. In this sense it is used not to express but impress. With such doublespeak, the act of smelling something becomes "organoleptic analysis," glass becomes "fused silicate," a crack in a metal support beam becomes a "discontinuity," conservative economic policies become "distributionally conservative notions."

Lawyers, for example, speak of an "involuntary conversion" of property when discussing the loss or destruction of property through theft, accident, or condemnation. If your house burns down or if your car is stolen, you have suffered an involuntary conversion of your property. When used by lawyers in a legal situation, such jargon is a legitimate use of language, since lawyers can be expected to understand the term.

However, when a member of a specialized group uses its jargon to communicate with a person outside the group, and uses it knowing that the nonmember does not understand such language, then there is doublespeak. For example, on May 9, 1978, a National Airlines 727 airplane crashed while attempting to land at the Pensacola, Florida airport. Three of the fifty-two passengers aboard the airplane were killed. As a result of the crash, National made an after-tax insurance benefit of $1.7 million, or an extra 18¢ a share dividend for its stockholders. Now National Airlines had two problems: It did not want to talk about one of its airplanes crashing, and it had to account for the $1.7 million when it issued its annual report to its stockholders. National solved the problem by inserting a footnote in its annual report which explained that the $1.7 million income was due to "the involuntary conversion of a 727." National thus acknowledged the crash of its airplane and the subsequent profit it made from the crash, without once mentioning the accident or the deaths. However, because airline officials knew that most stockholders in the company, and indeed most of the general public, were not familiar with legal jargon, the use of such jargon constituted doublespeak.

Third Kind of Doublespeak

A third kind of doublespeak is gobbledygook or bureaucratese. Basically, such doublespeak is simply a matter of piling on words, of overwhelming the audience with words, the bigger the words and the longer the sentences the better. Alan Greenspan, then chair of President Nixon's Council of Economic Advisors, was quoted in *The Philadelphia Inquirer* in 1974 as having testified before a Senate committee that "It is a tricky problem to find the particular calibration in timing that would be appropriate to stem the acceleration in risk premiums created by falling incomes without prematurely aborting the decline in the inflation-generated risk premiums."

Nor has Mr. Greenspan's language changed since then. Speaking to the meeting of the Economic Club of New York in 1988, Mr. Greenspan, now Federal Reserve chair, said, "I guess I should warn you, if I turn out to be particularly clear, you've probably misunderstood what I've said." Mr. Greenspan's doublespeak doesn't seem to have held back his career.

Sometimes gobbledygook may sound impressive, but when the quote is later examined in print it doesn't even make sense. During the 1988 presidential campaign, vice-presidential candidate Senator Dan Quayle explained the need for a strategic-defense initiative by saying, "Why wouldn't an enhanced deterrent, a more stable peace, a better prospect to denying the ones who enter conflict in the first place to have a reduction of offensive systems and an introduction to defensive capability? I believe this is the route the country will eventually go."

The investigation into the Challenger disaster in 1986 revealed the doublespeak of gobbledygook and bureaucratese used by too many involved in the shuttle program. When Jesse Moore, NASA's associate administrator, was asked if the performance of the shuttle program had improved with each launch or if it had remained the same, he answered, "I think our performance in terms of the liftoff performance and in terms of the orbital performance, we knew more about the envelope we were operating under, and we have been pretty accurately staying in that. And so I would say the performance has not by design drastically improved. I think we have been able to characterize the performance more as a function of our launch experience as opposed to it improving as a function of time." While this language may appear to be jargon, a close look will reveal that it is really just gobbledygook laced with jargon. But you really have to wonder if Mr. Moore had any idea what he was saying.

Fourth Kind of Doublespeak

The fourth kind of doublespeak is inflated language that is designed to make the ordinary seem extraordinary; to make everyday things seem impressive; to give an air of importance to people, situations, or things that would not normally be considered important; to make the simple seem complex. Often this kind of doublespeak isn't hard to spot, and it is usually pretty funny. While car mechanics may be called "automotive internists," elevator operators members of the "vertical transportation corps," used cars "pre-owned" or "experienced cars," and black-and-white television sets described as having "non-multicolor capability," you really aren't misled all that much by such language.

However, you may have trouble figuring out that, when Chrysler "initiates a career alternative enhancement program," it is really laying off five thousand workers; or that "negative patient care outcome" means the patient died; or that "rapid oxidation" means a fire in a nuclear power plant.

The doublespeak of inflated language can have serious consequences. In Pentagon doublespeak, "pre-emptive counterattack" means that American forces attacked first; "engaged the enemy on all sides" means American troops were ambushed; "backloading of augmentation personnel" means a retreat by American troops. In the doublespeak of the military, the 1983 invasion of Grenada was conducted not by the U.S. Army, Navy, Air Force, and Marines, but by the "Caribbean Peace Keeping Forces." But then, according to the Pentagon, it wasn't an invasion, it was a "predawn vertical insertion."

DOUBLESPEAK THROUGHOUT HISTORY

Doublespeak is not a new use of language peculiar to the politics or economics of the twentieth century. In the fifth century B.C., the Greek historian Thucydides wrote in *The Peloponnesian War* that

> revolution thus ran its course from city to city. . . . Words had to change their ordinary meanings and to take those which were now given them. Reckless audacity came to be considered the courage of a loyal ally; prudent hesitation, specious cowardice; moderation was held to be a cloak for unmanliness; ability to see all sides of a question, inaptness to act on any. Frantic violence became the attribute of manliness; cautious plotting, a justifiable means of self-defense. The advocate of extreme measures was always trustworthy; his opponent, a man to be suspected.

Julius Caesar, in his account of the Gallic Wars, described his brutal and bloody conquest and subjugation of Gaul as "pacifying" Gaul. "Where they make a desert, they call it peace," said an English nobleman quoted by the Roman historian Tacitus. When traitors were put to death in Rome, the announcement of their execution was made in the form of saying "they have lived." "Taking notice of a man in the ancestral manner" meant capital punishment; "the prisoner was then lead away" meant he was executed.

In his memoirs, *V-2*, Walter Dornberger, commanding officer of the Peenemünde Rocket Research Institute in Germany during World War II, describes how he and his staff used language to get what they needed from the Bureau of Budget for their rocket experiments. A pencil sharpener was an "Appliance for milling wooden dowels up to 10 millimeters in diameter," and a typewriter was an "Instrument for recording test data with rotating roller." But it was the Nazis who were the masters of double speak, and they used it not just to achieve and maintain power but to perpetrate some of the most heinous crimes in the history of the human race.

In the world of Nazi Germany, nonprofessional prostitutes were called "persons with varied sexual relationships"; "protective custody" was the very opposite of protective; "Winter Relief" was a compulsory tax presented as a voluntary charity; and a "straightening of the front" was a retreat, while serious difficulties became "bottlenecks." Minister of Information (the very title is doublespeak) Josef Goebbels spoke in all seriousness of "simple pomp" and "the liberalization of the freedom of the press."

Nazi doublespeak reached its peak when dealing with the "Final Solution," a phrase that is itself the ultimate in doublespeak. The notice, "The Jew X.Y. lived here," posted on a door, meant the occupant had been "deported," that is, killed. When mail was returned stamped "Addressee has moved away," it meant the person had been "deported." "Resettlement" also meant deportation, while "work camp" meant concentration camp or incinerator, "action" meant massacre, "Special Action Groups" were army units that conducted mass murder, "selection" meant gassing, and "shot while trying to escape" meant deliberately killed in a concentration camp.

GEORGE ORWELL AND LANGUAGE

In his famous and now-classic essay, "Politics and the English Language," which was published in 1946, George Orwell wrote that the "great enemy of clear language is insincerity. When there is a gap between one's real and one's declared aims, one turns as it were instinctively to long words and exhausted idioms, like a cuttlefish squirting out ink." For Orwell, language was an instrument for "expressing and not for concealing or preventing thought." In his most biting comment, he observed that, "in our time, political speech and writing are largely the defense of the indefensible. . . . [P]olitical language has to consist largely of euphemism, question-begging and sheer cloudy vagueness. . . . Political language . . . is designed to make lies sound truthful and murder respectable, and to give an appearance of solidity to pure wind."

Orwell understood well the power of language as both a tool and a weapon. In the nightmare world of his novel, *1984*, Orwell depicted a society where language was one of the most important tools of the totalitarian state. Newspeak, the official state language in the world of *1984*, was designed not to extend but to *diminish* the range of human thought, to make only "correct" thought possible and all other modes of thought impossible. It was, in short, a language designed to create a reality that the state wanted.

Newspeak had another important function in Orwell's world of *1984*. It provided the means of expression for doublethink, the mental process that allows you to hold two opposing ideas in your mind at the same time and believe in both of them. The classic example in Orwell's novel is the slogan, "War Is Peace." Lest you think doublethink is confined only to Orwell's novel, you need only recall the words of Secretary of State Alexander Haig when he testified before a congressional committee in 1982 that a continued weapons build-up by the United States is "absolutely essential to our hopes for meaningful arms reduction." Or remember what Senator Orin Hatch said in 1988: "Capital punishment is our society's recognition of the sanctity of human life."

At its worst, doublespeak, like newspeak, is language designed to limit, if not eliminate, thought. Like doublethink, doublespeak enables speaker and listener, writer and reader, to hold two opposing ideas in their minds at the same time and believe in both of them. At its least offensive, doublespeak is inflated language that tries to give importance to the insignificant.

THE DOUBLESPEAK ALL AROUND US

Orwell was concerned primarily with political language because it is the language of power, but it is not just political language that is so misleading these days. Everywhere you turn you encounter the language with which Orwell was so concerned. It's not an economic recession but, according to the Reagan Administration, a "period of accelerated negative growth" or simply "negative economic growth." There's no such thing as acid rain; according to the Environmental Protection Agency, it's just "poorly buffered precipitation" or, more impressively, "atmospheric deposition of anthropogenetically-derived acidic substances." And those aren't gangsters, mobsters, the Mafia, or La Cosa Nostra in Atlantic City; according to the "New Jersey Division of Gaming Enforcement" (a doublespeak title that avoids the use of that dreaded word, "gambling") they're just "members of a career-offender cartel."

Military Doublespeak

Military doublespeak seems to have always been around. In 1947 the name of the Department of War was changed to the more pleasing if misleading Department of Defense. How much easier it is to spend hundreds of billions of dollars for defense instead of war. During the Vietnam War the American public learned that it was an "incursion" into Cambodia, not an invasion; a "protective reaction strike" or "a limited duration protective reaction strike" or "air support," not bombing.

When asked why U.S. forces lacked intelligence information on Grenada before they invaded the island in 1983, Admiral Wesley L. McDonald told reporters that "We were not micromanaging Grenada intelligence-wise until about that time frame." In today's armed forces it's not a shovel but a "combat emplacement evacuator," not a bullet hole but a "ballistically induced aperture in the subcutaneous environment."

Business Doublespeak

The world of business has produced large amounts of doublespeak. If an airplane crash is one of the worst things that can happen to an airline company, a recall of automobiles because of a safety defect is one of the worst things that can happen to an automobile company. In April of 1972, when the Ford Motor Company had to recall 423,000 1972 Torino and Mercury Montego models to correct "mechanical deficiencies," the company sent a letter to all those who had bought the defective cars. In its letter, Ford said that the rear axle bearings of the cars "can deteriorate" and went on to say "Continued driving with a failed bearing could result in disengagement of the axle shaft and adversely affect vehicle control." This is the language of nonresponsibility. What are "mechanical deficiencies"— poor design, bad workmanship? The rear axle bearings "can deteriorate," but will they deteriorate? If they do deteriorate, what causes the deterioration? Note that "continued driving" is the subject of the sentence, which suggests that it is not Ford's poor manufacturing that is at fault but the driver who insists on driving the defective car. Note, too, the expression "failed bearing," which implies that the bearing failed, not Ford. Finally, the phrase "adversely affect vehicle control" means simply that, because of the mechanical defect, the driver could lose control of the car and get killed.

If you ask the questions for examining language to see if it's doublespeak (who is saying what to whom, under what conditions and circumstances, with what intent, and with what results), you can quickly discover the doublespeak here. What Ford should be saying to its customers is that the car Ford sold them has a serious defect that should be corrected immediately, otherwise the customer runs the risk of being seriously injured or killed. But you have to find this message beneath the doublespeak that Ford has used to disguise its embarrassing and unpleasant message. We will never know how many customers didn't bring their cars in for repairs because they didn't understand from that letter just how serious the problem was and that they'd better get their car to the service department fast.

When it comes time to fire or lay off employees, business has produced more than enough doublespeak to deal with the unpleasant situation. Employees are, of course, never fired or laid off. They are "selected out," "placed out," "non-retained," "released," "dehired," or "non-renewed." A corporation will "eliminate the redundancies in the human resources area," assign "candidates for derecruitment" to a "mobility pool," "revitalize the department" by placing executives on "special assignment," "enhance the efficiency of operations," "streamline the field sales organization," or "further rationalize marketing efforts." The reality behind all this doublespeak is that companies are firing or laying off employees, but no one wants to acknowledge to the stockholders, public, or competition that times are tough, business is bad, and people have to go.

When the oil industry was hit hard by declining sales and a surplus of oil, after years of great prosperity and a shortage of oil, the doublespeak flowed thicker than crude oil. Because of "reduced demand for product," which results in "space refining capacity" and problems in "down-stream operations," oil companies have been forced to "re-evaluate and consolidate their operations" and take "appropriate cost-reduction actions," in order to "enhance the efficiency of operations," which has meant the "elimination of marginal outlets," "accelerating our divestment program," and the "disposition of low throughput marketing units." This doublespeak really means that oil companies have fired employees, cut back on expenses, and closed gas stations and oil refineries because there's a surplus of oil and people are not buying as much gas and oil as in the past.

One oil company faced with declining business sent a memorandum to its employees advising them that the company's "business plans are under revision and now reflect a more moderate approach toward our operating and capital programs." The result of this "more moderate approach" is a "surplus of professional/technical employees." To "assist in alleviating the surplus, selected professional and technical employees" have been "selected to participate" in a "Voluntary Program" providing "incentives" for employees who "resign voluntarily." What this memorandum means, of course, is that expenses must be cut because of declining business, so some employees will have to go.

Wall Street produces doublespeak right along with the junk bonds. It is rare to read in a trade publication that the stock market "fell." Others might say the stock market fell, but those who work on Wall Street prefer to say that the stock market "retreated," "eased," made a "technical adjustment" or a "technical correction," or perhaps that "prices were off due to profit taking," or "off in light trading," or "lost ground." In October 1987, when the stock market collapsed, losing billions of dollars, one brokerage house called the collapse a "fourth quarter equity retreat." As a side note, it is interesting to observe that the stock market never rises because of a "technical adjustment" or "correction," nor does it ever "ease" upward. Stock prices always "climb," "advance," "move forward," "edge up," or "surge."

Business magazines, corporate reports, executive speeches, and the business sections of newspapers are filled with words and phrases such as "marginal rates of substitution," "equilibrium price," "getting off margin," "distributional coalition," "non-performing assets," and "encompassing organizations." Much of this is jargon or inflated language designed to make the simple seem complex, but there are other examples of business doublespeak that misleads or is designed to avoid a harsh reality. What should you make of such expressions as "negative deficit" or "revenue excesses" (i.e., profit), "invest in" (spend money or buy something), "price enhancement" or "price adjustment" (price increase), "shortfall" (mistake in planning), or "period of accelerated negative growth" or "negative economic growth" (recession)?

Business doublespeak often attempts to give substance to pure wind (to use Orwell's term), to make ordinary actions seem complex. Executives "operate" in "timeframes" within the "context" of which a "task force" will serve as the proper "conduit" for all the necessary "input" to "program a scenario" that, within acceptable "parameters," and with the proper "throughput," will "generate" the "maximum output" for a "print out" of "zero defect terminal objectives" which will "enhance the bottom line."

Education Doublespeak

Politicians, members of the military, and businesspeople are not the only ones who use doublespeak. People in all parts of society use it. Education has more than its share of doublespeak. On some college campuses, what was once the Department of Physical Education is now the "Department of Human Kinetics" or the "College of Applied Life Studies." You may have called it Home Economics, but now it's the "School of Human Resources and Family Studies." These days, you don't go to the library to study; you go to the "Learning Resources Center."

Those aren't desks in the elementary school classroom, they're "pupil stations." Teachers, who are "classroom managers" applying an "action plan" to a "knowledge base," are concerned with the "basic fundamentals," which are "inexorably linked" to the "education user's" "time-on-task." Students don't take simple tests; now it's "criterion-referenced testing" that measures whether a student has achieved the "operational curricular objectives." A school system in Pennsylvania, making absolutely no mention of whether the student learned anything, uses the following grading system on its report cards: "no effort, less than minimal effort, minimal effort, more than minimal effort, less than full effort, full effort, better than full effort, effort increasing, effort decreasing."

B. W. Harlston, president of City College in New York, said in 1982 that some college students in New York come from "economically nonaffluent" families, while a spokesperson at Duke University said in 1982 that coach Red Wilson wasn't being fired, "He just won't be asked to continue in that job." An article in a scholarly journal suggests teaching students three approaches to writing to help them become better writers: "concretization of goals, procedural facilitation, and modeling planning."

In its August 3, 1981 issue, *Newsweek* magazine reported that the prestigious National Bureau of Economic Research published a working paper by Brown University economist Herschel I. Grossman entitled "Familial Love and Intertemporal Optimality." Professor Grossman reached this conclusion about family love: "An altruistic utility function promotes intertemporal efficiency. However, altruism creates an externality that implies that satisfying the conditions for efficiency does not insure intertemporal optimality."

A research report issued by the U.S. Office of Education in 1966 contains this sentence: "In other words, feediness is the shared information between toputness, where toputness is at a time just prior to the inputness." At times, doublespeak seems to be the primary product of educators.

DEADLY DOUBLESPEAK

There are instances, however, where doublespeak becomes more than amusing, more than a cause for a laugh. At St. Mary's Hospital in Minneapolis in 1982, an anesthetist turned the wrong knob during a Cesarean delivery, giving a fatal dose of nitrous oxide which killed the mother and unborn child. The hospital called it a "therapeutic misadventure." In its budget request to Congress in 1977, the Pentagon called the neutron bomb "an efficient nuclear weapon that eliminates an enemy with a minimum degree of damage to friendly territory." The Pentagon also calls the expected tens of millions of civilian dead in a nuclear war "collateral damage," a term the Pentagon also applies to the civilians killed in any war. And in 1977 people watching the Dick Cavett show on television learned from former Green Beret Captain Bob Marasco that during the Vietnam war the Central Intelligence Agency created the phrase "eliminate with extreme prejudice" to replace the more direct verb "kill."

President Reagan and the Doublespeak of Politics

Identifying doublespeak can at times be difficult. For example, on July 27, 1981, President Ronald Reagan said in a speech televised to the American public that "I will not stand by and see those of you who are dependent on Social Security deprived of the benefits you've worked so hard to earn. You will continue to receive your checks in the full amount due you." This speech had been billed as President Reagan's position on Social Security, a subject of much debate at the time. After the speech, public opinion polls revealed that the great majority of the public believed that the president had affirmed his support for Social Security and that he would not support cuts in benefits. However, only days after the speech, on July 31, 1981, an article in the *Philadelphia Inquirer* quoted White House spokesperson David Gergen as saying that President Reagan's words had been "carefully chosen." What President Reagan had meant, according to Gergen, was that he was reserving the right to decide who was "dependent" on those benefits, who had "earned" them, and who, therefore, was "due" them.

The subsequent remarks of David Gergen reveal the real intent of President Reagan as opposed to his apparent intent. Thus, the criteria for analyzing language to determine whether it is doublespeak (who is saying what to whom, under what conditions and circumstances, with what intent, and with what results), when applied in light of David Gergen's remarks, reveal the doublespeak of President Reagan. Here, indeed, is the insincerity of which Orwell wrote. Here, too, is the gap between the speaker's real and declared aim.

Doublespeak and Political Advertisements

During the 1982 congressional election campaign, the Republican National Committee sponsored a television advertisement that pictured an elderly, folksy postman delivering Social Security checks "with the 7.4% cost-of-living raise that President Reagan promised." The postman then adds that "he promised that raise and he kept his promise, in spite of those sticks-in-the-mud who tried to keep him from doing what we elected him to do." The commercial was, in fact, deliberately misleading. The cost-of-living increases had been provided automatically by law since 1975, and President Reagan had tried three times to roll them back or delay them but was overruled by congressional opposition. When these discrepancies were pointed out to an official of the Republican National Committee, he called the commercial "inoffensive" and added, "Since when is a commercial supposed to be accurate? Do women really smile when they clean their ovens?"

Again, applying the criteria for identifying doublespeak to this advertisement reveals the doublespeak in it, once you know the facts of past actions by President Reagan. Moreover, the official for the Republican National Committee assumes that all advertisements, whether for political candidates or commercial products, do not tell the truth; in his doublespeak, they do not have to be "accurate." Thus, the real intent of the advertisement was to mislead, while the apparent purpose of the commercial was to inform the public of President Reagan's position on possible cuts in Social Security benefits. Again there is insincerity, and again there is a gap between the speaker's real and declared aims.

Alexander Haig and Doublespeak

One of the most chilling and terrifying uses of doublespeak in recent memory occurred in 1981 when then Secretary of State Alexander Haig was testifying before congressional committees about the murder of three American nuns and a

Catholic lay worker in El Salvador. The four women had been raped and then shot at close range, and there was clear evidence that the crime had been committed by soldiers of the Salvadoran government. Before the House Foreign Affairs Committee, Secretary Haig said:

> I'd like to suggest to you that some of the investigations would lead one to believe that perhaps the vehicle the nuns were riding in may have tried to run a roadblock, or may accidentally have been perceived to have been doing so, and there'd been an exchange of fire and then perhaps those who inflicted the casualties sought to cover it up. And this could have been at a very low level of both competence and motivation in the context of the issue itself. But the facts on this are not clear enough for anyone to draw a definitive conclusion.

The next day, before the Senate Foreign Relations Committee, Secretary Haig claimed that press reports on his previous testimony were "inaccurate." When Senator Claiborne Pell asked whether the secretary was suggesting the possibility that "the nuns may have run through a roadblock," he replied, "You mean that they tried to violate . . . ? Not at all, no, not at all. My heavens! The dear nuns who raised me in my parochial schooling would forever isolate me from their affections and respect." Then Senator Pell asked Secretary Haig, "Did you mean that the nuns were firing at the people, or what did 'an exchange of fire' mean?" The secretary replied, "I haven't met any pistol-packing nuns in my day, Senator. What I meant was that if one fellow starts shooting, then the next thing you know they all panic." Thus did the secretary of state of the United States explain official government policy on the murder of four American citizens in a foreign land.

Secretary Haig's testimony implies that the women were in some way responsible for their own fate. By using such vague wording as "would lead one to believe" and "may accidentally have been perceived to have been doing so," he avoids any direct assertion. The use of the phrase "inflicted the casualties" not only avoids using the word "kill" but also implies that at the worst the killings were accidental or justifiable. The result of this testimony is that the secretary of state has become an apologist for rape and murder. This is indeed language in defense of the indefensible; language designed to make lies sound truthful and murder respectable; language designed to give an appearance of solidity to pure wind.

THE DANGERS OF DOUBLESPEAK

These previous three examples of doublespeak should make it clear that doublespeak is not the product of carelessness or sloppy thinking. Indeed, most doublespeak is the product of clear thinking and is carefully designed and constructed to appear to communicate when in fact it doesn't. It is language designed not to lead but mislead. It is language designed to distort reality and corrupt thought. In the world created by doublespeak, if it's not a tax increase, but rather "revenue enhancement" or "tax base broadening," how can you complain about higher taxes? If it's not acid rain, but rather "poorly buffered precipitation," how can you worry about all those dead trees? If that isn't the Mafia in Atlantic City, but just "members of a career-offender cartel," why worry about the influence of organized crime in the city? If Supreme Court Justice William Rehnquist wasn't addicted to the pain-killing drug his doctor prescribed, but instead it was just that the drug had "established an interrelationship with the body, such that if the drug is removed precipitously, there is a reaction," you needn't question that his decisions might have been influenced by his drug addiction. If it's not a Titan II nuclear-armed intercontinental ballistic missile with a warhead 630 times more powerful than the atomic bomb dropped on Hiroshima, but instead, according to Air Force Colonel Frank Horton, it's just a "very large, potentially disruptive reentry system," why be concerned about the threat of nuclear destruction? Why worry about the neutron bomb escalating the arms race if it's just a "radiation enhancement weapon"? If it's not an invasion, but a "rescue mission" or a "predawn vertical insertion," you won't need to think about any violations of U.S. or international law.

Doublespeak has become so common in everyday living that many people fail to notice it. Even worse, when they do notice doublespeak being used on them, they don't react, they don't protest. Do you protest when you are asked to check your packages at the desk "for your convenience," when it's not for your convenience at all but for someone else's? You see advertisements for "genuine imitation leather," "virgin vinyl," or "real counterfeit diamonds," but do you question the language or the supposed quality of the product? Do you question politicians who don't speak of slums or ghettos but of the "inner city" or "sub-standard housing" where the "disadvantaged" live and thus avoid talking about the poor who have to live in filthy, poorly heated, ramshackle apartments or houses? Aren't you amazed that patients don't die in the hospital anymore, it's just "negative patient-care outcome"?

Doublespeak such as that noted earlier that defines cab drivers as "urban transportation specialists," elevator operators as members of the "vertical transportation corps," and automobile mechanics as "automotive internists" can be considered humorous and relatively harmless. However, when a fire in a nuclear reactor building is called "rapid oxidation," an explosion in a nuclear power plant is called an "energetic disassembly," the illegal overthrow of a legitimate government is termed "destabilizing a government," and lies are seen as "inoperative statements," we are hearing doublespeak that attempts to avoid responsibility and make the bad seem good, the negative appear positive, something unpleasant appear attractive; and which seems to communicate but doesn't. It is language designed to alter our perception of reality and corrupt our thinking. Such language does not provide us with the tools we need to develop, advance, and preserve our culture and our civilization. Such language breeds suspicion, cynicism, distrust, and, ultimately, hostility.

Doublespeak is insidious because it can infect and eventually destroy the function of language, which is communication between people and social groups. This corruption of the function of language can have serious and far-reaching consequences. We live in a country that depends upon an informed electorate to make decisions in selecting candidates for office and deciding issues of public policy. The use of doublespeak can become so pervasive that it becomes the coin of the political realm, with speakers and listeners convinced that they really understand such language. After awhile we may really believe that politicians don't lie but only "misspeak," that illegal acts are merely "inappropriate actions," that fraud and criminal conspiracy are just "miscertification." President Jimmy Carter in April of 1980 could call the aborted raid to free the American hostages in Teheran an "incomplete success" and really believe that he had made a statement that clearly communicated with the American public. So, too, could President Ronald Reagan say in 1985 that "ultimately our security and our hopes for success at the arms reduction talks hinge on the determination that we show here to continue our program to rebuild and refortify our defenses" and really believe that greatly increasing the amount of money spent building new weapons would lead to a reduction in the number of weapons in the world. If we really believe that we understand such language and that such language communicates and promotes clear thought, then the world of *1984*, with its control of reality through language, is upon us.

Alan Siegel

The Plain English Revolution

Alan Siegel, a leading consultant in the Plain English movement, is the Chairman and Chief Executive Officer of Siegel & Gale, a New York firm specializing in language simplification.

> *LOUIS XVI: C'est une grande révolte?*
>
> *DUC DE LA ROCHEFOUCAULD— LIANCOURT: Non, Sire, c'est une grande révolution. (When the news arrived at Versailles of the Fall of the Bastille, 1789)*

Three-and-a-half years ago I wrote an article . . . on the movement away from legalese toward forms and documents that can be understood by ordinary citizens.* At that time, the movement consisted of isolated revolts against legalistic gobbledygook and bureaucratese. A handful of business documents—mainly insurance policies and loan notes—had been simplified by forward-looking companies. In government, the scattered advocates of clear communication shared the lament of Alfred Kahn, then head of the Civil Aeronautics Board: "There's nothing I can do but cry. I feel so lonely and futile."

Now the plain language movement is becoming a revolution. . . . But the Bastille has not fallen, yet. Part of the legal community remains resistant to change. Some supposed practitioners of Plain English have confused simplicity with simplemindedness. And the greatest obstacle has yet to be fully overcome— a misunderstanding of what Plain English is all about.

That misunderstanding was dramatized when New Jersey's Plain English Bill, passed by the legislature, expired in February 1980, through a pocket veto by

---

*"To lift the curse of legalese—Simplify, Simplify," *Across the Board*, June 1977.

Gov. Brendan Byrne, who said: ". . . although regulating the protection of consumers is a worthy goal, legislating the style of a society's prose is another thing."*

But the movement has nothing to do with "legislating the style of a society's prose." We are not trying to turn English into one-syllable words, or to translate Saul Bellow into baby-talk. We are not even trying to do away with professional jargon, though that is a tempting target. Let lawyers talk to lawyers, or accountants to accountants, as they please. So long as they understand one another, that is just fine.

Well, what is the Plain English movement all about? *Its aim is to make functional documents function, whether they are put out by business or by government. If a consumer is expected to abide by a formal document—an insurance policy, a mortgage, a lease, a warranty, a tax form—then the consumer should be able to understand the document.* Carl Felsenfeld, vice president and counsel for Citicorp N.A., expresses the concept in legal terms: "There is growing dissatisfaction with contracts where consumers merely 'sign here' and can't, under any reputable system of contract law, be deemed to have agreed to all the printed verbiage." In social terms, I would say that people are learning a new right: the right to understand.

It's been argued that the fault lies with consumers, or with the educational system that does not train them to cope with legalese, or accounting, or insurance terminology. But that argument is neither reasonable nor realistic. In fact, documents meant for consumers have been made much more difficult than they really need to be. Redressing the balance—making sure that the documents consumers are expected to understand are made understandable—is a matter of simple fairness, and simple efficiency. In the long run it's in the interests of business as well as the consumer.

Five years of experience have proved that consumer contracts and forms of all kinds can be made much more understandable without sacrificing legal effectiveness. The very first plain language loan note, which I helped write for Citibank of New York, was introduced in 1975. Since then, Citicorp counsel Felsenfeld notes, "We've lost no money and there has been no litigation as a result of simplification."

That loan note remains a good example of just what plain language means. Among other things:

- A personal tone was used throughout—"I" and "me" rather than "the undersigned" or "Borrower," "you" and "yours" rather than "the Bank."
- Language was radically simplified. For example, "To repay my loan, I promise to pay you . . ." instead of "For value received, the undersigned (jointly and severally) hereby promise(s) to pay . . ."
- When unfamiliar terms couldn't be eliminated, we added explanatory phrases. For instance, ". . . if this loan is refinanced—that is, replaced by a new note— you [the bank] will refund the unearned finance charge, figured by the rule of 78—a commonly used formula for figuring rebates or installment loans."

---

*Rewritten and reintroduced in June, the bill was passed again. This time Gov. Byrne signed it (on October 10th), out of deference to its "unusually large" number of sponsors—65 of the Assembly's 80 members.

- We shortened the sentences wherever possible and even used contractions ("I'll pay this sum . . ."). To enhance clarity, we chose active instead of passive verb forms where possible.
- Improvements in design included the use of larger (12-point) type printed in green on light brown stock. Compared with the previously intimidating format, this visually appealing approach suggests immediately that the document is supposed to be read.

The Citibank note taught us a fundamental lesson about simplifying language. Consumer contracts have traditionally been adapted from mercantile contracts, which feature verbose protective clauses accumulated over the years in an effort to cover all possible contingencies. Many of these provisions have no practical value in the consumer marketplace. So the first task in simplifying a document is not rewriting it in Plain English, but identifying clauses taken from commercial contracts that can be eliminated from the consumer contract without jeopardizing its validity. The secret to doing this lies in analyzing actual business experience to see which provisions are really being used.

In the case of the Citibank note, the provisions describing the lender's protections in case of default proved to be the biggest challenge. The traditional note listed a string of contingencies more than 180 words long:

Before

In the event of default in the payment of this or any other Obligation or the performance or observance of any term or covenant contained herein or in any note or other contract or agreement evidencing or relating to any Obligation or any Collateral on the Borrower's part to be performed or observed; or the undesigned Borrower shall die; or any of the undersigned become insolvent or make an assignment for the benefit of creditors; or a petition shall be filed by or against any of the undersigned under any provision of the Bankruptcy Act; or any money, securities or property of the undersigned now or hereafter on deposit with or in the possession or under the control of the Bank shall be attached or become subject to distraint proceedings or any order of process of any court; or the Bank shall deem itself to be insecure, then and in any such event, the Bank shall have the right (at its option), without demand or notice of any kind, to declare all or any part of the Obligations to be immediately due and payable. . . .

But analysis of Citibank's business experience disclosed that the typical consumer loan transaction needs only one event of default—failure to pay. With one additional protection added, the following replaced all the fine print above:

After

Default   I'll be in default—

1. If I don't pay an installment on time, or
2. If any other creditor tries by legal process to take any money of mine in your possession.

A number of insurance companies, whose product, after all, is words on paper, continue to get good marks for simplification. The Massachusetts Savings Bank Life Insurance (SBLI) Whole Life Policy used to begin like this:

IN CONSIDERATION OF THE APPLICATION for this policy (copy attached hereto) which is the basis of and a part of this contract and of the payment of an annual premium as hereinafter specified for the basic policy as of the Date of Issue as specified herein and on the anniversary of such date in each year during the continuance of this contract until premiums have been paid for the number of years indicated in the POLICY SPECIFICATIONS or until the prior death of the Insured . . .

By contrast, the cover of their simplified policy begins with this message:

Please take the time to read your SBLI policy carefully. Your SBLI representative will be glad to answer any questions. You may return this policy within ten days after receiving it. Deliver it to any SBLI agency. We'll promptly refund all premiums paid for it.

The insurer's promise to pay the consumer used to be phrased in these forbidding terms:

The Bank Hereby Agrees upon Surrender of this Policy to Pay the Face Amount Specified Above, less any indebtedness on or secured hereunder . . . upon receipt of due proof of the Insured's death to the beneficiary named in the Application herefor or to such other beneficiary as may be entitled thereto under the provisions hereof, or if no such beneficiary survives the Insured, then to the Owner or to the estate of the Owner.

In Plain English, policyholders can understand what they are buying:

We will pay the face amount when we receive proof of the Insured's death. We will pay the named Beneficiary. If no Beneficiary survives the Insured, we'll pay the Owner of this policy, or the Owner's estate.

Any amount owed to us under this policy will be deducted. We'll refund any premiums paid beyond the month of death.

Perhaps the most ambitious simplification program so far has been that undertaken by the St. Paul Fire and Marine Insurance Company, of St. Paul, Minnesota. In 1975 they became one of the first large insurers to begin simplifying their policies. As a pilot project, the company produced an easy-to-read "personal liability catastrophe policy" that used personalized examples in colloquial English:

Before

a. Automobile and Watercraft Liability:

1. any Relative with respect to (i) an Automobile owned by the Named Insured or a Relative, or (ii) a Non-owned Automobile, provided his actual operation or (if he is not operating) the other actual use thereof is with the permission of the owner and is within the scope of such permission, or
2. any person while using an Automobile or Watercraft, owned by, loaned or hired for use in behalf of the Named Insured and any person or organization legally responsible for the use thereof is within the scope of such permission.

After

We'll also cover any person or organization legally responsible for the use of a car, if it's used by you or with your permission. But again, the use has to be for the intended purpose.

You loan your station wagon to a teacher to drive a group of children to the zoo. She and the school are covered by this policy if she actually drives to the zoo, but not if she lets the children off at the zoo and drives to her parents' farm 30 miles away.

By 1978, St. Paul had reduced the number of policy forms in its commercial business package from 366 to 150. Plans now call for most of its commercial insurance policies to be rewritten and reprinted in a more readable format by 1982. The company has appointed a Manager of Forms Simplification, with his own staff and a detailed style manual, to oversee the effort.

The financial field is a particularly promising new area for Plain English efforts. A case in point: Sanford C. Bernstein & Co. is a New York broker-dealer advising clients on investments as well as trading for them. It is subject to regulation under the Federal Securities Exchange Act of 1934 and the Investment Advisers Act of 1940, as well as regulation by the New York Stock Exchange, the National Association of Securities Dealers, the Board of Governors of the Federal Reserve System, and state securities laws in various jurisdictions in which it does business or has clients. These various regulations extend to every aspect of the firm's business, from the content of advertising and soliciting materials to client reports, statements, and the consents that must be obtained before the firm can effect certain types of transaction on clients' behalf. Nonetheless, the firm was able to replace jargon with Plain English, while satisfying legal and technical requirements.

Before

It is the express intention of the undersigned to create an estate or account as joint tenants with rights of survivorship and not as tenants in common. In the event of the death of either of the undersigned, the entire interest in the joint account shall be vested in the survivor or survivors on the same terms and conditions as theretofore held, without in any manner releasing the decedent's estate from the liability provided for in the next preceding paragraph.

After

Other signers share your interest equally. If one of you dies, the account will continue and the other people who've signed the agreement will own the entire interest in it.

Where obscure terms could not be eliminated, the new forms explain them. The meaning of the term "clearing agents," for instance, is described in this way:

Clearing agents. We can use other broker/dealers as clearing agents to execute transactions, to hold securities in custody on your behalf and to perform other routine procedures in connection with your account. These clearing agents will act at our direction, and they won't have any part in investment decisions.

Especially complex ideas are not only explained but accompanied by illustrative examples:

Buying on margin. In managing your account, we may buy securities on margin. This means that we'll loan you part of the cost of the securities and charge you interest on the loan. By buying on margin, we can purchase more securities on your behalf than we could

Plain Language Scorecard

- Thirty-four states have laws or regulations setting standards for clear language in insurance policies. In 1978, New York became the first state to require that business contracts "primarily for personal, family, or household purposes" be plainly written in everyday language. Since then, New Jersey, Maine, Connecticut, and Hawaii have passed similar laws requiring consumer contracts to be understandable to the public, and a score of other states are considering such laws.

- Hundreds of corporations across the country are simplifying their documents—employee benefit manuals, brokerage account agreements, trusts, the notes to financial statements, customer correspondence, billing statements, and internal communications, ranging from corporate policy manuals to the humble memo.

- President Carter, following up a promise for "regulations in plain English for a change," issued two unprecedented Executive Orders to Federal agencies telling them to simplify paperwork and eliminate gobbledy-gook from regulations. Agencies that have started to do so include the Environmental Protection Agency and the Department of Health and Human Services (formerly HEW). The Civil Aeronautics Board, where Alfred Kahn started a push for Plain English, is rewriting the notices it requires to be posted at airline counters and printed on the backs of tickets. The Federal Acquisition Regulation Project (FARP), in the Office of Federal Procurement Policy (Department of Management and Budget), offers great potential; FARP is rewriting and recodifying the massive Federal regulations on procurement of good and services.

- Major law firms, such as Shearman & Sterling in New York, have launched programs to train their young lawyers in clear legal drafting.

- The Internal Revenue Service committed over one million dollars in an all-out effort to simplify the Federal income tax forms. Simplified prototypes were presented to the Congress in November.

- The 1978 Amendments to the Constitution of the State of Hawaii include a provision that, "Insofar as practicable, all governmental writing meant for the public . . . should be plainly worded, avoiding the use of technical terms."

- On April 15, 1980, Gov. Hugh Carey of New York issued an Executive Order directing all state agencies to write their forms and regulations in plain language.

- Last fall, New York City's Department of Consumer Affairs proposed simplified language for more than 40 of its regulations covering deceptive and unconscionable business practices.

- The prestigious Practising Law Institute held a program and published a course book on "Drafting Documents in Plain Language." Another such program is planned for 1981.

- A lobbying group for the movement, Plain Talk, Inc., has been established in Washington, D.C.

- The National Institute for Education, a Federal research agency, has funded the Document Design Project, which is being run by the American Institutes of Research in conjunction with Siegel & Gale and Carnegie-Mellon University. The project's purpose is to study and encourage the use of simplified public documents.

—A.S.

using only the cash and securities actually available in your account at the time. Because it increases the amount we can invest on your behalf, buying on margin can increase the return on your investment. However, it can also increase the amount of loss on your account.

> Example: Assume that the value of A Manufacturing Corp. stock in your portfolio has increased from $5,000 to $10,000. Because of changed market conditions, we decide to sell it and buy Z Industries common stock for your portfolio. If, for instance, the margin requirement is 50%, in theory we could purchase $20,000 worth of Z Industries stock for your account by using the $10,000 proceeds from the sale of your A Manufacturing stock and loaning you the additional $10,000.

In government, the promise—and difficulty—of the Plain English movement is illustrated by the case of the Food Stamp program. That program, begun in 1961, was substantially revised by the Food Stamp Act of 1977. As a result, the forms that had been used by the states were made obsolete. There was a bewildering variety of these forms to begin with. Wisconsin's, for instance, was 37 pages long. To help the states design simple, client-oriented forms reflecting the changes in the law, the Food and Nutrition Service (FNS) of the Department of Agriculture commissioned my firm to come up with new forms to offer to the states as models.

Our approach is illustrated by the new titles we gave our forms:

Old	New
Notice of Expiration	Continuing Your Food Stamps
Notice of Eligibility, Denial or Pending Status	Action Taken on Your Food Stamp Case
Tax Dependency Form	Student Tax Report

Significantly, the new forms ran into a special kind of opposition. First, *we* discovered that not all state governments agreed that the forms should *help* people to use the program, as the law intended. Food Stamp recipients are seen as one measure of a state's poverty, so some officials were reluctant to "encourage" their use by simplifying the paperwork. Some states had actually sought to discourage applicants by using forms that needlessly demanded embarrassing information, such as whether the applicant was a drug user or had a criminal record. Some state officials also felt that the language of the new forms, which included words such as "please," "thank you," and "sincerely," was too nice to people wanting money from the government.

Second, we found that state caseworkers themselves often appeared threatened by new, simplified forms. Some felt that their authority was being undermined. For example, people whose applications for food stamps are turned down or who are cut off are allowed by law to request a fair hearing. We embodied this feature of the law in the forms themselves: letters bearing bad news to Food Stamp clients included a perforated tear-off portion that could be sent back to appeal the decision. Some caseworkers objected heatedly: they felt the tear-offs would cast doubts on their competence and would generate "unnecessary" hearings, increasing their already heavy workload.

Plain Research Is Needed

The writing of Plain English documents is still a developing field. Researchers have discovered that the reader's comprehension has as much to do with what is inside his or her head as it has to do with what is on the printed page. Diagnostic research can help clarify what information must be explained at greater length.

A study conducted by law professor Jeffrey Davis, described in *Virginia Law Review* 63: No. 6, 1977, provides an excellent example. Davis prepared a simplified installment purchase contract for a refrigerator, showed it to shoppers from various backgrounds, and then asked them questions about its contents. The contract included this definition of default: "I will be in default if I fail to pay an installment on time or if I sell or fail to take proper care of the collateral." Yet most of the readers—even the well educated—failed to understand that they would be in default if they carelessly damaged the refrigerator. They persisted in the common but mistaken belief that they could not be in default if they made all their payments on time.

—A.S.

After more than two years, the objections seem to have been overcome. Our model forms are used now by about half the states, and one FNS official commented earlier this year: "I've never heard a critical comment about the forms themselves." In fact, as in other simplifications projects, researchers got a strong impression that staff objections had as much to do with the fact the forms were new and unfamiliar as with their style or content.

Earlier in this piece I mentioned that some Plain English efforts have confused simplicity with simplemindedness. A major source of this confusion is the quick-and-dirty "readability formula" that equates clarity with short words and sentences. Here is a rider to a health insurance policy obviously written to such a formula. It is both uncommunicative and insulting to the reader's intelligence.

THE DRUG SPECIAL ENDORSEMENT

The DRUG SPECIAL is a small Endorsement. It goes with your [name of company] Contract. It depends on that other Contract to say who's covered. It is added coverage, with big extra benefits.

The DRUG SPECIAL gives you special help in paying for DRUGS . . .

Some attempts at simplification produce a kind of black humor. One life insurance company offers a "simplified" policy requiring that, "The Insured must die while this policy is in force." In a grisly parody that might be called "Dick and Jane Become Underwriters," the policy continues:

The Insured's death may be caused by accidental Bodily Injury. If so the Beneficiary may be paid an additional amount. It will be equal to the Face Amount. Three things must all happen. (1) Death must have been caused by Accidental Bodily Injury. Not by sickness. Not by anything else. . . .

Another misuse of the movement is the way some lawyers are cynically using "simplified" language, notably in apartment leases, to misinform consumers and to mislead them by playing on their ignorance. The nation's first plain language law went into effect in New York State in November 1978. One year later, the State Consumer Protection Board found that most revised lease forms "force tenants to surrender nearly every right they have under law." Rosemary S. Pooler, the board's executive director, referring to the lease prepared by the New York City Bar Association, commented: "It is ironic that tenants were in some respects better off with the 'legalese' of the 1965 lease since the . . . 'plain language' lease seems to have taken almost every opportunity to resolve legal issues in favor of landlords, sometimes at the expense of existing statutory and decisional law."

To illustrate her point, Mrs. Pooler gave, among others, the following examples:

RENT PAYMENT PROVISION

Pre-plain language version (1965): "The tenant will pay the rent as herein provided."	Plain language version (1978): "Tenant will pay the rent without any deductions, even if permitted by law."

The new document flies in the face of New York State's Multiple Dwelling Law (Sec. 302-a); Real Property Law (Sec. 235-2); Real Property Actions and Proceedings (Sec. 756 and Article 7-A); and several recent court decisions, all of which give tenants the right to reduce or withhold rent under certain circumstances—for instance, when these are serious building code violations.

ASSIGNMENT AND SUBLETTING

Pre-plain language version (1965): "(Tenant) will not assign this lease or underlet the leased premises or any part thereof without the landlord's written consent, which landlord agrees not to withhold unreasonably."	Plain language version (1978): "Tenant shall not assign this lease or enter into a sublease unless it is allowed by a law of the State of New York."

Since 1975, New York's Real Property Law (Sec. 226-b) has given tenants the right to request their landlord's permission to sublet or assign their lease. If the landlord withholds permission unreasonably, he must let the tenant break the lease. The 1965 version is a better statement of existing law than the plain language version, which implies that subletting is allowed by state law only in very special situations.

INSTALLATION OF LOCKS, CHAINS, ETC.

Pre-plain language version (1965): None.	Plain language version (1978): "Tenant will not, without landlord's written approval: . . . 3. Put in any locks or chain guards or change any lock-cylinders on the doors of the Apartment."

But New York State's Multiple Dwelling Law (Sec. 51-C) gives tenants of multiple dwellings the right to install additional locks without the landlord's permission, provided that the landlord who requests a key is given one.

Though many lawyers have seen the light, some of them—and their corporate clients—misguidedly try to protect themselves by insisting on too-precise standards for compliance. Such traditionalists do not like the Plain English laws that have been passed in New York, Maine, and Hawaii. Those laws simply require that each document for a residential lease or for a loan, property or services of less than $50,000 for personal, family or household purposes be:

1. Written in a clear and coherent manner using words with common everyday meanings;
2. Appropriately divided and captioned by its various sections.

Ironically, some lawyers and executives who are quick to decry regulatory minutiae in other areas object to this general approach to plain language legislation. They yearn for the false security of the simplistic "readability formula" approach, which ignores less easily quantified elements such as grammar, logic, and organization. They fear that without an exact definition of "clear, coherent, everyday language," compliance will elude them. Lawyers such as Wilbur H. Friedman, chairman of the New York County Lawyers' Association Special Committee on Consumer Agreements, direfully predicted that New York's law would create "upheaval" among businesses and an "absolutely staggering" burden on the courts.

Safe English

The *New York Times* (Aug. 31, 1980) reported the first case under the Plain English law on contracts:

"The New York State Attorney General was puzzled by this sentence.

'The liability of the bank is expressly limited to the exercise of ordinary diligence and care to prevent the opening of the within-mentioned safe deposit box during the within-mentioned term, or any extension or renewal thereof, by any person other than the lessee or his duly authorized representative and failure to exercise such diligence or care shall not be inferable from any alleged loss, absence or disappearance of any of its contents, nor shall the bank be liable for permitting a colessee or an attorney in fact of the lessee to have access to and remove the contents of said safe deposit box after the lessee's death or disability and before the bank has written knowledge of such death or disability.'

"Saying, 'I defy anyone, lawyer or lay person, to understand or explain what that means,' Attorney General Robert Abrams sued the Lincoln Savings Bank in New York City . . . demanding that it simplify a customer agreement on safe-deposit boxes.

"The case is settled . . . The former 121-word sentence now says: 'Our liability with respect to property deposited in the box is limited to ordinary care by our employees in the performance of their duties in preventing the opening of the box during the term of the lease by anyone other than you, persons authorized by you or persons authorized by the law.'"

But these fears were unfounded: there has been no flood of lawsuits. The arguments of the technically minded still have their appeal, however, as demonstrated by Connecticut's plain language law. It lists two alternate sets of criteria for plain language, with nine and eleven different tests respectively, including these:

1. The average number of words per sentence is less than twenty-two; and
2. No sentence in the contract exceeds fifty words; and
3. The average number of words per paragraph is less than seventy-five; and
4. No paragraph in the contract exceeds one hundred fifty words; and
5. The average number of syllables per word is less than 1.55; and . . .

Cookbook detail like this only guarantees that the spirit of plain language legislation will be lost in attempts to follow it to the letter: the clarity of a sentence becomes less important than seeing that it has "less than 1.55" syllables per word. And what would happen to interstate commerce if one of Connecticut's neighbors were to require less than 20 words per sentence instead of 22, or 160 words per paragraph instead of 150?

Keeping plain language laws simple, like the law in New York State, is essential. The courts should be left free to judge particular cases according to a general standard, as they do now with the concept of the "reasonable man." Lawyers who hang back, in favor of the traditional approach to legal language, should remember the traditional result: as Harold Laski put it, in every revolution the lawyers lead the way to the guillotine.

University of Wisconsin–Extension Equal Opportunities Program Office and Department of Agricultural Journalism

Guide to Nonsexist Language

With a little thought, you can use accurate, lively, figurative language in your classrooms, publications, columns, newsletters, workshops, broadcasts and telecommunications—and still represent people fairly. Breaking away from sexist language and traditional patterns can refresh your style.

You can follow two abbreviated rules to check material for bias: Would you say the same thing about a person of the opposite sex? Would you like it said about you? That's the bottom line. Use your own good sense on whether a joke, comment or image is funny—or whether it unfairly exploits people and perpetuates stereotypes.

Most fairness rules improve communication. Use explicit, active words; give concrete examples, specifics and anecdotes to demonstrate facts; present your message in context of the "big picture"; draw your reader in; use parallel forms; present a balanced view; and avoid clichés and generalizations that limit communication. Balanced language rules also guide you to choose or create balanced visual images.

Good communication respects individual worth, dignity, integrity and capacity. It treats people equally despite their sex, race, age, disability, socioeconomic background or creed. And it expresses fairness and balance. To communicate effectively, use real people, describe their unique characteristics and offer specific information. Using stereotypes or composites stifles communication and neglects human potential.

Routinely using male nouns and pronouns to refer to all people excludes more than half the population. There have been many studies that show that when the generic "he" is used, people in fact think it refers to men, rather than men *and* women. Making nouns plural to ensure plural pronouns can help you avoid using the singular "generic" male pronoun.

Many professional titles and workplace terms exclude women and unfairly link men with their earning capacity, while others patronize and subordinate women. Such nongeneric titles reinforce assumptions restricting women and men to stereotypical roles, inaccurately identify people, and give false images of people and how they live and work.

As a communicator, teacher or illustrator, you can help correct and eliminate irrelevant and inaccurate concepts about what it means to be male or female, black or white, young or old, rich or poor, healthy or disabled, or to hold a particular belief.

Editing, publishing and style manuals recognize the need for creating accurate, quality messages without slighting anyone and are beginning to prescribe standards for writing and evaluating manuscripts that represent people without stereotyping them.

Colleges and universities may want to develop their own booklets on nonsexist communication. For example, Franklin and Marshall College (PA) and Michigan State University have developed their own materials on bias-free language.

PRONOUNS: Each Person, to the Best of Her or His Ability

1. *Address Your Reader*
   No. If he studies hard,
   a student can make the honor roll.

   Yes. If you study hard, you can make the honor roll.

2. *Eliminate the Pronoun*
   No. Each nurse determines
   the best way she can treat a patient.

   Yes. Each nurse determines the best way to treat a patient.

3. *Replace Pronouns with Articles*
   No. A careful secretary consults
   her dictionary often.

   Yes. A careful secretary consults a dictionary often.

4. *Use Plural Nouns and Pronouns*
   No. Teach the child to walk
   by himself.
   He is expanding his operation.
   Everyone needs his own space.

   Yes. Teach children to walk by themselves.
   They are expanding their operation.
   All people need their own space.

5. *Alternate Male and Female Pronouns Throughout Text*
   No. The baby tries to put
   everything he finds in his mouth.

   Yes. The baby tries to put everything she finds in her mouth.

6. *Use Both Pronouns and Vary Their Order*
   No. A worker with minor
   children should make sure his
   will is up to date.

   Yes. A worker with minor children should make sure her or his will is up to date.

7. *Use Specific, Genderless Nouns*
   No. The average man on the
   street speaks his mind on the issues.

   Yes. The average voter speaks out on political issues.

8. *Substitute Job Titles or Descriptions*
   No. He gave a test on Monday.

   Yes. The professor gave a test on Monday.

9. *Repeat the Noun or Use a Synonym*
   No. The professor who gets published frequently will have a better chance when he goes before the tenure board.

   Yes. The professor who gets published frequently will have a better chance when faculty tenure is granted.

(*Note:* We don't recommend using "their" to refer to a singular noun.)

(*Note:* Nations, battleships, gas tanks and other objects have no gender.)

Titles: People Working
Replace Language Stereotyping Men

No	Yes
Businessman/men	Business person/people, people in business, executive, merchant, industrialist, entrepreneur, manager
Cameraman	Camera operator, photographer
Chairman	Chairperson, chair, moderator, group leader, department head, presiding officer
Congressmen	Members of Congress, Representatives, congressmen and congresswomen
Craftsman	Craftsperson, artisan
Deliveryman/boy	Delivery driver/clerk, porter, deliverer, courier, messenger
Draftsman	Drafter
Fireman	Fire fighter
Foreman	Supervisor
Guys	Men, people
Headmaster	Principal
Kingpin	Key person, leader
Lumberman	Wood chopper, tree/lumber cutter
Male nurse	Nurse

Manhole/cover	Sewer hole, utility access/cover
Man-hours	Labor, staff/work hours, time
Man-made	Manufactured, handbuilt, hand made, synthetic, simulated, machine-made
Night watchman	Night guard, night watch
Policeman	Police officer, detective
Pressman	Press operator
Repairman, handyman	Repairer (Better: plumber, electrician, carpenter, steam fitter's apprentice)
Salesman/men	Salespeople, salesperson(s), sales agent(s), sales associate(s), sales representative(s), sales force
Spokesman	Representative, spokesperson, advocate, proponent
Sportsman	Sports/outdoor enthusiast (Better: hunter, fisher, canoer)
Sportsmanship	Fair play
Statesman	Political leader, public servant, diplomat
Statesmanship	Diplomacy
Steward/stewardess	Flight attendant
Weatherman	Weather reporter, meteorologist
Workmen	Workers

Replace Titles Stereotyping Women

No	Yes
Authoress	Author
Aviatrix	Pilot, aviator
Career girl/woman	Professor, engineer, mathematician, administrative assistant
Coed	Student
Gal, Girl, Girl Friday	Woman, secretary, assistant, aide (Better: full name)

Housewife, lady of the house	Homemaker, consumer, customer, shopper, parent
Lady/female doctor, lawyer	Doctor, lawyer
Little lady, better half	Spouse, partner
Maid, cleaning lady	Houseworker, housekeeper, custodian
Poetess	Poet
Sculptress	Sculptor
Usherette	Usher
Waitress	Wait person, waiter
Working wife/mother	Worker

Replace Stereotypical Adjectives and Expressions

No	Yes
Act like a lady and think like a man	Act and think sensitively and clearly
Act like a gentleman/man	Be polite, brave, keep your chin up
Dear Sir	Dear Madam or Sir, Dear Personnel Officer/Director, Dear Executive/Manager (Better: name)
Fatherland	Homeland, native land
Founding fathers	Pioneers, colonists, patriots, forebears, founders
Gentleman's agreement	Informal agreement, your word, oral contract, handshake
Ladylike, girlish, sissy, effeminate	Tender, cooperative, polite, neat, fearful, weak, illogical, inactive (Both male and female characteristics)
Lady luck	Luck
Layman, layman's terms	Lay, common, ordinary, informal, nontechnical
Maiden name	Birth name
Maiden voyage	First/premiere voyage
Male chauvinist	Chauvinist

Male ego	Ego
Man-sized	Husky, sizable, big, large, voracious
Man-to-man defense/talk	Player-to-player, person-to-person, face-to-face, one-to-one
Manly, tomboy	Courageous, strong, vigorous, adventurous, spirited, direct, competitive, physical, mechanical, logical, rude, active, messy, self-confident (Both female and male characteristics)
Mother doing dishes, father reading the paper	Men and women doing dishes, women and men reading the paper (Note: Also applies to visual images)
Mother Nature, Father Time	Nature, time
Mothering, fathering	Parenting, child-rearing
Motherly	Protective, supportive, kind
Unwed mother	Mother
Woman did well for a woman/ as well as a man	Woman did well, woman performed competently
Woman's/man's work	Avoid (Too broad; use specifics)
Women's page	Lifestyle, living section

Part 3

Business and Technical Correspondence

Traditionally, business and technical correspondence has taken two forms: the letter and the memo. More recently, it has assumed a third electronic form, e-mail. This electronic innovation has brought with it a myriad of issues, most of which unfortunately remain unresolved. There has been little discussion of, let alone agreement about, the etiquette of e-mail.

In the past, business and technical correspondence *seemed* easier on some levels: letters went to people outside of the company; memos, to people inside the company. Letters were advertisements for writers and their companies. Memos provided a record of decisions made and actions taken within a department of a company.

Letters and memos announce or reaffirm policies, confirm decisions and conversations, and send or request information. Some letters and memos are routine; others concern pressing issues. In either case, letters and memos require careful writing.

Given the volume of correspondence many companies produce, it is easy for business and technical writers to be tempted to use shortcuts when writing. Looking for a shortcut can, however, court danger if the shortcut involves relying on form responses or copies of previous—though not necessarily good examples of—similar correspondence.

There is nothing wrong with form letters or memos, as long as such correspondence is appropriate to the given writing situation. A letter reminding customers of bills past due or a memo transmitting attached data is often an effective and time-efficient way to communicate—provided such letters and memos are appropriate to the situations and audiences with which they are used.

The danger with form letters and memos is that they can become a crutch that writers depend on even in situations in which they are inappropriate. Such misuse of form letters and memos can create new problems for writers. Writers may have to send a follow-up letter or memo to fix what they tried to do in the first piece of correspondence. Or worse, they might have to send *several* follow-up messages.

A process approach to business and technical correspondence will help determine when form letters and memos are appropriate and when something more original is required. The careful use of the writing process can also make follow-ups unnecessary.

The advent of electronic correspondence would seem to offer a boon to business and technical writers, but every silver lining has its cloud. The ease with which electronic correspondence can be transmitted has led some writers to become more casual—if not careless and sloppy—in their electronic correspondence than they would be in producing correspondence in the more traditional forms of letters and memos.

Electronic correspondence can be sent within and outside companies to multiple audiences across the globe. A misstatement, mistake, or unintended slight can suddenly be transmitted to a very large audience, making the writer look foolish or worse to a large body of readers. As I indicated earlier, there is yet no established or generally adopted etiquette for electronic correspondence. In the absence of such etiquette, writers would be wise to be conservative in such correspondence, adhering closely to the rules and principles that work so well for letters and memos. Whether the product is a letter, a memo, or an electronic message, a process approach is the safest approach to take.

The selections in this section of *Strategies* follow a decidedly process-oriented approach to business and technical correspondence. Readers in the world of work are impatient with people who waste their time, and David V. Lewis offers some suggestions on how to write to people effectively by showing business writers how to save their readers time while also selling themselves through their correspondence.

Writers for the Royal Bank of Canada agree with Lewis, arguing that *all* letters can be letters that sell. They suggest writers use formulas for their correspondence to achieve this goal. Of course, when the purpose is to say "no," writers face special difficulties. As Allan A. Glatthorn suggests, writers can say "no" in their letters and still keep or make friends. Harold K. Mintz offers a formula for writing better memos. John S. Fielden and Ronald E. Dulek suggest a model for efficient writing, and Heidi Schultz concludes this section of *Strategies* with a discussion of the advantages and problems of that seemingly ubiquitous new form of business communication, electronic mail.

David V. Lewis

Making Your Correspondence Get Results

When he wrote this essay, David V. Lewis was in-house consultant in sales and management training with Western Company of North America in Fort Worth, Texas.

If you turn out an average of five letters a day, you'll produce nearly twice as many words during the year as the typical professional writer.

These letters reflect *your* attitude—and obviously your organization's—toward customers. Realizing this, many progressive organizations train key people in the art of writing readable, results-getting letters. For example, the New York Life Insurance Company produces more than a million letters a year and uses its correspondence as a public relations tool.

"Anyone who writes a letter for New York Life holds a key position in our organization," says Nathan Kelne, vice president of the company. "By the letters our people write, they help determine how the public feels toward our company, and toward the life insurance business as a whole. Since they are instrumental in shaping the personality of the company, they are in a very real sense public relations writers."

For example, before New York Life launched its companywide letter-writing courses, here's how one of its executives tried to explain to a beneficiary the way the death claim was to be paid on a $3,000 policy:

> The monthly income per $1,000 under option 10 years certain is $7.93 per $1,000 total face amount of insurance, which total is $3,000.

Fortunately, the beneficiary was able to cut through the jargon. He replied:

> As I understand your letter, you seem to be saying that I should receive a monthly check for the amount of $23.79 for at least 10 years.

But another policyholder's response to a similarly bewildering letter is more typical:

Please tell me what you want me to do, and I'll be glad to do it.

Some universally proven principles that can help you sell yourself and your company to the public will now be examined.

WRITE FOR HIM, NOT TO HIM

General Foods is an organization that believes strongly in creating a favorable image through its correspondence. The public relations people came across this letter signed by a marketing executive and ready for mailing:

```
Dear Sir:
    Enclosed please find a questionnaire which we are sending to
all our retail contacts in this state.
    Will you please answer this as soon as possible? It's very
important that we have an immediate reply. We're delaying final
plans for our retail sales program in this area until we get
answers to the questionnaire.
    With thanks in advance, we are,

                                        Gratefully yours, . . .
```

The letter is clear and to the point; it *does* communicate readily. But there's a major flaw. It points out benefits the company will receive instead of suggesting how the program will help the recipient. The letter is writer- rather than reader-oriented.

Psychologists say that each of us is basically interested in himself or herself. We want to know "what's in it for me?" Once you routinely approach letter writing from this point of view, you'll find yourself telling your readers in specific terms how the letter will benefit them.

The questionnaire letter was rewritten this way:

```
Dear Sir:
    Enclosed is a questionnaire on our proposed retail sales pro-
gram. Your advance opinion on how this new plan would help you
and others will be very useful in our evaluation of the program.
    If you will fill out and return this questionnaire as soon as
possible, we can let you and other retail contacts in this area
know promptly of changes in the present program. Thank you.

                                        Very truly yours, . . .
```

The revised version asks for your advance opinion, suggests how quick action might help you and *others,* and promises to let you know promptly of changes. There's also another improvement. The tired and outdated "With thanks in advance" has been replaced by a modern "Thank you."

The best way to persuade your readers to your point of view is to show them that it will be worth their while to do so. This rule holds true for just about every successful letter, sermon, sales presentation, or advertisement. A good sales letter, for example, is much like a good ad in that it attempts to dramatize benefits to the reader.

What does it take to get—and hold—a reader's attention? The most powerful letters appeal to basic needs and emotions rather than to purely logical reasons. Mainly, management people want to know how to save time, money, and effort. Show them how your product or service can help them do one or more of these things, and you're likely to get their attention.

With effort, almost any letter can be oriented to the reader, even the usually hard-to-write collection letter. Here's such a specimen, written almost entirely from the writer's point of view:

```
Dear Sir:
    Our records show that you are three months delinquent in pay-
ment of your bill for $37.50.
    Perhaps this is an oversight on your part. Otherwise we cannot
understand why you have not taken care of this obligation.

                                    Very truly yours, . . .
```

Keeping the reader's self-esteem in mind, the letter could have been rewritten this way:

```
Dear Sir:
    We know you'll want to take care of your small past-due
account for $37.50. This will help you to maintain the fine
credit you have built up with us over the years.

                                    Sincerely, . . .
```

To help develop the "you" attitude, put yourself in your reader's shoes, then write from his or her point of view. Once you've developed this attitude, you'll automatically start telling your readers what's in it for them. Instead of saying, "I wish to thank you," you'll write, "Thank you." Instead of writing, "We'd like to have your business," you'll write, "You'll find our service can help your business in many ways."

PERSONALIZE YOUR LETTERS (THERE'S POWER IN PRONOUNS)

Some years ago, many would have considered this letter to have been perfectly proper and very effective:

```
Gentlemen:
    Enclosed herewith are the subject documents which were
requested in yours of the 10th. The documents will be duly
reviewed and an opinion rendered as to their relevancy in the
involved litigation.

                                    Very truly yours, . . .
```

The letter *was* all right—back in the horse-and-buggy days! Phrases like "enclosed herewith," "duly received," and "involved litigation" would have marked the writer as learned. But the executive who makes a habit of writing like that today is generally regarded as an anachronism.

Current usage calls for clear, to-the-point letters, written mostly in conversational language. Like good conversation, your letters should generally be friendly, filled with personal references, and almost always informal in tone and language. Here's how the horse-and-buggy letter might have been rewritten by the modern executive:

> Dear Sam:
>
> Here are the documents you asked for. I'll look them over and let you know if we can use them in our lawsuit.
>
>   Cordially, . . .

When experts write a letter, even a form letter, they generally try to make it sound as personal as possible. Almost always, it contains a sprinkling of personal pronouns. In orienting your letter to the reader, fill it with "you," "your," and "yours." Use "I" and "me" sparingly.

Here's a case in point, a letter sent out by a mortgage company (emphasis added by the author):

> Dear Mr. Jones:
>
> *We* want to thank you for your query about *our* new mortgage insurance plan.
>
> *We* are enclosing a pamphlet which outlines benefits of *our* new policy and gives testimonials from some of *our* policyholders.
>
> *We* would like very much to enroll you within the next 30 days, since we are offering a special low premium rate as *our* introductory offer.
>
>   Very truly yours, . . .

The repeated use of *we* and *our* clearly shows the letter is written with the mortgage company's interests at heart. But notice how substituting *you* and *yours* orients the letter to the reader's interest. (Emphasis has again been added by the author.)

> Dear Mr. Jones:
>
> Thank *you* for *your* recent inquiry about our new mortgage insurance plan.
>
> *You'll* find that the enclosed pamphlet outlines benefits of the policy in detail. *You'll* also probably be interested in the testimonials furnished by some policyholders.
>
> If *you* sign up within the next 30 days, *you'll* be able to take advantage of special premiums we're offering on an introductory basis. Thank you.
>
>   Cordially, . . .

The word is out now in enlightened business circles. Companies are telling their executives to regard every letter as a personal contact: to write your own way, to develop your own style (within bounds), to make your letter distinctively *you*.

MASTERING TONE (YOUR PERSONALITY IN PRINT)

Writing a "rejection letter" that leaves the recipient's self-esteem intact and preserves his or her goodwill is a difficult task. It requires tact, diplomacy, and empathy—all of which must be effected through appropriate *tone*. For example, here's the way one banker turned down a builder, a long-time customer:

```
Dear Mr. Jones:
   We regret to inform you that your request for an additional
loan in the amount of $250,000 must be rejected. It was the
judgment of the loan committee that, with your present commit-
ment, such a loan would present too much of a risk for us.
   Naturally, we look forward to doing business with you on your
existing commitment.


                              Sincerely, . . .
```

What's wrong: Tone, mainly—the *way* the rejection is phrased. It deflates the reader's ego and makes it virtually certain the builder will do his banking elsewhere in the future.

True, the rejection was "justified." But why not let the reader down more gently and leave the door open for future business? For example:

```
Dear Mr. Jones:
   One of the most distasteful tasks we have is to turn down a
loan application, particularly when it is from a regular and
respected customer like you.
   We know you'll be disappointed, and so are we. We share your
excitement about your plans to expand the Roseland Project, and
hope that circumstances will later warrant our working with you
on this and other projects.
   However, the current economic outlook, combined with your
delinquent status on your existing loan, makes your new loan
application a questionable venture for us at the present time.
   We will be glad to work with you in any way we can to resolve
your current financial problems, and we look forward to helping
you meet your future financial commitments.


                              Sincerely, . . .
```

The two letters say the same thing, but in vastly different ways. Suffice it to say that tone is as important to a letter as good muscle tone is to an athlete. A negatively phrased letter can have the same effect as an abrasive personality. As one New York Life executive put it:

By its very nature, the "no" letter is a turndown and leaves the reader unsatisfied. Yet there are ways to soften the blow, and one of the best is simple candor. So, if you must say "no," state the reason first: "Currently there are no vacancies in that department. Therefore, I cannot offer you a position with the company at this time."

A "no" response often requires positive alternatives if they are appropriate (for example, "Although, for the reasons mentioned, I cannot do as you ask, may I suggest that . . ."). You have to be more diplomatic and more sensitive if you're to have any chance of salvaging your reader's goodwill or ego.

And that really is your goal in an effective "no" letter—to tell your readers something they don't want to hear in a way that compels understanding and, ideally, acceptance.

Unfortunately, tone requirements vary not only from person to person, but from one situation to another. For example, you would ordinarily use fast-paced, persuasive prose for a sales-promotion letter. In writing to a highly regarded law firm, you might use a slightly more "dignified" approach. If you're writing for a service-type organization, you might use a middle-of-the-road approach, as the government does.

Government letter writers are told to strive for a tone of "simple dignity," to make letters brief and to the point, and to avoid gobbledygook. "Don't act as if you're the only game in town," they're told. "On the other hand, don't bow and scrape just because you're a service organization."

Poor letters start with poor thinking on the part of the writer. Negative thoughts lead to negative words—and before you know it, there goes the old ball game. Here's a letter from a manager of a department store to a customer who complained that an appliance she had recently bought didn't work. (Negative words have been italicized by the author.)

> Dear Sir:
>
> I am in receipt of your letter in which you *state* that the hair dryer you purchased from us recently *failed* to meet the warranty requirements.
>
> You *claim* that the dryer *failed* to do the things *you say* our salesman promised it would do.
>
> *Possibly you misunderstood* the salesman's presentation. Or perhaps *you failed to follow instructions* properly. We *positively* know of no other customer who has made a similar complaint about the dryer. The feeling is that it will do all that is stated *if properly used.*
>
> However, we are willing to make some concessions for the *alleged* faulty part. We will allow you to return it; however, *we cannot* do so *until* you sign the enclosed card and return it to us.
>
> Very truly yours, . . .

The tone is unmistakably negative. "State" and "failed to meet" in the opening paragraph imply there's some doubt that the claim is valid. In the second paragraph, "you say" suggests the salesman didn't make the statement at all.

Such phrases as "you misunderstood," "you failed to follow instructions," and "if properly used" tell you in so many words that you're not too bright. And to wrap it up, the writer uses the negative "we cannot . . . until" instead of the positive "we will . . . as soon as."

Most letters aren't this negative, of course. But it doesn't take much to offend. Any of these negative words or phrases, in themselves, could have spoiled the tone of an otherwise effective letter.

If you've been guilty of taking a negative approach, study the following examples. In each case, the negative thought (emphasis added by the author) has been converted into a positive one.

Negative	Since you *failed* to say what size you wanted, we cannot send you the shirts.
Positive	You'll receive the shirts within two or three days after you send us your size on the enclosed form.
Negative	We *cannot* pay this bill in one lump sum as you requested.
Positive	We can clear up the balance in six months by paying you in monthly installments of $20.
Negative	We're sorry we *cannot* offer you billboard space for $200.
Positive	We can offer you excellent billboard space for $300.
Negative	We are *not* open on Saturday.
Positive	We are open from 8 A.M. to 8 P.M. daily, except Saturday and Sunday.

Negative words aren't the only cause of poor letter tone. Many letters are made more or less "neutral" by mechanical, impersonal, or discouraging language. Here's an example of each fault with a preferred alternative:

Interpersonal	Many new names are being added to our list of customers. It is always a pleasure to welcome our new friends.
Personal	It's a pleasure to welcome you as our customer, Mr. Jones. We will make every effort to serve you well.
Mechanical	This will acknowledge yours of the 10th requesting a copy of our company's annual report. A copy is enclosed herewith.
Friendly	Thanks for requesting a copy of our annual report, which is enclosed. We hope you will find it helpful.
Discouraging	Since we have a shortage of personnel at this time, we won't be able to process your order until the end of the month.
Encouraging	We should have more help shortly, which will enable us to get to your order by the end of the month.

Ideally, the tone of your letter should reflect the same ease in conversation that you enjoy when talking about your favorite hobby, business, or pastime.

HOW TO WRITE (MORE) THE WAY YOU TALK

This letter was sent out by a large department store. Imagine yourself as the recipient.

```
Dear Sir:
   We are in receipt of yours of July 10 and contents have been
duly noted.
```

As per your request, we are forwarding herewith copies of our new fall brochure. Thanking you in advance for any business you will be so gracious as to do with us,

Yours truly, . . .

Conversational? Of course not. Who uses such language as "We are in receipt of . . . ," "duly noted . . . ," and "as per your request"? Practically no one. People just don't talk that way. Most of these phrases went out with the Model-T—or should have.

Face to face, the writer probably would have said, "Here's the fall brochure you requested. Let me know if we can serve you." It says the same thing, in less space, and without all the fuss.

Some business and professional people still feel it isn't quite proper to write conversationally, mainly because they have seen so much stilted business writing. But many progressive companies are telling their people to communicate in plain English, using only those technical terms that are absolutely necessary. As one executive of a major company said, "The best letters are more than just stand-ins for personal contact. They bridge any distance by the friendly way they have of talking things over person to person." And this from a U.S. Navy bulletin: "At best, writing is a poor substitute for talking. But the closer our writing comes to conversation, the better our exchange of ideas will be."

The consensus clearly is that informal, natural business writing is *in;* stilted business writing is *out.*

One word of caution about writing (more) the way you talk. Since World War II, readability experts have urged business people to "write more the way they talk" or "write as they talk."

Detractors have soft-pedaled the idea, claiming that most conversation is rambling, often incoherent, and frequently a bit too earthy. These are valid objections, but they miss the point. You're not being asked to write *exactly* the way you talk. Rather, you're being asked to bridge the gap between the spoken and the written word—to narrow the difference in the way you would *give* an order verbally and the way you would *write* that same order in a memo. Naturally, writing requires more restraint than speech, and the writer must normally use fewer words. But you can do these things and still capture the tone and cadence of spoken English.

The first step in making your correspondence more conversational is to rid your vocabulary of worn-out business phrases. Here are some of the more flagrant offenders. They were stylish once, but they've done their duty and need to be honorably discharged.

Old Hat	Conversational
At a later date	Later
If this should prove to be the case	If this is the case
This will acknowledge receipt of	Thank you for
Attached herewith please find	Here is; Enclosed is
We shall advise you accordingly	We'll let you know

Old Hat	Conversational
Due to the fact that	Because
With regard to	About
Please notify the writer as to	Please let me know
Enclosed please find a stamped envelope	I've enclosed a stamped envelope
We are submitting herewith a duplicate copy	Here is a copy
In compliance with your request	Here is
We are submitting herewith our check in the amount of $75	Here is our check for $75
We beg to advise (acknowledge) that	[Begging is unnecessary]
The information will be duly recorded	We'll record the information
The subject typewriter	This typewriter
In compliance with your request	As you requested
We will ascertain the facts and advise accordingly	We'll let you know
The writer wishes to state	[Just say it]

This list is far from complete but you get the idea. Once you're mentally geared to writing more conversationally, you'll detect many other clichés. Try to eliminate them.

Contractions will also make your writing more conversational. They play a part—a very large part—in almost everyone's everyday conversation. Even your most learned associate doesn't say, "We shall endeavor to be there at eight o'clock"; he's more likely to say, "We'll try to be there at eight o'clock." Instead of saying, "I am going to the ball game," he'll probably say, "I'm going to the ball game." It takes less effort to say, "You needn't bother to call," than to say, "You need not bother to call." Contractions tend to make the spoken words flow more smoothly. That's why they make writing appear more natural.

Probably the most common contractions are here's, there's, where's, what's, let's, haven't, hasn't, hadn't, won't, wouldn't, can't, couldn't, mustn't, don't, doesn't, didn't, aren't, isn't, and weren't. Then there are the pronoun contractions: I'll, I'm, I'd, I've, he's, he'll, he'd, and so forth.

Using these and other contractions when they facilitate the flow of words will do much to give your writing a quality of spontaneity and warmth. But they must be used with discretion. Using too many contractions can sometimes make your writing too informal. And used in the "wrong" place, they might not "sound right." Indeed, the key is whether the contraction sounds right when the sentence is read.

Take this section from the Gettysburg Address: "But in a larger sense, we cannot dedicate, we cannot consecrate, we cannot hallow this ground." The passage would undoubtedly have lost its historic tone if *can't* had been used instead of cannot.

On the other hand, the advertising slogan "We'd rather fight than switch" would lose some of its punch if phrased, "We would rather fight than switch." Appropriate usage depends on how the contraction makes the sentence sound when it's read aloud.

Next time you get ready to write a letter, ask yourself, "How would I say this if I were talking to the person?" Then go ahead and write in that vein.

The Royal Bank of Canada

Letters That Sell

Everyone writes letters that sell, and every letter has as its purpose the selling of something: goods, services, ideas or thoughts.

Someone may say that a family letter has no such purpose, but consider this: a letter telling about the children seeks to promote a favourable impression of their welfare and happiness; a letter telling about illness is designed to gain sympathy; the letter that says nothing but "I hope you are well" is selling the idea "I am thinking of you."

Family letters are usually rambling letters. They would be improved both in their readability and their informativeness if they adopted some of the principles that are used to sell goods and services. Business building letters, on the other hand, could with advantage incorporate some of the friendly informality of family letters.

Salesmanship of any kind is basically a person moving goods by persuading another person that he needs them, or winning that person's support or approval of an idea or a plan.

Some noncommercial type sales letters are those that champion good causes, such as community welfare or health standards or national unity. They seek to influence the thinking of individuals or groups.

It is not a simple task to compose a letter designed to sell. Like any other product of value, it calls for craftsmanship. There are techniques to be learned, techniques of conveying ideas, propositions, conclusions or advice appealingly and purposefully.

IN THE BEGINNING

In creating a letter to sell something we need to begin by thinking about the person to whom we are writing. A lawyer studies his opponent's case just as sharply as his client's; the manager of a baseball or hockey team analyses the qualities, good and bad, of members of the opposing team.

The writer must anticipate and answer in his letter questions that will occur to the reader: What is this about? How does it concern me? How can you prove it? What do you want me to do? Should I do it?

People buy goods or services because these will give them a new benefit or extend or protect a benefit they already have, so the writer needs to translate what he offers into owner benefits.

The proffered benefits must be accessible and adapted to the reader's position, environment and needs. No letter is likely to sell sun-bonnets to people who live beyond the Arctic Circle or baby carriages to bachelors. We may classify a potential customer as a man, woman, company or institution that will have use for a product or service, has sufficient money to pay for it, and in whom a desire for possession may be created.

The reader's interest: that is the guiding star in sales letter writing. See his interests, his angle, and accommodate your stance to them. A simple precaution against sending a letter to the wrong person is to ask yourself what use you would have for the commodity if you were in the reader's place.

It is a good rule to spend more time thinking about the reader than about what you have to say. Otherwise you may become wrapped up in the virtues of your product so that you forget that the decision to buy rests with your prospect.

The self-interest of the person to whom you write is a major factor to consider in successful sales communication. When you remember it, you give the impression that you have singled out this reader as being an important individual, and that is an excellent introduction.

It is not to be expected that the writer of letters that sell will know every person to whom he writes, but he must know certain facts: approximate income and age, occupational level, his business, and things like that. Then he is able to slant his sales points accurately toward the reader's needs, interests and purchasing power.

KNOW YOUR PRODUCT

The reader's attention should be attracted to the product or service, not to the grand style or picturesque phraseology of your letter. When you catch a person's attention you are focusing his consciousness on something. Concentrate on your commodity. The best magnet to draw and hold attention is what you say about the product, showing it to be useful and the means of fulfilling a desire.

It is no small accomplishment to analyse and marshal into order the facts about a product so as to win the thoughtful consideration of a person who has plenty of other things on his mind.

In purchasing almost any sort of commodity, the buyer has a choice between what you are offering and what others are selling. Your sales job is to show the superiority of your product. Tell why what you offer is necessary or desirable, what it will accomplish in your reader's business, and how it can be fitted into his present layout and his plans. Do not content yourself with telling about the article as it sits on display: picture it in use in the reader's home or factory.

Your letter needs to convey the assurance that you are telling the truth about your goods. It is not a sensational offer that makes a letter convincing, but the feeling that the reader can depend upon what is said. He should feel assured that he will be buying what he thinks he is buying. Customer dissatisfaction caused by misleading sales talk can cause shock waves that affect the whole selling organization.

LET YOUR PERSONALITY SHOW

Make your letter sound friendly and human: put your personality on paper. Your letter is you speaking. Some of the features in your personality that you can display are: friendliness, knowledge, keen-mindedness, trustworthiness and interest in the prospect's welfare.

What you have in your mind about the good quality, appearance and usefulness of your product has to be communicated to your reader so as to arouse his interest, create a desire to possess, and induce him to buy.

Communication is not the easiest thing in the world to attain in writing, in art, or in music. Dr. Rollo May wrote in *Man's Search for Himself:* "We find in modern art and modern music a language which does not communicate. If most people, even intelligent ones, look at modern art without knowing the esoteric key, they can understand practically nothing."

It is not enough to write something so that it can be read. The degree to which communication occurs depends upon the degree to which the words represent the same thing for the reader as they do for the writer.

The recipient of a letter that is not clear is likely to blame its opacity on the lack of intelligence of the writer.

The art of composing sales letters is not one to be mastered by minds in which there is only a meager store of knowledge and memories.

The art consists in having many mental references and associating them with new thoughts. Consider a poem. Its theme will likely have arisen from a single event, but the images used in its construction will have been drawn from the total life experience of the poet.

Put some flavour into your letters so that they taste good. Your letter will not be like anyone else's. That is a virtue, just as being an individual is a virtue in conversation. Who wishes to be a carbon copy of a textbook letter or to parrot phrases that other people use?

Practise talking on paper as if you were on the telephone. First write down the imagined questions asked by the person on the other end of the line and then your answers, given in simple, direct and pleasing words. To humanize your letters in this way with the natural idiom of conversation does not mean that you use cheap slang or clever verbal stunting.

SHOW SOME STYLE

The style in which you write is not a casual feature of your letter. It is vital to your reader's understanding of what you are saying to him. It is not your job to please the reader's sense of the aesthetic, but to tell and explain plainly what is necessary to introduce your goods or your idea to his favourable attention. This may be done in a way that has grace and comeliness.

Never "talk down" to a reader. Make him feel that he knows a great deal, but here is something he may have missed. There is a big difference, when trying to build business, between making a suggestion and preaching a sermon.

It is highly important in writing a letter to sell something that it should be appropriate. Whatever your writing style may be, it will fit the occasion if it gives this particular correspondent information that will be useful to him, conveys to him a feeling of your interest in him and his business, and assures him of your goodwill.

Besides being grammatically correct, language should be suitable. At one extreme of unsuitability is the language that is too pompous for its load, and at the other is the language of the street which belittles the receiver's intellectual level.

Your words should be the most expressive for their purpose that the language affords, unobstructed by specialty jargon, and your sentences should be shaken free of adjectives—the most tempting of forbidden fruit to a person describing something.

Properly chosen words will convey your appreciation of the addressee as a person, and such friendliness is contagious. Some people are afraid to be friendly in their letters. They fear they will be thought of as "phonies" who have disguised themselves as Santa Claus for the occasion. Being friendly and showing it should not raise this scarecrow. It would be a grave mistake, indeed, for any of us to indulge in flowery language foreign to our natural talk; but it is no mistake at all to incorporate in our letters the warm, personal language that comes naturally to us in person-to-person social contacts.

Letter writing invites us to use the same etiquette as we use in courteous conversation. We look at the person with whom we are talking, converse on his level of understanding, speak gently, and discuss matters he considers important or interesting.

What the reader of your letter will notice is not its normal courtesy, but the extra touch that demonstrates care and understanding, a genuine interest in the reader's wants, a wish to do what is best for him, and the knowledge you show of how it can be done.

Everyone who writes a letter has a moral as well as a business reason to be intelligible. He is placing his reader under an obligation to spend time reading the letter, and to waste that time is to intrude upon his life plan.

There is an eloquence of the written as well as of the spoken word. It consists in adapting a statement to the receptive system of the reader so that he will have maximum help against confusion, against mistaking what is incidental from what is fundamental. A familiar device to use in this effort is to relate the new commodity you are offering to something that is familiar.

USE SUITABLE FORMULAS

There are formulas you may wish to make use of. Your letter must conform in some respects to what letters are expected to be. This does not mean pouring all letters into the same mould. Within the accepted pattern you are free to develop your talent for expression.

Skill is needed in the use of formulas. A form letter reveals itself to the reader and gets short shrift. It is possible to make use of the form as a guide to what points to cover, and then speak your piece on paper in a natural way.

Here are three formulas for letters. The first may be called the sales formula, the second the logical formula, and the third the rhetorical formula.

1. Get attention, provoke interest, rouse desire, obtain decision. Attention is curiosity fixed on something; interest is understanding of the nature and extent of what is new and its relationship to what is old; desire is the wish to take advantage of the proffered benefits; decision is based on confidence in what the writer says about his goods.
2. This is summarized: general, specific, conclusion. You start with a statement so broad and authoritative that it will not be disputed; you show that the general idea includes a specific idea; the conclusion is that what has been said about the general idea is also true of the specific.
3. This is very simple: picture, promise, prove, push. You write an attractive description of what you are selling; you promise that it will serve the reader well in such-and-such a way; you give examples of the commodity in use, proving that it has utility and worth; you urge the reader to take advantage of the promised values.

SELLING NEEDS IDEAS

Selling is done with ideas, so never throw away an idea even if it is of no use at the moment. Put it into your idea file where it will rub against other ideas and perhaps produce something new. The file is like an incubator. Thoughts and fancies you put into it will hatch out projects and plans.

Imagination helps in this operation. A correspondent of ordinary ability may never write anything that is not absolutely accurate and yet fail to interest his readers. This is a real weakness: to be perfect as to form but lacking in imagination and ideas.

Imagination should be given priority over judgment in preparing your first draft of a letter designed to sell. Then put reason to work: delete what is unnecessary; marshal your sentences into logical form so that your ideas advance in an orderly way; revise your words so that your thought is conveyed exactly as you wish it to be.

When you tell the advantages of your product or service or idea, and show how it will fill a need in the reader's life or job, in clear, truthful words placed in easily understood sentences brightened by ideas and imagination, you have done a good job of writing a sales letter.

Desire of the reader to do what you want done is created, just as in conversation, by both rational and emotional means, by proof and by persuasion, by giving reasons. Some goods and some buyers need nothing more than facts. An office manager buying pencils or pens for his staff will respond to an informative, factual, statistical sales letter. He is already sold on the idea of using pencils and pens, so you do not have to coach him about their usefulness; in fact, you may lose a sale if you give the impression of "teaching grandmother to suck eggs." What is needed is to catch his attention, give pertinent information about your product, and show him why buying from you will be profitable to him.

Try to make the information you give really enlightening. Comparing something unknown with something already known makes it possible to talk about the unknown. The analogy (like that between the heart and a pump) can be used as an aid in reasoning and in explaining or demonstrating.

THE SOFT SELL

The tone of a letter designed to sell something should be persuasive rather than insistent. It should seek to create a feeling of wanting, or at least an urge to "let's see."

People do not want to be told how to run their affairs, but anyone who shows them how to do things more economically or faster or better will find keen listeners. Soft sell gives the prospect credit for knowing a good thing when it is shown him, and acknowledges his right to make up his own mind.

The soft sell is recognition of the Missouri mule in human nature. Try to push a mule and he lashes out with his heels. Try to pull him by the halter rope and he braces his legs and defies you to budge him. "In the old cavalry," says A. C. Kemble in *Building Horsepower into Sales Letters,* "they said all it took to get a mule working for you was to recognize that he was an individualist who hated nagging and needed a chance to make up his own mind about things."

One hears a lot in advertising circles about "appeal." It is, according to the dictionary, "the power to attract, interest, amuse, or stimulate the mind or emotions."

Obviously, when you wish to influence someone you must take into account the kind of person you are addressing and what you want him to do. Your appeal must touch his feelings, needs and emotions. It strengthens your position if you can relate your own experience to that of the person you are addressing and write your message around the overlap of that experience.

The sort of mistake to be avoided with great care is slanting your appeal in a way that runs counter to the feelings of those whom you wish to influence.

It has been found in recent years that the advertising messages addressed to older people *as* older people did not win the desired response. In travel, for example, only a very small minority who want or need a sheltered situation are attracted by the semicustodial "trips for the elderly."

The Swiss Society for Market Research decided that "to sell anything to the over–65 age group it is important to keep one concept in mind: most senior citizens are vigorous and independent. Don't try to reach them with a head-on approach to the senior market. It probably won't work."

Writing a letter that pleases the recipient is not enough: it must be designed to lead to action. Do not fear to be explicit about what you want. Coyness in a letter is not attractive, and it exasperates the reader. Answer the reader's questions:

"What has this to do with me?" and "Why should I do what this person is asking me to do?"

You may answer these questions and encourage a purchase by appealing to emotional motives like pride, innovation, emulation, or social prestige; or to rational motives like money gain, economy, security, time-saving or safety.

READ YOUR LETTER CRITICALLY

Imagine your letter to be your garden upon your return from vacation. You have to get into it and prune, clean up, tie up, and trim the edges.

Read the letter as if you were the recipient. How does it strike you? What can be added to attract attention? Is there anything irrelevant in it? Read the letter aloud to capture the conversational rhythm.

If you are not satisfied, do not crumple up the paper on which your draft is written. Try rearranging the paragraphs, the sentences, the words. Give the letter a new twist. Change the shape of your appeal. Delete anything that is distracting.

Be careful when trying to shorten a letter that seems to be too long. While a letter should be as short as possible, consistent with clearness and completeness, it is not the length that counts, but the depth. Since clearness and brevity sometimes get in the way of each other, remember that the right of way belongs to clearness. It will make a good impression if you find occasion to write: "I can be quite brief because this letter deals with a topic already well known to you."

The end of your letter, like the end of your pencil, should have a point. It should answer the reader's natural question: "So what?"

FOLLOW THROUGH

Do not let your customer forget you. When you produce a piece of copy that hits the bull's-eye, that is not the time to sit back and take things easy. It is a time to imagine what you would do if you were in your competitor's chair . . . and then do it first.

Competition is a fact of life. Wherever there are two wild animals trying to live on the same piece of land or two persons depending upon the same source of sustenance, there is competition. The customer who was a prospect to you before he bought your goods is now a prospect to your competitor. With the proper follow-through attention, he will turn to you when he needs up-dating of equipment or new goods.

Writing a letter that sells goods or ideas, and following through so as to retain the customer, requires just as much specialized talent and mental ability as any other kind of advertising, if not more.

When you run into difficulties, composition of the sales or follow-up letter may give you a feeling of confusion. You may feel like throwing up your hands in despair of finding the exactly right slant or the perfect array of words. That is not unnatural. Nietzsche, the German philosopher, said in *Thus Spake Zarathustra:* "I tell you, one must still have chaos in one, to give birth to a dancing star."

The effort is worthwhile. When you set yourself to snap out of the depressing pedestrian type of letter that is so commonplace, you are raising yourself and your firm to a place where people will sit up and pay attention. As a student of sales letter writing you will generate ideas, as a philosopher you will assess the letters as to their purpose and usefulness, and as a writer you will energize them.

To summarize: the backbone of the principles of writing letters that sell is made up of these vertebrae—know why you are writing and what about; believe in what you are writing; be tactful and friendly and truthful; base your appeal on the prospect's interests. . . . and check your letter and revise it.

Allan A. Glatthorn

"I Have Some Bad News for You"

A consultant in management communications and the author of more than thirty books on writing, Allan A. Glatthorn is retired Senior Faculty Member in the Teaching of Writing Program at the University of Pennsylvania.

One of the most difficult letters or memos to write is the one containing bad news. A subordinate requests a salary increase which you cannot grant. A community organization asks for a contribution which you do not wish to make. An unqualified applicant asks for an appointment to discuss a position with your company, and you do not wish to take the time. You decide to have to dismiss a good worker because declining business mandates cutbacks. In each case you have some unpleasant news to deliver—but you wish to deliver it in a way that does not offend or alienate. How do you accomplish this difficult task? Let me offer some general guidelines and then explain the specific techniques.

The first guideline is to remind you that the successful manager is people-sensitive, able to empathize with others. When you receive a request, you stand in the shoes of the petitioner. You realize, to begin with, that from the standpoint of the asker there are no foolish requests. People ask for things that they need, that they believe they are entitled to, that they hope to receive. And you also realize that there is really no good way to break bad news. No matter how empathic and tactful you may be, your bad news will inevitably cause disappointment and will often cause distress.

The second guideline derives from the first: remember that bad news is best delivered face to face. People getting bad news have some strong needs. They want an opportunity to express their negative feelings. They want the chance to press the request or appeal the decision. They want to explore the reasons more fully to be sure that there is no hidden message. These needs can best be met in a face-to-face meeting. So if you have bad news to give to an employee or a valued client, the best method is to confer with the individual, deliver the bad news in person, and then, if desirable, follow up with a written statement for the record. Don't avoid the unpleasant confrontation by writing a memo, hoping that the petitioner will go away quietly.

Finally, remember that everyone values honesty and forthrightness, especially when being disappointed. Typically, the receiver of bad news is inclined to be cynical, often resenting your attempts to be tactful and politic. You therefore should be sure that your desire to soften the blow does not beguile you into distorting the truth. Explain as tactfully as possible the real reasons for the bad news, instead of offering lame excuses.

With these general guidelines in mind, you should be able to send two kinds of "bad news" messages: the indirect and the direct.

THE INDIRECT BAD NEWS MESSAGE

The indirect message of bad news uses the soft and gentle approach. It tries to cushion the blow by burying the bad news in the middle of the letter or memo, surrounded by positive expressions of appreciation. You send the indirect message under one or more of these circumstances:

- You want further contact with the petitioner.
- You want to project the image of a caring individual.
- You believe that the petitioner won't be able to handle a more direct statement.

The formula is a simple one: THANKS . . . BECAUSE . . . SORRY . . . THANKS.

Thanks

You begin with a positive statement. You express appreciation for the idea offered, the interest expressed, the petition received. Here's how an indirect bad news letter from an insurance company begins:

> Thank you very much for your recent inquiry about our automobile insurance coverage. We are pleased that you have considered transferring your account to our company.

Because

You then continue with the reasons for the bad news. The strategy here is that stating the reasons first cushions the shock, preparing the reader for the bad news to come. Remember to use reasons that will make sense to the petitioner and that will at least project an image of sincerity. If at all possible, state reasons that will depersonalize the rejection, as the next paragraph of that letter does:

> In order to keep our premiums as low as possible, we find it necessary to accept only a small number of new accounts. And in fairness to our present customers, we accept

applications only from those whose claims records are comparable to those of drivers we presently insure.

Sorry

You next present the bad news itself, but you state it in a positive fashion. If you can suggest an alternative, do so. If you can find a way to leave the door open, make that clear—but do not give false hopes. Notice how the insurance letter continues:

> Even though we would like to include you among our insured drivers, we will not be able to do so at the present time. It seems to us that your best choice at this time is to remain with your present company. If your claims record improves during the coming year, we would be happy to have you reapply for coverage with us.

Thanks

The indirect message of bad news closes with another expression of appreciation—and ends on a positive note, as this example illustrates:

> We do appreciate your considering our company. And we hope you will be able to reapply under more positive circumstances.

Figure 1 shows another example of this formula at work.

THE DIRECT MESSAGE OF BAD NEWS

The direct message is a tough no-nonsense statement. While courteous, it gets right to the point and does not try to bury the bad news. It would be used under one or more of these circumstances:

- You want to slam the door shut, discouraging any other request from that petitioner.
- You want to project an image of toughness and directness.
- You are addressing an individual who prefers forthrightness and equates indirect-ness with softness or dishonesty.

The formula for the direct message is THANKS . . . SORRY . . . BECAUSE . . . THANKS. It uses the same ingredients as the indirect message, but it changes the order. And the language is less subtle and more direct.

Thanks

You begin with a courteous expression of appreciation, since courtesy is expected even in the most direct communication. So a firm and direct letter rejecting an applicant might begin like this:

Thank you for your letter and resume. We appreciate your interest in joining Mutual Life.

Sorry

The direct message moves quickly to the bad news, to be sure that the message is heard as intended. The bad news is stated forthrightly—but without offending. So the rejection letter continues:

Unfortunately, we do not have a position available that would match your qualifications, and it seems unlikely that such a position will develop in the future.

Because

Now the reasons come, after the bad news has been delivered. The reasons are stated directly but not offensively, as in this example:

We have found that our most successful middle-level managers are those who have had the benefit of working with Mutual in a variety of nonmanagerial positions. We therefore tend to promote from within and do not encourage applications for managerial positions from those who have not previously worked with us.

```
To: Bill Harkins
From: Joanne Clemens
Subject: Flex-Time Proposal
Date: February 19,1985     File: A-342

    Bill, thanks very much for forwarding your proposal for a
flex-time schedule for your department. I appreciate the cre-
ative energy—I can always count on getting good ideas from you.
    Our data suggest that flex time would result in an uneven
distribution of worker hours at a time when uniformity seems
desirable. As you are aware, the recent increase in customer
demand is presenting us with the right kind of problem: our
people and equipment are working at maximum capacity. My expe-
rience with other companies using flex time indicates that our
productivity would suffer if we instituted it now.
    And my concern for maintaining productivity makes me reluc-
tant to implement your excellent suggestion now. However, I
have asked our personnel department to review the data. Because
of my respect for your leadership, I want to give all of your
ideas the most careful consideration.
    Thanks again for taking the time to share your ideas with
me. I appreciate your developing a very sound proposal at a
time when I know you are quite busy.
```

FIGURE I   The Indirect Bad News Message

Thanks

The message ends courteously—but the door is firmly closed. There must be no mistake about it: the news is bad. So the rejection letter ends like this:

> We hope you will be able to find employment with a company that can make use of your excellent qualifications, and we do appreciate your interest in applying with us.

The successful manager projects both people-sensitivity and toughness. The indirect message emphasizes sensitivity; the direct message affirms the toughness.

Harold K. Mintz

How to Write Better Memos

Harold K. Mintz was Senior Technical Editor for the RCA Corporation when he wrote this article.

Memos—interoffice, intershop, interdepartmental—are the most important medium of in-house communication. This article suggests ways to help you sharpen your memos so that they will more effectively inform, instruct, and sometimes persuade your coworkers.

Memos are informal, versatile, free-wheeling. In-house they go up, down or sideways.* They can even go to customers, suppliers, and other interested outsiders. They can run to ten pages or more, but are mostly one to three pages. (Short memos are preferable. Typed single-space and with double-space between paragraphs, Lincoln's Gettysburg Address easily fits on one page, and the Declaration of Independence on two pages.) They can be issued on a one-shot basis or in a series, on a schedule or anytime at all. They can cover major or minor subjects.

Primary functions of memos encompass, but are not limited to:

- Informing people of a problem or situation.
- Nailing down responsibility for action, and a deadline for it.
- Establishing a file record of decisions, agreements and policies.

Secondary functions include:

- Serving as a basis for formal reports.
- Helping to bring new personnel up-to-date.
- Replacing personal contact with people you cannot get along with. For example, the Shubert brothers, tyrannical titans of the American theatre for 40 years, often refused to talk to each other. They communicated by memo.

---

*We will return to this sentence later.

- Handling people who ignore your oral directions. Concerning the State Department, historian Arthur Schlesinger quoted JFK as follows: "I have discovered finally that the best way to deal with State is to send over memos. They can forget phone conversations, but a memorandum is something which, by their system, has to be answered."

Memos can be used to squelch unjustified time-consuming requests. When someone makes what you consider to be an unwarranted demand or request, tell him to put it in a memo—just for the record. This tactic can save you much time.

ORGANIZATION OF THE MEMO

Memos and letters are almost identical twins. They differ in the following ways: Memos normally remain in-house, memos don't usually need to "hook" the reader's interest, and memos covering a current situation can skip a background treatment.

Overall organization of a memo should ensure that it answer three basic questions concerning its subject:

- What are the facts?
- What do they mean?
- What do we do now?

To supply the answers, a memo needs some or all of the following elements: summary, conclusions and recommendations, introduction, statement of problem, proposed solution, and discussion. Incidentally, these elements make excellent headings to break up the text and guide the readers.

In my opinion, every memo longer than a page should open with a summary, preferably a short paragraph. Thus, recipients can decide in seconds whether they want to read the entire memo.

Two reasons dictate placing the summary at the very beginning. There, of all places, you have the reader's undivided attention. Second, readers want to know, quickly, the meaning or significance of the memo.

Obviously, a summary cannot provide all the facts (see Question 1, above), but it should capsule their meaning, and highlight a course of action.

When conclusions and recommendations are not applicable, forget them. When they are, however, you can insert them either right after the abstract or at the end of the memo. Here's one way to decide: If you expect readers to be neutral or favorable toward your conclusions and recommendations, put them up front. If you expect a negative reaction, put them at the end. Then, conceivably, your statement of the problem and your discussion of it may swing readers around to your side by the time they reach the end.

The introduction should give just enough information for the readers to be able to understand the statement of the problem and its discussion.

LITERARY QUALITIES

A good memo need not be a Pulitzer Prize winner, but it does need to be clear, brief, relevant. LBJ got along poorly with his science adviser, Donald Hornig, because Hornig's memos, according to a White House staffer, "were terribly long and complicated. The President couldn't read through a page or two and understand what Don wanted him to do, so he'd send it out to us and ask us what it was all about. Then we'd put a short cover-memo on top of it and send it back in. The President got mad as hell at long memos that didn't make any sense."

Clarity is paramount. Returning to the asterisked sentence in the second paragraph of the introduction, I could have said: "Memoranda are endowed with the capability of internal perpendicular and lateral deployment." Sheer unadulterated claptrap.

To sum up, be understandable and brief, but not brusque, and get to the point.

Another vitally important trait is a personal, human approach. Remember that your memos reach members of your own organization; that's a common bond worth exploiting. Your memos should provide them with the pertinent information they need (no more and no less) and in the language they understand. Feel free to use people's names, and personal pronouns and adjectives: you-your, we-our, I-mine. Get people into the act; it's they who do the work.

Lastly, a well-written memo should reflect diplomacy or political savvy. More than once, Hornig's memos lighted the fuse of LBJ's temper. One memo, regarded as criticizing James E. Webb (then the head of NASA), LBJ's friend, infuriated the President.

Another example of a politically naive memo made headlines in England three years ago. A hospital superintendent wrote a memo to his staff, recommending that aged and chronically ill patients should not be resuscitated after heart failure. Public reaction exploded so overwhelmingly against the superintendent that shock waves even shook Prime Minister Wilson's cabinet. Result? The Health Ministry torpedoed the recommendation.

Two other courses of action would have been more tactful for the superintendent: make the recommendation orally to his staff or, if he insisted on a memo, stamp it "private" and distribute it accordingly.

Literary style is a nebulous subject, difficult to pin down. Yet if you develop a clear, taut way of writing, you may end up in the same happy predicament as Lawrence of Arabia. He wrote "a violent memorandum" on a British-Arab problem, a memo whose "acidity and force" so impressed the commanding general that he wired it to London. Lawrence noted in his *Seven Pillars of Wisdom* that, "My popularity with the military staff in Egypt, due to the sudden help I had lent . . . was novel and rather amusing. They began to be polite to me, and to say that I was observant, with a pungent style. . . ."

FORMAT OF THE MEMO

Except for minor variations, the format to be used is standard. The memo dispenses with the addresses, salutations, and complimentary closes used in letters. Although format is a minor matter, it does rate some remarks.

To and From Lines—Names and departments are enough.

Subject—Capture its essence in ten words or less. Any subject that drones on for three or four lines may confuse or irritate readers.

Distribution—Send the memo only to people involved or interested in the subject matter. If they number less than say, ten, list them alphabetically on page 1; if more than ten, put them at the end.

Text—Use applicable headings listed after the three questions under "Organization."

Paragraphs—If numbering or lettering them helps in any way, do it.

Line Spacing—Single space within paragraphs, and double space between.

Underlines and Capitals—Used sparingly, they emphasize important points.

Number of Pages—Some companies impose a one-page limit, but it's an impractical restriction because some subjects just won't fit on one page. As a result, the half-baked memo requires a second or third memo to beef it up.

Figures and Tables—Use them; they'll enhance the impact of your memos.

CONCLUSIONS

Two cautions are appropriate. First, avoid writing memos that baffle people, like the one that Henry Luce once sent to an editor of *Time*. "There are only 30,000,000 sheep in the U.S.A.—same as 100 years ago. What does this prove? Answer???"

Second, avoid "memo-itis," the tendency to dash off memos at the drop of a pen, especially to the boss. In his book, *With Kennedy,* Pierre Salinger observed that "a constant stream of memoranda" from Professor Arthur Schlesinger caused JFK to be "impatient with their length and frequency."

John S. Fielden and Ronald E. Dulek

How to Use Bottom-Line Writing in Corporate Communications

When they wrote this article, John S. Fielden and Ronald Dulek were professors of management communications at the University of Alabama. Jointly, they also authored a series of books on effective business writing.

Every top executive complains about "wordy" memos and reports. From Eisenhower to Reagan, stories have circulated about their refusal to read any memo longer than one page. The CEO of one of the largest companies in the United States actually demands reports so short they can be typed on a three-by-five card. And J.P. Morgan is reputed to have refused audience to anyone who could not state his purpose on the back of a calling card.

"Don't be wordy!" "Be brief, brief, brief!" "Be succinct!" One writing expert after another exhorts business people with these slogans. And who will disagree?

We do.

As a result of an in-depth study of the writing done at the division headquarters of a very large and successful company, we are absolutely convinced that advice such as "Be brief!" is not only useless, it does not even address itself to the real writing problem.

What causes trouble in corporate writing is not the length of communications (for most business letters, memos, and reports are short), but a lack of efficiency in the organizational pattern used in these communications. And, as you will see, it is for the most part a lack of organizational efficiency on the part of writers that is often deliberate, or, if not consciously deliberate, so deeply ingrained in their behavioral programming that it causes an irresistible impulse to beat around the bush.

Put simply, people organize messages backwards, putting their real purpose last. But people read frontwards and need to know the writer's purpose immediately. That purpose is what we eventually came to call the message's bottom line.

At the beginning of the study we wondered, why do people write backwards? One possibility is that they are writing histories of their mental processes as they think their way through a problem. Since their conclusion could only be arrived at after analysis, the report therefore would state its conclusions last. But if that were the reason, it would only be analytical memos and reports that would be organized backwards.

Such was not the case in the study we did. Almost *all* memos and reports put their purpose last. Why? We determined to study the problem to see if we could design a cure for such blatantly inefficient writing.

COMPREHENSION IS THE KEY

In the division headquarters we worked with, 9,000,000 (internal and external) messages of all types are distributed annually. Our study began with an intense analysis of a sample of 2,000 letters, memos, and reports randomly drawn from company files. The typical communication was one page. Only in rare cases did any memo or report exceed three pages. These various documents were well-written in the sense of being above average in terms of mechanical correctness and aptness of word choice.

Yet only one in twenty of these communications was organized efficiently. Below is one of the actual memos we analyzed (disguised, of course). Look at your watch before you read it. Keep a record of how many times you have to read it before you really *comprehend* its message, and how long it takes to understand the report's purpose.

Memo A

The Facilities people have been working on consolidating HQ marketing Functions into the new building at Pebble Brook. As presently envisioned, Marketing Research will remain in its current location but be provided with additional space for expansion. The following functions will be moved into the new facility—Business Analysis, Special Applications, and Market Planning. It is expected that Public Sector will be relocated in a satellite location. The above moves will consolidate all of Marketing into the Pebble Brook location with the exception noted above.

Attached is a preliminary outline of the new building by floor and whom it will house. I am interested in knowing if this approach is in agreement with your thoughts.

This memo seems brief on the surface. It contains only 115 words. But let's measure brevity not by words but by the length of time it takes a reader to comprehend a message.

We feel that if you were really the addressee, you would have had to read this memo twice. Why? Because you didn't know *why* you were reading it until the last paragraph. Once you discover the purpose—that you are being asked to approve a plan—you want to reread the memo to see if you do, in fact, agree with the moves. The memo suddenly (and, unfortunately, at its end) informed you that you were on the hook. Obviously, there is a big difference between the way you will read a memo containing information of general (and casual) interest and one which requires you to make a decision involving the physical moving and reshuffling of hundreds of powerful and sensitive people. Yet organizationally this memo as presented does not show even a foggy awareness of this difference.

Now read the revision. Notic+e how your comprehension time would have dropped significantly had you received this revision instead of the original.

Memo B

Attached is a preliminary outline—by floor—of the new building at Pebble Brook and a statement of whom it will house. I am interested in knowing if this approach is in agreement with your thinking.

Our suggestion is that we make the following changes:

1. Business Analysis, Special Applications, and Market Planning will. . . .
2. Public Sector will be. . . .
3. Marketing Research will remain. . . .

In terms of comprehension, the revision lets you know right away that you are expected to make a decision. Whether or not you would mull over that decision, we cannot tell. But we do know this: in terms of actual time expended in comprehending what is being asked of the reader, Memo B can be comprehended in one-third the time required by Memo A. And, if we measure brevity in terms of comprehension time, rather than number of words, Memo B produces a 66 percent savings in comprehension time.

We learned immediately in our study that comprehension time drops dramatically when a memo states its purpose—why it is being written for the reader—at the very beginning.

Confirm this point by reading Memo C, another disguised memo drawn from the division headquarters. Time your comprehension as before.

Memo C

The first of a series of meetings of the Strategic Marketing planning group will be held on Thursday, September 7, from 1 to 4 P.M. in Conference Room C. These important meetings are for the purpose of monitoring and suggesting changes in overall market strategies and product support. Attached is a list of those managers who should attend on a regular basis. These managers should specifically be prepared to review alternative strategies for the new product line. The purpose of this reminder is to ask your help in encouraging attendance and direct participation by your representatives. Please have them contact Frank Persons for any further information and to confirm their attendance.

Now read Memo D and compare comprehension time once again.

Memo D

Please encourage those of your managers whose names are listed on the attachment to attend regularly and directly participate in the meetings of the Strategic Marketing Planning group.

The next meeting is to be held on Thursday, September 7, from 1 to 4 P.M. in Conference Room C.

Please have your representative(s):

1. Contact Frank Persons for any further information and to confirm attendance.
2. Be prepared specifically to review alternative strategies for the new F–62 line.
3. Be ready to discuss changes in overall market strategies and product support.

Memo D is obviously more efficient. Why? Not only because it begins by stating its purpose but also because it itemizes the actions requested of the reader, organizing them in an easy-to-digest checklist. The original gives extensive background about the meetings but buries the requested action in a fat paragraph. Most readers would have to read the original at least twice just to be able to ferret out exactly what is being asked of them.

The time being saved, of course, seems insignificant on communications as short as Memos A and C. But consider how significant the savings would be corporate-wide if every manager's comprehension time in reading all messages could be reduced by even a small percentage.

HIGH COST OF COMPREHENSION

A recent study done by International Data Corporation states that managers in information industries spend an astonishing 60 percent of their time reading and writing; professionals spend 50 percent.[1] If cost accounted, how much would, say, a 20 percent to 30 percent savings in reading and writing time amount to for companies in this industry alone? The possibilities are arresting.

We made some cost estimates for the communications undertaken by the division headquarters we studied. The 9,000,000 messages distributed annually by the division headquarters included, of course, all sorts of mailings and multiple copies of such things as new product announcements, price changes, and the like, often running into the thousands of copies. Therefore, it was not fair to assume all 9,000,000 messages mailed were individually composed. Instead, we determined through conservative estimates that 12 percent of the 9,000,000 mailings were individual communications. And for each of these we will assume, for the purpose of this article, the ridiculously low figure of $10 to be the cost of creating, typing, and distributing. Based on these estimates, the minimum total composition cost for this one divisional headquarters would be $10,800,000 a year (see Table 1).

TABLE I   Estimated Division Headquarters Writing Costs

Number of messages sent annually	9,000,000
Percent individually composed	12%
Total individually composed	1,800,000
Composition cost (per message)	$10
Minimum total composition cost	$10,800,000

---

[1] *Automated Business Communications: The Management Workstation.* (Framingham, Mass.: International Data Corporation, 1981): 21.

But writing time is, of course, only part of the story. What about the cost of comprehending all 9,000,000 of these messages? For while we estimate that only 12 percent of these messages were individually composed (that is, not copies), all 9,000,000 messages were presumably intended to be read. Again, in an attempt to dramatize through understatement, we will use a low salary figure: $20,000 per year. You can, of course, substitute the actual salary and other figures for your own company and determine for yourself at least roughly the magnitude of your company's reading costs.

As Table 2 shows, we approximated the division's minimum reading costs to be over $4,500,000. Of course, this figure is bound to be far below actual costs. Not only are our salary estimates unrealistically low, but we haven't taken into account the fact that many of the documents were read by multiple readers. In fact, many memos urged recipients to pass information on to colleagues and subordinates.

TABLE 2   Estimated Division Headquarters Reading Costs

Low-median salary	$20,000
	$10/hour
	17¢/minute
Average* reading time	3 minutes
Reading cost per document	$0.51
Reading cost for 9,000,000 messages	$4,590,000

* Assumed that some messages read in depth; some not given more than a glance; some barely looked at.

USING DIRECT PATTERNS

While these dollar figures were somewhat astonishing and the possibilities of dollar savings enticing, the company under study evidenced the greater concern about the waste of productivity involved in such inefficient communications having become the norm. The specter of hard-working employees' time being wasted by an inundation of inefficiently organized memos and letters was distressing. The company asked us to teach people how to report in a "bottom-line" fashion. Therefore, we taught people to tell readers immediately what was their purpose in writing and what they expected of the reader, if anything. If people had no purpose in writing, they probably shouldn't write in the first place. If they didn't expect anything of the reader and were just offering possibly useful information, we told them to say so right away.

In short, we were teaching people to use a direct organizational pattern. We were urging them to eschew the circuitous pattern in which writers, because of some sensitively (real or imagined), withhold their purpose and do not let their readers know why they are being written to and what is being asked of them until their minds have been conditioned to accept the points the writers are trying to get across.

Obviously, there is nothing wrong with a circuitous organizational pattern in certain circumstances. But in this company, and we suspect in many other companies across the country, the circuitous pattern has become the norm for all types of communication in all situations.

Just ask yourself: how sensible is it to always write backwards, in a way that is just the opposite of how people comprehend information? Look at one more illustrative letter from our study:

Memo E

This is in reference to the letter sent you by Joe Smith of ABC Materials, Inc.

Mr. Smith requested information available from Product Analysis Reports (PAR's). As soon as information was made available to me from this source, I orally relayed the response to Mr. Smith.

Making use of the Planning Application Model, I was able to respond to Mr. Smith's request for further information about potential new products of possible interest to ABC. It was not until I received the copy of Mr. Smith's letter that I was aware that the data provided for him was not sufficient.

I have used all the resources that I am aware of to resolve Mr. Smith's concerns. Mr. Smith has informed me he is more than satisfied with the work done and considers the project completed. He has also announced an intention of doing further business with us.

Attached are copies of all the requests that I've been asked to submit during the six months that I have been assigned to this account. Also attached is a copy of an ABC analysis, submitted by my predecessor, which related to one of the items referenced in his letter. Upon request, I will forward copies of all of the relevant analysis that are in my files.

Where's the bottom line? What's this writer trying to get across? Isn't it the following?

Memo F

I have reviewed and acted upon the letter sent to you by Joe Smith of ABC Materials. Mr. Smith has informed me he is more than satisfied with the work done and considers the project completed. He has also announced an intention of doing further business with us.

Here in some detail are the steps I have taken for Mr. Smith. . . .

Since what this writer is reporting is good news, the communication is not sensitive. There's no need to report this information as circuitously as we might well be tempted to do if we had bungled the situation with Mr. Smith and had lost his business.

What percentage of all communications would you estimate to fall into the sensitive category that may call for a circuitous organizational pattern? Ten percent is the outer limit of possibility, unless one has a specialized job such as handling complaints, writing sales letters, dealing with shareholders, or the like. Why, then, upon analyzing these 2,000 sample letters, memos, and reports from this corporate division, did we find that almost all documents were organized circuitously? Why, in this extremely well-managed and successful company, was it the exceedingly rare letter or memo that did not bury its purpose somewhere in the third or fourth paragraph, and most frequently in the last sentence?

At the time, we had no idea of the etiology of the disease, but we felt we had a simple cure. We would tell people how to organize their thoughts so that the bottom line of their message would be immediately highlighted and promptly presented in everything they wrote. To facilitate this goal, we invented a series of bottom-line reporting principles which, if followed, would enable writers to communicate in a direct, straightforward, no-nonsense fashion in all situations that were not fraught with sensitivity. This would, we thought, save writing time and expense (see Table 3).

TABLE 3   Principles of Bottom-Line Reporting

Principle 1:	State your purpose first unless there are overriding reasons for not doing so.
Principle 2:	State your purpose first, even if you believe your readers need a briefing before they can fully understand the purpose of your communication.
Principle 3:	Present information in order of its importance to the reader.
Principle 4:	Put information of dubious utility or questionable importance to the reader into an appendix or attachment.
Principle 5:	In persuasive situations, where you do not know how your reader will react to what you ask for, state your request at the start in all cases except: a. Those where you don't (or barely) know the reader, and to ask something immediately of a relative (or absolute) stranger would probably be perceived as being "pushy." b. Those where the relationship between you and your reader is not close or warm.
Principle 6:	Think twice before being direct in negative messages upward.

And readers, too, would benefit—instead of having to search through a memo to find out what purpose the writer had in writing to them and what the writer wanted or expected them to do, they could look at the first paragraph and see the answer to these vital questions. Moreover, if the purpose and topic seemed irrelevant to their interests or needs, they could reject reading it, or merely give it a glance. Significant reading time and dollar savings should certainly result.

The program was instituted throughout the division. Did it work? Yes, in terms of getting the principles across and in terms of getting intellectual acceptance of these principles.

But getting emotional commitment to these principles was quite another story. We sensed in discussions that writers' commitment to being circuitous was not merely a bad habit. It was something else, something that was so ingrained that forcing personnel to be direct actually caused disquiet in many people. What could have been the reason?

PROGRAMMING FOR INEFFICIENCY

Obviously, people who work in large organizations were not born there. They have come to those organizations programmed by their social upbringing and by their educational experiences. And both of these earlier programmings strongly contribute to resistance to bottom-line reporting.

Social Upbringing

People seldom are conscious of how their social upbringing programs them to be indirect. Yet almost every sensitive social situation reinforces the wisdom of being circuitous, of not being direct. Aren't most brief answers to sensitive questions regarded as brusqueness or curtness, as being short with someone?

It begins early. The children are asked, "Do you want to go to Aunt Alice's house?" The children answer, honestly and directly, "No!" Unacceptable! The children are scolded and soon learn to beat around the bush the next time, all the while searching for some plausible excuse to forestall the visit. It is not surprising that as adults, the same children, when asked by the boss, "What do you think of my new plan? Think it'll work?" think twice before responding, "No!"

Educational Programming

Having been thoroughly programmed by their families that being direct is being impolite, the children now go to school. Here they soon learn that a twenty-page term report gets a high grade; a two-page report gets a low grade. A five-page answer to a test question is good; a one-paragraph answer is bad. Regardless of what teachers may profess, they invariably give extra credit for "effort." And effort is most easily measured by numbers of words or by pounds of pages. A premium is placed on long-windedness, and long-windedness is achieved by being circuitous rather than direct.

Indoctrination into Anxiety

On their first jobs in a large organization, young people are naturally nervous. They are very concerned that whatever they write or say not make people upset. They are also very concerned about "getting good grades." Therefore, they fall back upon the same behavior that was rewarded in school. They are going to do everything possible not to look lazy. They are going to be thorough in everything they write. Every chance to write a report to a superior is a chance to write that blockbuster of a term paper that could not fail to impress the boss. They are going to get that "A."

The fact that young people enter organizations at the bottom provides a final step in their programming for being circuitous and indirect. Everybody knows that writing *up* in an organization is far different from writing *down*. When young people enter an organization, the only direction they *can* write is up. Therefore, all the early experiences received in corporations consist of writing situations where they have to write information to people who are in fact, or may someday be, their superiors. Naturally, they become very uneasy.

Now let's suppose the company institutes a program to encourage personnel to be more direct. Imagine yourself as that newly hired young person in the organization. Are you going to believe any program suggesting that you be blunt and direct in your upward or lateral communications, when your entire lifetime programming has proved to you over and over again that bluntness is all too frequently suicidal? No chance!

Young people may give lip service to such a program, but in any real-life situation in which they feel threatened (in actuality, almost all situations) they will avoid coming to the point with an almost religious passion. And in negative or sensitive situations, the last thing they are going to do is state their purposes and requests directly.

By contrast, their higher level superiors, having enjoyed years of power positions in the hierarchy (from whence they could write down to anyone in any fashion they pleased), take a far different view of writing. The superiors now pride themselves on directness and bewail long-windedness on the part of their subordinates. But the subordinates' desires for self-preservation (reinforced by all their preorganizational programming) force them to give lip service at best to corporate attempts to "get to the point."

And, let's face it, in the corporate pyramid almost everybody is somebody's subordinate and, perhaps because of files, one never knows who is going to read what has been written. Therefore, circuitous writing is partly the habit of a lifetime and partly CYA.

WHAT YOUR ORGANIZATION CAN DO

Is a cure then impossible? Is inefficient, circuitous writing simply to be endured and its costs in lost productivity and wasted dollars merely written off?

No; a cure for inefficient writing is possible. But a thorough organization-wide cure requires that:

- People recognize and reject their social and educational programming for being circuitous in all non-sensitive writing situations. This deprogramming is the responsibility of the individual.
- People learn to write efficiently; that is, learn to organize their messages in such a way as to make it easy (and fast) for readers to comprehend the message. Teaching the bottom-line principles will impart this skill. But implementing the bottom-line principles requires strong high-level management support.
- People must develop the self-confidence necessary to send bottom-line messages upward in nonsensitive (or slightly sensitive) messages. A long-range cure depends to a great extent on attitude, on reducing the tensions inherent in superior-subordinate communications. Most writing insecurities stem from real, not imagined, failures on the part of superiors to communicate clearly and unequivocally their willingness to accept bottom-lined messages from subordinates.

Higher level executives have to appreciate how threatening directness can be to subordinates. Superiors, therefore, need to be persuaded not only to have the following credo taped to the wall above their desks but also communicated to all subordinates with whom they relate:

The Superior's Credo

1. I will ask all subordinates to be direct in their messages to me and I will not become angry if subordinates do so politely—even when those thoughts run counter to mine.
2. I will recognize and appreciate subordinates' attempts to conserve my time (and other readers' time) in all memos and reports they write to me, or for my signature.
3. I will work out with subordinates some general understanding of how much detail I require in various circumstances.
4. I will make clear to subordinates that I judge their communications not by length and weight, but by directness and succinctness.
5. And when I myself report up in the organization, I will be as direct as I expect my subordinates to be.

Once senior executives have adopted and put this credo into practice, all subordinates should recognize that there is now no excuse for them not to live by the following credo:

The Subordinate's Credo

1. I will have the courage (in all but the most sensitive or negative situations) to state at the beginning of messages my purpose in writing, exactly what information I am trying to convey, and/or precisely what action(s) I want my reader to take (if possible, itemized in checklist form for easy comprehension).
2. My readers, especially if they are my superiors, are extremely busy. I must not waste their time by making them read unnecessary undigested detail any more than I would waste their time chattering on in a face-to-face interview.
3. I will make a judgment as to how much my readers need to know in order to take the action required by the communication.
4. If I am in doubt as to whether specific information is necessary to my readers, I will either put this information in summary form in attachments, or tell readers that I stand ready to offer more information if so requested.
5. I will avoid the arsenal of the con man. If I want something of a superior, I will ask for it forthrightly.

Heidi Schultz

Electronic Mail

Heidi Schultz is a member of the faculty at Kenan-Flagler Business School of the University of North Carolina at Chapel Hill.

During the past 15 years or so, the number of e-mail users has increased 100-fold! Some 35 million of us spend much of our time sitting at computers, pounding out more than 100 million e-mail messages every day. Our computers are linked through the Internet or intranets, which allow us to send and receive urgent business messages, convey important medical and legislative information, and simply stay in touch with scattered friends and relatives around the community, the country, and the world.

The widespread use of e-mail has changed much about the way we communicate, breaking social hierarchies, geographical boundaries, and time barriers. E-mail also has challenged the traditional channels of communication among members within organizations. Closed doors no longer translate into inaccessible CEOs, journalists, lawyers, physicians, professors, teachers, or upper management. Simply put, everyone with e-mail has access to everyone else with e-mail— and that's a lot of accessibility! People who live or work down the hall, up the street, across campus, or over the continental divide are as reachable as those folks who share your school, your office, or your house. Communication with your Japanese relatives, South African business partners, or Norwegian academic contacts is no longer restricted by time zone considerations. You'd never call someone in Europe at 6:00 p.m. mountain time. But with e-mail, any time is the right time. Given the convenience of e-mail, we have once again become prolific message writers, which may seem like a blessing—until you open your e-mail to discover 50 messages that require your attention!

Because e-mail sidesteps traditional communication channels, it is primarily an informal communication tool. As does its accessibility, e-mail's informality brings challenges to electronic communication. It has an interesting way of getting people

141

to drop their inhibitions because of a *perceived* anonymity some writers tend to adopt. Without the accountability that comes with face-to-face conversations, e-mail users sometimes write what they would never say to a recipient's face or write in a formal letter. For example, a manager from a large organization wrote the following e-mail message to a technology-support staff person, who had been busily exchanging the organization's 15-inch monitors for 17-inch monitors:

> NEVER, NEVER touch my machine without telling me first! At best, it's rude. At worst, it's unprofessional!

Thinking he was doing the manager a favor *by* exchanging the smaller monitor for a larger one, the technology-support person said, "I couldn't believe it. She's always been real nice to my face. But her reaction was a surprise, if I've ever had one in this job."

Because it's so easy to send e-mail messages, we send lots and lots of them. An article published in *The Baltimore Sun* revealed that President Clinton received 2.7 million e-mail messages from 1993 to 1998, when he started to get 2,500 e-mail messages every day. And a recent survey revealed that e-mail messages interrupt the typical worker three times every hour to respond to and send an average of 190 messages each day.

This voluminous exchange of information creates yet another challenge with e-mail. Connecting yourself to e-mail—whether at work, at home, or at school—can lead to something called *communication enslavement,* in which you feel as though you're always on call because e-mail writers expect instant responses to their messages. Moreover, e-mail's quick pace causes many people to check messages regularly or have e-mail programs automatically alert them by beeping each time a new message arrives. (In fact, as I was writing the sentence about beeping, *my* e-mail beeped, indicating a new message had arrived! Despite the fast-approaching deadline for this book, I stopped writing to check the message—but it wasn't anything important.) While curiosity is a powerful emotion, I have no doubt that continual e-mail interruptions can reduce anyone's productivity.

Given these challenges, you'll want to ask and answer some questions before sending any e-mail message. So let's do that now.

WHY SEND E-MAIL?

Despite its challenges, e-mail certainly has advantages. Once you've written your message, you don't have to address a separate envelope, seal the message in the envelope, put a stamp on the envelope, and drop it in a mailbox. Furthermore, you can send the same message to lots of people without taking time to make photocopies. With a few simple keystrokes, you can send your message to 1 or 100 people. So, e-mail is extremely convenient.

E-mail is also the fastest way to send a message. When you send an e-mail to someone in your organization, your message stays within your organization's mail server. Because your electronic message doesn't have to travel on outside servers, your message is transmitted in seconds. If, however, you send an e-mail message outside your organization—for example, across the country or halfway around the world—your message must move from one server to another. While this process slows your message's transmission, 99% of the time, your message arrives in less than a minute.

E-mail is also inexpensive, when you consider paper, envelope, and postage expenses. Sure, messages travel over expensive computer networks. But because those networks serve additional purposes, we can't blame those expenses exclusively on e-mail.

WHEN IS E-MAIL APPROPRIATE?

Even though e-mail is convenient, fast, and inexpensive, you'll need to determine its appropriateness in relation to your specific communication situation. E-mail is appropriate in the following instances:

- When you don't need to speak to someone face to face
- When the recipient of your message is not immediately available
- When the information you're sending is not sensitive or personal

E-mail is also appropriate when you need to convey detailed information that would be cumbersome to deliver over the phone, on voice mail, or even in a face-to-face conversation. For example, when your message includes names and addresses or long account and telephone numbers, e-mail allows you to deliver that information not only efficiently but also accurately. One word of caution: If you are sending account numbers, credit card numbers, or social security numbers, you need to be certain that the information is protected from unintended eyes. Knowing *when* to use e-mail instead of another communication medium is as important as what information is contained in the message.

WHEN IS E-MAIL INAPPROPRIATE?

Never write or send an e-mail message when the information you are conveying is personal, confidential, or inflammatory. Simply stated, e-mail is *not* private. Even if you delete personal, confidential, or inflammatory e-mail messages from your computer, they may still be on your server's system indefinitely. Internet service providers, government agencies, educational institutions, and corporations all maintain copies of your messages long after you delete them from your e-mail account. For instance, the now-infamous *Starr Report* revealed that Monica Lewinsky's "deleted e-mail [had been] recovered from her home computer and her Pentagon computer." In October 1998, at the start of the Department of Justice's antitrust trial against software giant Microsoft Corporation, government lawyers read dozens of CEO Bill Gates's private e-mail messages. In both cases, the information conveyed in e-mail messages was personal, confidential, inflammatory, or some combination of all three.

Here's another example: In late 1998, the U.S. commander overseeing the no-fly zone in northern Iraq sent the following inflammatory e-mail message:

> It's a good day for bombing.

While Air Force Brigadier General David Deptuala admitted he should have chosen his words more carefully, he learned a very important lesson the hard way: that the air force's systems administrator has the technological ability to monitor all e-mail activity.

What does all this mean? It means you should never, *never* write anything in an e-mail message at home, at work, or at school that you wouldn't write on the back of a postcard, want to see on a billboard, or be ashamed to show your Great Aunt Elly.

E-mail is also inappropriate in the following situations:

- When facial expressions, gestures, or speech patterns are important
- When a message requires your actual signature
- When you need to send mass mailings but fail to limit your recipients to a specific target audience. . . .

Writer's Role

Because your relationship to your audience affects your choice of words and the way you organize your documents, it's important that you have a clear understanding of *your role* in the communication process. For example, you may be in the role of subordinate if you're writing to your supervisor. In that case, you may start your e-mail message with a salutation followed by your supervisor's name and title. You may also buffer your ideas with hedges such as *I believe* and *I think*. If, however, you're writing an e-mail message to a close friend, you'll probably forgo the salutation and complimentary close and begin directly.

When you are sensitive to your role—based on your relationship to your audience—you make organizational and stylistic choices that are appropriate to each rhetorical situation.

Purpose

E-mail can have a number of useful purposes. It can help you schedule meetings efficiently, find former college roommates, or disseminate copies of your most recent proposal. Generally speaking, the purposes associated with e-mail messages include (but are not limited to) the following:

- To entertain
- To provide information
- To seek information
- To persuade
- To transmit attached documents
- To confirm plans

If you simply ask yourself, What do I want this e-mail message to accomplish? and you provide a thoughtful answer, you'll have a clear purpose to guide the organization of your message. Recently, I wanted to know the origin of the phrase *I'm in a pickle,* so I e-mailed the Mt. Olive Pickle Company with my question. Because my purpose was to seek information from someone I didn't know, my message's organization was direct, my style somewhat formal, and my tone respectful.

Upon learning that I'd soon need new glasses, a friend who lives some distance from me sent the following e-mail message:

Clearly, my friend's purpose was to provide a chuckle, to entertain me. Because his style and tone were informal and casual, his organization was direct and his message was short.

Audience

Once you've established a clear purpose, you'll also need to consider your audience, or readers, and your relationship to them. Rhetoricians call this step in the communication triangle *audience analysis.* Whether you're communicating with your friends, your colleagues, or even strangers, taking a moment to create a mental profile of your readers and their preferences will guide you in determining the content and organization of your e-mail message. Of course, if you know your readers well and the context is a personal rather than a professional one, most of your responses about audience analysis will be intuitive. Briefly, you need to consider these questions for your audience analysis:

- Who are your readers, and what are their preferences?
- What do they currently know, and what do they need to know?

Let's take a closer look at these questions. When determining who your readers are, you may want to consider their average educational level, experience, gender, age, socioeconomic status, and anything else that might help you characterize them. For example, the more education and experience your readers have, the more likely they are to understand complex concepts. The younger they are, the more likely you'll have to fill in their knowledge gaps.

Considering who you readers are allows you to formulate some ideas not only about the content of your e-mail message but also about the format. For example, which of the following considerations do your readers prefer?

- Lists, headings, or highlighted text?
- Shorter messages or longer messages?
- The good news or the bad news first or last?

Related to this issue of preference, it can also help to ask, How will my readers use the information conveyed in my e-mail message? Once again, your answer may be less structured if you are e-mailing information to friends or relatives, but if you are e-mailing business associates, teachers, government officials, or medical professionals, for example, you'll need to think carefully about your answer to this question. Most professionals appreciate receiving e-mail messages that won't waste their time and will help them do their jobs better. Will your readers use your message to make decisions, perform duties, learn something about the organization, or fulfill job-related requirements?

ORGANIZING AN E-MAIL MESSAGE

In addition to considering your role, your readers, and your purpose, you'll also need to consider the most effective way to organize your message. Even though e-mail is primarily an informal communication medium, specific conventions govern organization if your goal is for that message to be effective. More than 3,000 years ago, Greek rhetorician Aristotle referred to this part of rhetoric as *logos*. For Aristotle—and for today's writers—the logic of any message is the one component over which a writer has the most control. After all, you may not be able to change the way your readers perceive you. For better or worse, your status or your convictions reflect who you are and can color the way readers interpret your message. Likewise, your readers have their own biases, which you may or may not be able to change.

But you do have control over the way you organize the information in your electronic messages. You can choose a direct or indirect structure, depending on your purpose and your relationship to your readers. (As you can see, the rhetorical components of all messages are interdependent.) Let's examine these two structures—*direct* and *indirect*—as they relate to the most common types of e-mail messages you'll write: good news, bad news, and persuasive e-mail messages.

The most effective organization for a good news message follows a direct structure. That is, you want to convey the good news first and follow at with supporting details and a positive, forward-looking close. Compare the following two good news examples:

Direct Good News Message

Dear Jo:

Congratulations! You've won $100 for your suggestion submitted to our "Save the City Money" contest.

The accounting department will issue checks on September 6, and you should receive your award a few days later in the mail.

We're excited by your suggestion. Thank you for participating!

Sincerely,

Chair, Suggestion Committee

Indirect Good News Message

Dear Jo:

As you know, the city has solicited suggestions for its "Save the City Money" contest. It's been exciting to receive so many good ideas. Out of the hundreds of suggestions offered, we've chosen five.

I'm happy to report we adopted your suggestion. We will implement your suggestion at the start of the next fiscal year. You can be sure that your idea will make a difference to the city's budget.

To thank you for your participation in the contest, we will be sending you a check for $100. The accounting department will issue checks on September 6, and you should receive your award a few days later.

We're excited by your suggestion. Thank you for participating.

Sincerely,

Chair, Suggestion Committee

If the committee had selected *your* suggestion, what would you want to know first? If you're like most people, you'd want to know what you'd won and when you'd receive your prize. So what does that mean? Direct organization is usually the best choice for conveying good news.

But let's say you need to send bad news to someone. What's the best choice in that situation? To answer this question, think about a time *you* received a message conveying bad news—perhaps rejecting your job, school, or mortgage application. How did those messages start? Immediately with the bad news? Probably not. Most bad news messages usually cushion the blow with what's called a *buffer:* a short, neutral, non-confrontational statement that appears at the start of your bad news message. Buffers can show appreciation, as in "Thank you for your mortgage application." Buffers can also demonstrate fairness, as in "We have been reviewing the absentee rate for all employees." Writers who want to convey a sense of cooperation might construct a buffer such as "I'm sure we both agree the curriculum here is quite demanding."

Once the writer creates a buffer that's relevant to the topic, he or she follows it with the reasons and the bad news. Here's an example:

Thank you for your mortgage application [appreciation buffer]. Because your income does not meet the minimum requirements for the loan you seek [reasons], we are unable to approve your mortgage request [bad news].

Typically, after conveying the reasons and bad news, the writer closes the message with a positive, forward-looking statement, such as "If your income level changes, you are welcome to reapply. We look forward to serving your future banking needs."

If the writer had started the message directly with "We are unable to approve your mortgage request," the bank could have alienated the customer permanently. While the indirect approach cannot guarantee the customer will be happy to receive the news, the message's indirect organization is less likely to alienate the customer. Of course, there are times when the direct approach to bad news is preferable. For instance, if the message stays within an organization or if the writer and reader have professional relationship, conveying bad news directly is acceptable. Direct bad news messages are more efficient because they are more succinct than indirect messages and thus take the reader less time to get to the point.

Persuasive messages can also follow direct or indirect patterns. In the direct pattern, the writer states the conclusion early and then follows it with the reasons. This approach is useful if you anticipate writing your e-mail message to a receptive audience. When applying the indirect pattern to e-mail messages, the writer first lists and explains the reasons before presenting the conclusion. This approach is effective when your audience is hostile or closed to your ideas. By first providing detailed information in support of your conclusion, you methodically prepare your audience to accept it.

Let's say you want to persuade your organization—your college, your place of work, or your parents—to replace some out-of-date computers. Because your place of work, for instance, may have a limited budget, you conclude that your readers will be closed to your suggestion. Knowing your readers' biases, you'll probably be more successful if you present your reasons before stating your conclusion that new computers are needed. Here's one way to organize this persuasive e-mail message:

Dear Budget Department VP:

We both know how quickly technology evolves. As GMS consultants depend more and more on internet access to remain competitive, it's important for us to access that information quickly.

While staying technologically up to date can be expensive at the outset, the increased efficiency of newer processors will no doubt increase productivity. In fact, just 6 months ago, MILDA Consulting spent $3.2 million dollars to upgrade employees' systems. And they've already doubled that investment due to increased contracts. That's not a bad investment. Given MILDA's success and our needs, I'd like to suggest that we upgrade all consultants' CPUs in the next two months.

Sincerely,

Marianne Reiher, Consultant

As you can see, the first paragraph sets up a general truth—that keeping technology up to date can be expensive. The second paragraph, however, provides two logically persuasive tools—figures and history—to argue that GMS can recover the initial expense. First, figures (specifically, the cost and financial return) support the argument about MILDA's success. And second, using history (MILDA as an example from the recent past) makes an argument in favor of GMS updating its computers. By providing these reasons, the writer has prepared the reader for the conclusion.

If you are e-mailing your parents from college and want them to update your computer, you might approach the same message directly rather than indirectly. If your parents are responsive to the legitimate needs associated with your education, you can consider them a receptive audience. In this case, your e-mail message might take this organizational structure:

> Hope you're OK, Mom and Dad. Will you transfer $2,000 to my checking account?
> I need to update my computer. With all the work that's piled on, I need a faster
> computer, so I can access the Internet and my e-mail account more quickly. What
> do you say?
>
> See ya!
>
> Andreas

In this letter home, Andreas presents his point immediately. After the greeting to his parents, he asks for the money directly.

So, what's the moral of this discussion about organization? If you consider the purpose for your e-mail message and your reader's likely reaction to it, you'll be prepared to decide whether direct or indirect organization is more effective.

FORMATTING CONVENTIONS

Now that you know how to complete a situational analysis and effectively organize your e-mail messages, let's talk a bit about what goes into a strategically formatted e-mail message.

Headings

When you create a new e-mail message, you'll see the following *headings*—"To," "Cc," "BCc," and "Subject"—which appear at the tops of new e-mail messages. . . . On the "To" line, simply type your reader's e-mail address. While e-mail addresses are not case sensitive (meaning you don't need to pay attention to typing upper- and lowercase letters), you should be very careful to insert underlines and other symbols accurately.

The "Subject" line typically needs the most thought. Because numerous e-mail messages crowd our electronic mail boxes every day, they compete with one another for attention. To this end, the subject line is your first defense against your reader's "Delete" key. Unless your position in an organization is important enough to guarantee your e-mail messages will be read, create a subject that reveals the *specific focus* of each message in as few words as possible.

Sometimes, it helps to write the e-mail message first and then go back and fill in the subject line. A vague subject line, such as "Basketball tickets," won't grab your reader's attention because you don't provide enough information. Starting with a noun and then adding a verb makes any subject line more specific. For example, you might start with the subject line "Basketball tickets." To make it more specific, add a verb; then "Basketball tickets." becomes "Basketball tickets *offered*." Adding appropriate modifiers can make a subject line even more specific. "Basketball tickets offered" becomes "*Free Ohio University-Ohio State* basketball tickets offered." If you're a fan of either team, you can see how much more enticing the specific subject line is compared to the first version.

Here's one more example of converting a general subject line to a specific one:

W–2 information → W–2 information provided → Important W–2 information provided

Another heading that requires your attention is the "Cc" line. Most e-mail software programs allow you to send copies of messages to secondary recipients. The abbreviation "Cc" is a carryover from the "carbon copy" days. While "Ec," or "electronic copy," is not only more current but also more accurate, tradition prevails, so "Cc" is the heading you'll usually see in e-mail software programs. If you want to forward an electronic copy of your message to another reader, then you'll simply need to include that reader's address on the "Cc" line.

The "BCc," or "blind carbon copy," line allows you to send an electronic copy of your message to a secondary recipient without letting the primary recipient know. For example, let's say you're a middle manager. You want to send an e-mail message to one of your employees, and you want your supervisor to know about it without letting your employee know you sent the second copy. To do so, you'd include your supervisor's e-mail address in the "BCc" line. Your employee, the primary recipient, will be "blind" to the fact that you've forwarded the same message to a secondary recipient, your supervisor.

Consider another example: Suppose a middle manager in the human resources division of a major automaker is responsible for evaluating several employees semi-annually. On one occasion, a furious employee challenges his evaluation, which is "below average" in four out of nine categories. E-mailing the employee her justification for the evaluation, this middle manager also should send a "BCc" to the VP of human resources to document this potentially volatile situation.

While all the usual reasons for sending blind copies of paper correspondence continue to be valid for electronic mail, the "BCc" line has yet another function. If you want to be sure the hardware and software sent your message and that your recipient received your message, send a "BCc" to your own e-mail address. When that message shows up in *your* "in" box, you'll know the transmission occurred successfully.

Salutations

Because e-mail messages combine format conventions from both letters and memos, the issue of whether to use a salutation becomes interesting. You may remember that letters include salutations and memos do not. So, should you include a salutation in an e-mail message?

The answer depends on your relationship to the recipient. If your message is going to friends, family, or colleagues whom you know well, either omit the salutation or simply use the person's first name. If you use the person's first name, you may write "Dear Kurt" or curtly write "Kurt." If your message is going to people whom you do not know well or do not address on a first-name basis, then include the salutation and appropriate title. If you're still unsure of what to do, adhere to standard formalities and simply create a salutation that accurately reflects the way you would address that individual in person.

Length

Keep your e-mail message to one screen or one page, if at all possible. While information overload may seem like the primary reason for this guideline, there is also a psychological reason. When you sit down to read articles in magazines or chapters in books, if you're like most people, you flip to the end to determine the number of pages. It's human nature to do so. We want to see what we're in for. And knowing how long a particular piece of writing is tells us how long we'll have to concentrate to read it.

The same thing holds true for an e-mail message. Before we read it, we want to know how long it is. But we can't physically handle an e-mail message or quickly flip to the end of it. So, do your readers a favor: Let them see the end onscreen by keeping your e-mail message short. If your message must be longer than one page, consider creating an executive summary or sending a "Heads up; it's coming" e-mail message. Then forward a hard copy of the longer message by regular, overnight, or interdepartmental mail.

Also consider the length of each line in your e-mail message, especially if you're sending it outside your business's, school's, or institution's network or if you're sending a message from home that may go through several different Internet service providers. Some terminals can't handle text longer than 80 characters per line. If your e-mail software *word wraps* each line automatically— in other words, it breaks the text into consecutive lines—at a specific character limit that doesn't exceed 80 characters, then you'll be OK. Either count the characters in one of your typical e-mail messages, or check with your technical support person. If your e-mail software doesn't automatically word wrap at 80 characters, then you'll have to remember to hit the "Enter" key manually before reaching 80 characters. While this manual procedure may be annoying, it's necessary if you want your message to appear to your recipient the way it appears to you.

You'll also need to signal the end of your message clearly. This is important even when your reader can see the physical end of a short message. By ending with an obvious closing, you won't risk leaving your readers wondering whether you forgot to include something important or whether a snag in the transmission clipped off the message. If the message is long enough, signal the end by briefly reviewing its content. If your message requires specific action from your reader, succinctly remind him or her of a deadline, quickly reiterate a request, or concisely repeat a question. If the message is somewhat complicated, end by concisely reviewing its contents.

Once you've ended your message, add your name or initials. You also have the option to include a quick complimentary close, such as *Sincerely, Thanks, Cheers,* or *See you at lunch on Tuesday.* Any of these options will make it clear to your reader that no more information follows.

Signatures

Current technology won't allow you to sign your e-mail in your own handwriting, which means your e-mail signature is always typed. And given the hybrid nature of e-mail messages—somewhere on the scale between a memo and a letter—much of the information about salutations applies to electronic signatures.

Typically, your relationship to the recipient of an e-mail message will determine two things: (1) whether your message should contain a signature and (2) if it should, what form that signature should take. Keep in mind, however, that your position in relation to your reader changes continually if you send e-mail messages on a regular basis. You may, for example, send messages to friends, colleagues, relatives, subordinates, and superiors—all in one day.

If you correspond frequently with someone whom you know well, it's OK to include only your first name or initials without the complimentary closing. However, if you want to maintain a formal or businesslike relationship with the recipient, it's a good idea to create a signature that reflects what you would include in a hard-copy signature. When your readers don't know you well or not at all, include a complimentary closing and your name along with your credentials. Knowing your title and position will likely cause your reader to take your communication more seriously.

Several other factors will help you decide whether to end your e-mail message with a complimentary closing and formal signature with credentials. If your message originates from company equipment, your signature should include not only your e-mail address and name but also your title and your organization's name. Doing so clearly signifies that you're using the organization's hardware and software for official communication. When it becomes common to send and receive e-mail messages on letterhead, you'll only include the signature information not included on the letterhead.

Because some organizations use abbreviations, numbers, or a combination of both when they assign e-mail addresses, the message you send might be identified cryptically in the "From" line once your message reaches its electronic destination. If you don't have an alias that clearly identifies you by name, your reader may not link the "From" line in your e-mail message to you. . . .

To be sure your recipient knows that *you* sent the message, make it a habit to include at least your name or initials in all e-mail messages, regardless of your relationship to the recipient. Doing so avoids any possible confusion about who sent the message. Moreover, a signature clearly signifies the end of the message.

Also consider including your e-mail address with your signature, especially if your e-mail message is going to someone outside your organization. You certainly could argue that your e-mail address is clearly indicated on the "From" line. You could also argue that the simple act of replying to a message eliminates your recipient's need for an explicit e-mail address. After all, if someone e-mailed you the first message, then he or she presumably has your e-mail address already.

But think of it this way: When you call someone and leave a voice mail message, you include your telephone number. If you're out of the office when that person calls you back, he or she leaves a message on your voice mail. Wouldn't life be easier if when you were ready to return the call, you were able to get the number from the voice mail message, instead of having to search for it? Surely, it would. So, do your readers a favor: Make e-mail life easy. Including your e-mail address with your signature ensures the address will be readily available, easily identifiable, and conveniently exchanged between e-mail writers.

The problem is that it can take a lot of time to include a signature with your name, title, credentials, organizational affiliation, and e-mail address in each message. To make this easier, your e-mail software gives you the option of creating a *signature file,* which automatically inserts all this information. When creating a signature file, just keep in mind your relationship to your readers. It's best to create a signature file that works in all rhetorical situations—from casual to informal to formal.

More and more e-mail writers are including quotations and characters—in addition to the items discussed above—in an attempt to personalize their signature files. For example, a development director for a not-for-profit organization ends all her e-mail messages with a quotation from Abraham Lincoln: "Determine that the thing can and shall be done, and then we shall find the way." I admire Abraham Lincoln; however, his ideas, as well as Indian proverbs and transcendental philosophies, are fascinating only in specific contexts—and e-mail is not one of those contexts!

Because most e-mail users suffer from information overload, include only relevant information in any e-mail signature file. After all, it's in your best interest to keep your message to the point. Ask yourself the following question: Would you include quotations or characters in *hardcopy messages* written to business associates, teachers and professors, or potential clients? If the answer is no (and it likely is), then avoid doing so in e-mail messages, too. If the answer is yes and you are compelled to include quotations and characters with your electronic signature, make sure they support the tone of your message. A serious message will be reduced by a light-hearted quotation, and a light-hearted message can't tolerate a serious quotation. Just be consistent in your tone, and you'll make the right choice when including quotations and characters with your signature.

```
        @@@@@@
      @@@@@@@@
      / ~      ~ \
     {(O) || (O)}
_____oOOOo_____oOOOo_____
```

One last point: If you include a quotation or a character with your signature, be sure to keep that information short and nonobtrusive. You want your reader to focus on your message, not on a long quotation or complex character.

```
_____.oooO_____Oooo. _____
            (   )        (   )
            \  (          )  /
             |__)        (__|
```

Emphasis Techniques

Earlier, we talked about the increasing number of e-mail messages most people receive each day. And we also talked about how quickly readers delete messages. If you create a message that enhances my ability to read it quickly and easily, that message is more likely to get my attention.

With this strategy in mind, let's look at some emphasis techniques that will protect your message from your reader's sweeping use of the "Delete" key. In general, observe the following suggestions:

- Present your main points quickly.
- Briefly preview the remaining content.
- Use bulleted lists, highlighted headings, and single-sentence paragraphs.

Presenting your main points quickly gets your information across to your reader more efficiently. Previewing the remaining content tells your reader what to expect. And offering information via bullets, highlighted text, and short paragraphs contributes to reading comprehension and efficiency. . . .

Style and Tone

Style and tone result, in part, from word choice, sentence length and patterns, figures of speech, and other linguistic choices, such as using all capital letters. The combined effect of these choices determines style and tone, which can be considered on a continuum that goes from formal to informal to casual. As with all writing, it's important to make intelligent choices about style and tone for e-mail messages.

Let's consider an example: A senior economics major with a GPA just under 3.0 wrote an e-mail message to an employment recruiter for one of the largest investment banks in the United States. The student's overly familiar tone made it sound as if he were chummy with the recipient of his e-mail message (whom, by the way, he had never met). He called one of this recruiter's colleagues "a very sweet person," and he included several smiley faces in the message. The recruiter responded by writing, "E-mail correspondence should not be treated as a sticky note." Needless to say, the applicant didn't get the job. So, what's the moral of this story? Don't assume just because e-mail is primarily an informal communication channel, style and tone don't matter. They do!

E-mail is a hybrid of written and spoken communication. As such, the conventions governing style and tone can be challenging. Should you adopt a casual style and tone appropriate to light-hearted circumstances? Should you adopt an informal style and tone appropriate to informal speaking situations? Or should you adopt a rigid style and tone appropriate to formal written contexts? The answer depends on your purpose for writing, your message, and your relationship to the audience who will read your message.

Having an appropriate style and tone clearly affect a message's effectiveness. For that reason, let's look at some general characteristics associated with each level of formality. Doing so will help you decide which style and tone are appropriate for your e-mail messages.

Levels of Formality

How would you characterize the style and tone of the following e-mail message sent to a company's stockholders?

> The impending liquidation of the company's assets will not take place until the end of next week, at which time the company's warehouse will be closed to everyone except registered participants. The company looks forward to a continued and prosperous relationship with its employees, its business partners, the community, and its investors and promises to devote resources that will help the company build up much of its remaining assets in the coming fiscal year.

Fairly formal, isn't it? A *formal* style and tone has these traits:

- Is serious
- Maintains a figurative distance between the writer and reader
- Avoids personal references to maintain an impersonal tone
- Avoids contractions
- Incorporates abstract language in long sentences and paragraphs

Most of your e-mail messages will avoid the stuffy, formal style and tone used in the preceding example, even when those messages contain attachments that require a high degree of formality.

The next level of formality can be labeled *informal*. Here's an example:

Hi, Keith,

Can you get the latest figures on the Performance case? I need them for our presentation at 2:00 p.m. Drop them on my desk. I'm heading out for a quick bite.

Thanks, Martha

As you can see from the preceding message, an informal style and tone means that your e-mail message has these qualities:

- Shows sentence variety
- Uses first- and second-person pronouns and contractions to achieve a more personal tone
- Demonstrates a slight figurative distance between the writer and reader

The style and tone of this book, for example, is informal. In fact, with a little effort, you'll be able to find all the characteristics on this very page associated with an informal style and tone. Keep in mind, too, that most of the e-mail messages you'll write for professional and business purposes will adopt an informal style and tone. Doing so maintains an approach that is appropriate without sounding stuffy. A word of caution though: Don't assume that using an informal style and tone gives you permission to ignore grammar, spelling, and word choice. It doesn't!

Next take at look at the *casual* style and tone of this e-mail message:

Hey, Bozo! What's happenin', dude? Matt and I are gonna hop down to Court Street and have us a brew. Wanna come? The libe will be there when we're done, man. We're leaving tonight at 11.

What do you notice about this e-mail message that makes its style and tone casual? Clearly, it's quite different from the previous two messages, which demonstrated a formal and informal style and tone. If you choose to adopt a casual style and tone, you'll typically create a message that can be described as follows:

- Uses short- to medium-length sentences (20 words or fewer)
- Has short paragraphs (3 to 5 sentences)
- Incorporates descriptive language using specific words in conjunction with whimsical topics
- Uses numerous personal pronouns and contractions

- Refers to people by first names, pet names, and nicknames
- Incorporates slang, jargon, and colloquialisms comfortably
- Minimizes the distance between the writer and reader

When corresponding with close friends and relatives, your e-mail messages can adopt a casual style and tone. However, don't allow casual writing to become obscene or otherwise indelicate. And be sure to reserve casual messages for those persons closest to you.

Flame Mail

One convention of electronic communication that contributes to an e-mail message's tone is called *flaming*. People who send flame mail, or *flames,* often send their unpleasant e-mail messages typed in all uppercase letters. While one or two words typed in capital letters is acceptable for conveying emphasis in a professional or personal e-mail message, you should avoid messages written in all caps. Doing so is the electronic equivalent of an ear-splitting cussing out.

Flames can be direct, as in "YOU ARE A STUPID NITWIT," or indirect, as in "I THINK THAT'S A DUMB IDEA." In both examples, the flames are directed at a person rather than at an idea or concept. While it's acceptable to disagree with a concept or an idea via an e-mail message, disagreeing with the source of that concept or idea—that is, the person who conveyed it—is inappropriate. Simply put, flames are unprofessional and impolite in all electronic communication contexts. So DON'T EVER, EVER SEND THEM! (Sorry I yelled.)

But, you might ask, what if I read an e-mail message that angers, irritates, or otherwise upsets me? Whatever you do, *don't respond immediately.* You might be tempted to flame someone. Instead, wait 24 hours to regain your composure. Then formulate a professional response, as if you were explaining politely and respectfully to your Great Aunt Elly (for the 103rd time) why the mold on the cheese she tried to serve you is unappetizing. Just remember: Don't hide behind your computer terminal with your "Caps Lock" button pushed.

Sometimes, you might be blamed for sending a flame even if you didn't intend to. If this happens, simply apologize by explaining that you didn't mean to convey that impression and you're sorry the message was misinterpreted.

Now you know how to avoid flaming someone else. But what techniques should you observe to avoid being flamed? First, don't offer negative comments about someone else's e-mail message. Many people still have the erroneous notion that e-mail tolerates a relaxed communication style and tone, in which attention to correctness is not valued. As a result, e-mail messages often contain more grammatical and mechanical mistakes than found in other more formal types of communication. Even so, never, ever express a view about someone else's grammar, mechanics, word choice, or writing style if you want to avoid getting a flame message in return.

Another way to avoid being flamed is by not using the "*ALL" (or "star all") function carelessly. Limit its use to *responsible* mass mailings. You can expect to get flamed if, for example, you offer your opinion on recent political developments to all the seniors at your high school, ask folks at the office to buy your daughter's Girl Scout cookies, or attempt to sell cosmetics or cookware or anything else via the hospital network. When in doubt, follow the "golden rule" of e-mail: E-mail unto others only what you would have them e-mail unto you!

Acronyms

E-mail has developed its own language filled with esoteric acronyms and cryptic symbols most likely developed to make e-mail convenient and efficient. Using acronyms allows us to type information quickly, IMHO. However, if acronyms obstruct rather than convey meaning to readers, then how truly efficient are they? Wouldn't you be more apt to understand "In My Humble Opinion" rather than "IMHO"? Certainly, incorporating acronyms in your messages will save you some time during the writing process, but it will cost your readers much time if they have to figure out what you mean. Since the purpose of any message is to get results, be as clear as you can.

So, how do you decide whether to include acronyms? The answer is simple. If the acronym you want to use is not standard in everyday English, avoid it. Besides, if you keep your e-mail messages succinct, then acronyms will be unnecessary. However, technical acronyms are useful for e-mail communication with systems administrators. Table 1 lists some of the technical acronyms that are becoming common among e-mail users. The point to stress here is *know your audience*. If you do, you'll use acronyms appropriately.

Emoticons

When you communicate face to face, your words convey only a portion of your message. Your body language—including posture, facial expressions, eye contact, and hand gestures—reveals whether you're interested or bored. And your voice—including articulation, inflection, pace, and volume—reveals whether

TABLE I   Technical Acronyms

Acronym	Full Term
ASCII	American Standard Code for Information Interchange
BBS	Bulletin Board Service or System
FAQ	Frequently Asked Question
FTP	File Transfer Protocol
GIF	Graphics Interchange Format
HTML	Hypertext Markup Language
HTTP	Hypertext Transfer Protocol
IRC	Internet Relay Chat
ISP	Internet Service Provider
JPEG	Joint Photographic Experts Group
URL	Uniform Resource Locator
WWW	World Wide Web

you're enthusiastic or secretive. A slow pace, for example, communicates deliberate emphasis, while a fast pace communicates excitement about your message. Taken together, the way you look and the way you sound give emotional significance to your words.

As is true of any written communication, an e-mail message cannot convey the meaning possible with a face-to-face conversation. Rather, in written communication, the meaning of your message depends exclusively on the words you've chosen. In an attempt to attach emotion to electronically communicated messages, e-mail has adopted symbols called *emoticons.*

The word *emoticon* was created from *emotion* and *icon.* Writers construct emoticons from keyboard characters, which are carefully placed after the specific words, sentences, or paragraphs to which the emoticons refer. The emoticons attempt to replicate what your readers hear and see when they communicate with you face to face.

For emoticons to make sense, you have to look at them sideways. For example, a colon followed by a hyphen and a closed parentheses creates an emoticon that conveys delight. A colon followed by a hyphen and an open parentheses creates an emoticon that conveys sadness. Some word-processing program can even convert these symbols into actual smiley and frowny faces:

:-) = ☺

:-(= ☹

Now that you know about emoticons, should you use them in your e-mail messages? The answer depends on your purpose for writing, the subject of your message, and your relationship to your readers. If you are writing within a professional, formal, or business context, avoid using emoticons. They add a level of informality that may trivialize any e-mail message. If you are writing within a personal, informal, or playful context, however, emoticons can be acceptable if you don't overuse them.

Just a few more words of caution: Too many emoticons can be distracting, intrusive, and annoyingly silly. Moreover, using emoticons can be risky. Even though the purpose of using them is to capture your emotion, your reader may interpret them differently. If you decide to use them, even in casual messages, be sure the emoticons hold the same meanings for your readers as they do for you.

And one more piece of advice: Never, ever substitute an emoticon for a carefully chosen word or phrase. Not every picture—certainly not a rudimentary picture constructed from keyboard characters—is worth a thousand words! In appropriate situations, if you stick to the emoticons in Table 2, you'll probably be OK in terms of successfully conveying your intended meaning.

Punctuation

As with emoticons, using too many unnecessary punctuation marks can be intrusive in an e-mail message. The key word here is *unnecessary.* I'm not suggesting that you eliminate necessary periods, semicolons, and commas. Rather, I'm sug-

TABLE 2 Common Emoticons

Emoticon	Name	Meaning
:-) or :)	Smiley	Expresses pleasure and delight, softens criticism, or hints at sarcasm
:-(or :(	Frowny	Expresses sadness or anger or hints at empathy or sympathy
;-)	Winky	Expresses "You know what I mean" or indicates "Just joking"
:-0	Shocked	Expresses shock or surprise

gesting that you not overuse exclamation points (known as *bangs* in computer circles) and question marks for emphasis. Again, your relationship to your audience, your subject, and your intended purpose for writing the e-mail message will determine how to use punctuation marks.

If you want to emphasize something, let your word choice, not your punctuation, communicate that meaning. If you are so anxious to get the answer to an e-mail question that you end the question with 20 question marks, you'd be better served to pick up the telephone and ask the question directly. If you are so hysterically agitated or excited that you need 20 exclamation marks to exorcise that emotion, you'd be better off to go to the top of the nearest uninhabited mountain and scream. Don't get into the habit of relying on punctuation marks to convey excessive inquisitiveness, impatience, or anxiety. Doing so will make it harder for you to shift from writing casual e-mail messages addressed to close friends to writing e-mail messages addressed to professional associates, teachers, and business contacts.

RESPONDING TO E-MAIL MESSAGES

We've spent a good deal of time talking about creating e-mail messages. Now let's discuss the proper way to respond to the e-mail messages that show up in your box. Because your response needs to be connected in some way to the original message, never create a new message if you're responding to an earlier e-mail message. Many of us tend to get multiple messages from the same person in a given day. So when you respond, it's important to help your reader know which message you're referring to.

Think of it this way: If you receive three questions from the same person via three separate telephone calls and each time you tell the caller you'll respond to the questions later that day, what will happen when you call back later and respond to question two without referring to that question? Your listener will be temporarily confused. An effective response, in any rhetorical situation, is sensitive to the entire communication context associated with the particular message. That always includes a reference to the original message and to the response.

Most e-mail software makes this requirement easy by automatically including a copy of the entire original message as part of a response. If you click on the "Reply" or "Respond" button in your e-mail program—which appears when you've opened any e-mail message sent to your "in" box—your response will appear just in front of the original message. In other words, when you hit the

"Send" button, your response and the entire original message will *both* go back to your reader, in that order.

When responding, sometimes it's better simply to quote the relevant parts of the original message and not the entire message. If, for example, the original message is long or includes several issues, quote only the single issue from the original message related to your response. Doing so helps the recipient put your response in a clearly focused context and allows your reply message to be more succinct.

Before we leave this section on responding to e-mail messages, I have one last comment: Not everyone or every organization can respond to e-mail in a timely manner. While businesses around the world accept e-mail from consumers, they have a hard time responding to those messages once they arrive. It's not just that e-mail messages are misrouted, lost, or responded to incorrectly. It's that the sheer volume of messages is overwhelming for many organizations. Consider a few examples:

- About a month ago, the German air carrier Lufthansa advertised that it would hold an auction for international airline tickets in late February. Because I wanted more information about the process, I e-mailed the company. By mid June, I had yet to hear from Lufthansa. I assume the auction happened without me.
- Earlier I told you that I e-mailed Mt. Olive Pickle Company because I wanted to know the origin of the phrase "I'm in a pickle." As with Lufthansa, I have yet to receive an e-mail from Mt. Olive.
- President Clinton receives 2,500 e-mail messages a day. Do you know how many responses are sent out in his name? Zero.

What's the moral of the story? E-mail may turn out to be less efficient than we thought. What's the other moral of the story? Don't neglect using the telephone or U.S. mail. . . .

Privacy, Legal, and Ethical Issues

In the late 1970s, as a result of the Watergate scandal, the federal government created the Electronic Communication Privacy Act (ECPA) to control excessive eavesdropping on telephone conversations. Since that time, the ECPA has evolved to address privacy issues related to electronic communication. Today, the ECPA prohibits unauthorized interception or disclosure of e-mail by *all* persons, businesses, and government not parties to the communication.

The first reported case concerning e-mail and privacy rights came about in 1996 in Pennsylvania and involved Michael Smyth, who worked for Pillsbury. Pillsbury's company policy assured employees that e-mail messages were confidential and the company could not use them as grounds for termination. However, when Smyth sent a biting message to his supervisor, characterizing other colleagues as "backstabbing bastards," the company fired him. Smyth sued Pillsbury in federal court, alleging wrongful discharge, but he lost the case. Under Pennsylvania law, the court said as that soon as Smyth mouthed off inappropriately in cyberspace, he lost any "reasonable expectation" for privacy. Smyth's case against Pillsbury illustrates what most lawyers currently believe about e-mail messages written on company systems: Unless your company explicitly gives you the right to privacy on the company e-mail system in *all* written contexts, you don't have that right.

The ECPA's protection can be somewhat nebulous because it contains exceptions. If a system administrator has reason to believe illegal activity is occurring over the system, he or she may disclose the information to legal authorities. Computer programmer Augustus Lindsey from Oklahoma learned this lesson the hard way when he sent the following message to a colleague: "That little sex kitten has been driving me wild. She's moaning and begging for it every minute. Last night I was afraid someone would hear, and we'd be thrown out of the building. But don't worry—all is arranged. Wednesday she gets the knife." Thinking a crime was about to occur, Lindsey's supervisor alerted the police, who arrested the computer programmer. Police released Lindsey the next morning, just in time for his veterinarian to spay his cat.

Despite the good intentions of the ECPA, e-mail is not private at home or at work. As I mentioned earlier in this chapter, some ISPs keep copies of your e-mail messages long after you've deleted them. And more and more businesses are keeping employee e-mail messages forever! It's becoming clear that computers, like elephants, never forget.

According to an American Management Association survey, 15% of employers store and regularly inspect company e-mail messages. But the actual number of U.S. businesses in industries such as banking, insurance, and telecommunications that eavesdrop on employees' e-mail messages is estimated to be higher than 25 percent. These companies want to monitor "cyber-lollygagging" to ensure employees aren't wasting valuable work time sending and receiving trivial messages. Moreover, companies are concerned with internal security and want to be sure employees aren't leaking company secrets. So systems administrators have the authority to intercept and monitor not only the flow but also the content of e-mail messages, often without employees' knowledge or consent.

How can you protect yourself? You can start by finding out the e-mail policies of your ISP and your employer. American Airlines, Federal Express, and United Parcel Service, for example, have e-mail systems that automatically tell employees the company reserves the right to monitor their e-mail. But today, corporations with such privacy policies are in the minority. In fact, only about one-third of companies currently have e-mail policies, even though the Electronic Messaging Association has, since the early 1990s, been urging organizations to make policies readily available to all employees.

To help you clarify appropriate e-mail use—whether at school, at work, or even at home—here are some questions to ask your e-mail system administrator:

- Is e-mail use exclusively for business, educational, or government use? Or are occasional personal e-mail messages acceptable?
- Does the company reserve the right to review, audit, and disclose my e-mail messages?

- What kinds of messages are considered derogatory, defamatory, obscene, or inappropriate?
- Does my deleting a message or file completely eliminate it from the system? If not, how long does the organization maintain the message?
- What monitoring policies does the company employ?

Even with answers to these questions, there is, of course, another issue: Regardless of the law, is it *ethical* for a system administrator to read someone's mail? Proponents say if the organization owns, operates, and repairs the equipment, then it's justified in reading employees' mail to ensure they are using the technology appropriately. Others argue that reading someone else's mail is simply wrong, and no organization is on sound ethical ground just because it tells people it may monitor their e-mail messages. No matter where you see yourself on this ethical continuum, remember that *e-mail is not private.* And even if it were, no e-mail message should contain anything that might cause your readers to feel threatened, discriminated against, or sexually harassed.

It's also important to remember that e-mail messages can sometimes go to the wrong audience. For example, while it is illegal to do so, hackers can intercept, read, and forward your e-mail messages. Not only that, e-mail software— even YOU—can malfunction, causing your message to go to an unintended reader. In one embarrassingly funny example, recently published by *The Phoenix Gazette,* an e-mail user sent the following message to thousands of online computer users rather than to his girlfriend: "I would love to kiss your hose-covered toes." While that message might make a hosiery company happy, it caused the boyfriend to run, high-tailing it out of the office for the rest of the day!

Certainly, organizations that transmit sensitive or confidential information— such as personnel actions, performance appraisals, and patient information—rely on encryption software to ensure the information they transmit via e-mail remains private. While encryption software can help protect the privacy of e-mail messages in the workplace, you should check with your organization's technology person before installing it on company hardware for your personal messages.

If you're worried about privacy on your personal computer at home, encryption software may protect you. Still, if you simply think of all your e-mail messages as public documents, you'll protect yourself from possible embarrassment, litigation, or discipline. Just remember: Never write anything in an e-mail message that you wouldn't write on a billboard. E-mail is not private.

SUMMARY

Benefits

- E-mail is convenient, fast, and inexpensive.
- E-mail is effective when you don't need to speak to your reader(s) face-to-face, when the recipient of your message isn't immediately available, or when the information you're sending isn't sensitive or personal.

Limitations

- E-mail cannot convey facial expressions, gestures, or speech patterns.
- E-mail cannot transmit your actual signature.
- E-mail isn't private. Familiarize yourself with privacy policies when using an organization's equipment.
- Because e-mail isn't private, don't send messages when the information is personal, confidential, or inflammatory.

"Nuts and Bolts"

- Before writing an e-mail message, determine your purpose and your relationship to your readers.
- E-mail addresses should be professional.
- Create a specific "Subject" line using a noun, verb, and modifiers.
- Salutations and signatures should reflect your relationship to your audience.
- Organize your message by presenting your main points quickly and previewing the remaining content.
- Keep your e-mail messages short.
- Include bulleted lists, highlighted headings, and single-sentence paragraphs.
- Use emoticons only in casual e-mail messages.
- Choose an informal style and tone for most messages.
- Don't send flames or spam.
- Use only those acronyms found in Standard English.
- Quote the original message or the appropriate part of it when responding.

Part 4

Reports and Other Longer Documents

Nearly everyone in business and industry writes reports. The word *report* is really just a generic term for a variety of documents that vary in form and purpose. Some reports are purely informative; others are persuasive or argumentative. Reports can be simple check lists, or they can take the form of interoffice memos, letters to clients, or more full-blown documents that are the results of weeks, if not months, of effort.

Business and technical writing practice sometimes distinguishes between formal and informal reports. Formal reports generally follow a multi-part format and are used primarily to present the results of a detailed project. Such a project often involves a considerable outlay of capital plus time and effort. The format for a formal report may mandate a cover letter or a memo of transmittal attached to a bound document consisting of an abstract, a table of contents, a glossary, an introduction, a detailed discussion of all aspects of the topic, a set of conclusions and recommendations, and pages and pages of attachments.

Informal reports tend to be shorter documents. Their formats are less complex, consisting only of such essential items as an introduction, a discussion, a set of conclusions, and, where appropriate, a list of recommendations.

To ensure that their reports are useful, writers should take the same kind of process approach that they would use when writing any other business or technical document. They should plan their reports carefully from the start so that they can clearly define and stick to their intended purpose whether that purpose be to provide information, to analyze information and draw conclusions, or to make recommendations.

Because they are often action-oriented, reports require writers to analyze their audience—or audiences—carefully. By virtue of their length, some reports can intimidate readers who need to know the information they contain.

A classification of the kinds of audiences that business and technical writing address, which Thomas Pearsall first delineated in his *Audience Analysis for Technical Writing* (1969), is especially relevant to the audiences for reports. Those audiences can include any, or all, of the following:

- the layperson
- the executive
- the expert
- the technician
- the operator.

Each brings a different background and a different set of needs to his or her reading of a report that writers must take into account if they hope to produce an effective document. In general, using abstracts and visual aids to supplement the contents of both formal and informal reports will make those reports more accessible to larger groups of readers of varying degrees of expertise.

Proposals are in some ways simply specialized reports, although they can be written in letter or memo form as well. The primary purpose of a proposal is to persuade readers to do something. Sales letters and requests for adjustments are two fairly simple proposals. Internal proposals, which are aimed at changing policies and procedures with an organization, and external proposals, such as documents seeking grants or funding, are more complex persuasive documents. Because proposals aim at convincing an audience to act in a way the writer wants, a process approach can help lead writers to a successful proposal.

This section of *Strategies* begins with an essay by J. C. Mathes and Dwight W. Stevenson that returns to a familiar theme in this anthology: the importance of audience analysis. Mathes and Stevenson offer a detailed examination of the problem of audience analysis and a solution to that problem designed to meet the needs of the several audiences that reports in particular, and other business and technical documents in general, address.

Underscoring the need for attention to audience, Robert W. Dodge reports on the results of a survey done at Westinghouse in the 1960s to determine the reading habits of busy managers. Those results still have relevance for business and technical writers today. Dodge concludes that, while the body of the report may have required the most time for the writer to produce, busy managers read this section of the report with considerably less frequency than they read the other sections, usually giving more attention to the abstract. Christian K. Arnold further underscores the importance of the abstract in meeting audience needs in his discussion of the key elements of an effective abstract. Vincent Vinci then offers what amounts to a checklist that business and technical report writers can use as a last step in an effective writing process.

Supposedly a picture is worth a thousand words. In creating and using illus-trations to supplement written text, business and technical writers need to make sure that those thousand words are the ones they intended. Walter E. Oliu, Charles T. Brusaw, and Gerald J. Alred offer easy-to-follow instructions for the effective use of the most widely used visual aids.

David W. Ewing then discusses persuasive writing, addressing the wider issue of the different strategies needed to persuade rather than inform readers. Ewing's general comments are then complemented by those of Philip C. Kolin about how to write effective proposals.

J. C. Mathes and Dwight W. Stevenson

Audience Analysis:
The Problem and a Solution

Both J. C. Mathes and Dwight W. Stevenson are professors of humanities in the College of Engineering at the University of Michigan.

Every communication situation involves three fundamental components: a writer, a message, and an audience. However, many report writers treat the communication situation as if there were only two components: a writer and his message. Writers often ignore their readers because writers are preoccupied with their own problems and with the subject matter of the communication. The consequence is a poorly designed, ineffective report.

As an example, a student related to the class her first communication experience on a design project during summer employment with an automobile company. After she had been working on her assignment for a few weeks, her supervisor asked her to jot him a memo explaining what she was doing. Not wanting to take much time away from her work and not thinking the report very important, she gave him a handwritten memo and continued her technical activities. Soon after, the department manager inquired on the progress of the project. The supervisor immediately responded that he had just had a progress report, and thereupon forwarded the engineer's brief memo. Needless to say, the engineer felt embarrassed when her undeveloped and inadequately explained memo became an official report to the organization. The engineer thought her memo was written just to her supervisor, who was quite familiar with her assignment. Due to her lack of experience with organizational behavior, she made several false assumptions about her report audience, and therefore about her report's purpose.

The inexperienced report writer often fails to design his report effectively because he makes several false assumptions about the report writing situation. If the writer would stop to analyze the audience component, he would realize that:

1. It is false to assume that the person addressed is the audience.
2. It is false to assume that the audience is a group of specialists in the field.
3. It is false to assume that the report has a finite period of use.
4. It is false to assume that the author and the audience always will be available for reference.
5. It is false to assume that the audience is familiar with the assignment.
6. It is false to assume that the audience has been involved in daily discussions of the material.
7. It is false to assume that the audience awaits the report.
8. It is false to assume that the audience has time to read the report.

Assumptions one and two indicate a writer's lack of awareness of the nature of his report audience. Assumptions three, four, and five indicate his lack of appreciation of the dynamic nature of the system. Assumptions six, seven, and eight indicate a writer's lack of consideration of the demands of day-by-day job activity.

A report has value only to the extent that it is useful to the organization. It is often used primarily by someone other than the person who requested it. Furthermore, the report may be responding to a variety of needs within the organization. These needs suggest that the persons who will use the report are not specialists or perhaps not even technically knowledgeable about the report's subject. The specialist is the engineer. Unless he is engaged in basic research, he usually must communicate with persons representing many different areas of operation in the organization.

In addition, the report is often useful over an extended period of time. Each written communication is filed in several offices. Last year's report can be incomprehensible if the writer did not anticipate and explain his purpose adequately. In these situations, even within the office where a report originated, the author as well as his supervisor will probably not be available to explain the report. Although organizational charts remain unchanged for years, personnel, assignments, and professional roles change constantly. Because of this dynamic process, even the immediate audience of a report sometimes is not familiar with the writer's technical assignment. Thus, the report writer usually must design his report for a dynamic situation.

Finally, the report writer must also be alert to the communication traps in relatively static situations. Not all readers will have heard the coffee break chats that fill in the details necessary to make even a routine recommendation convincing. A report can arrive at a time when the reader's mind is churning with other concerns. Even if it is expected, the report usually meets a reader who needs to act immediately. The reader usually does not have time to read through the whole report; he wants the useful information clearly and succinctly. To the reader, time probably is the most important commodity. Beginning report writers seldom realize they must design their reports to be used efficiently rather than read closely.

The sources of the false assumptions we have been discussing are not difficult to identify. The original source is the artificial communication a student is required to perform in college. In writing only for professors, a student learns to write for audiences of one, audiences who know more than the writer knows, and audiences who have no instrumental interests in what the report contains. The subsequent source, on the job, is the writer's natural attempt to simplify his task. The report writer, relying upon daily contact and familiarity, simply finds it easier to write a report for his own supervisor than to write for a supervisor in a different department. The writer also finds it easier to concentrate upon his own concerns than to consider the needs of his readers. He finds it difficult to address complex audiences and face the design problems they pose.

AUDIENCE COMPONENTS AND PROBLEMS THEY POSE

To write a report you must first understand how your audience poses a problem. Then you must analyze your audience in order to be able to design a report structure that provides an optimum solution. To explain the components of the report audience you must do more than just identify names, titles, and roles. You must determine who your audiences are as related to the purpose and content of your report. "Who" involves the specific operational functions of the persons who will read the report, as well as their educational and business backgrounds. These persons can be widely distributed, as is evident if you consider the operational relationships within a typical organization.

Classifying audiences only according to directions of communication flow along the paths delineated by the conventional organizational chart, we can identify three types of report audiences: *horizontal, vertical,* and *external.* For example, in the organization chart in Figure 1, *Part of organization chart for naval ship engineering center,*[1] horizontal audiences exist on each level. The Ship Concept Design Division and the Command and Surveillance Division form horizontal audiences for each other. Vertical audiences exist between levels. The Ship Concept Design Division and the Surface Ship Design Branch form vertical audiences for each other. External audiences exist when any unit interacts with a separate organization, such as when the Surface Ship Design Branch communicates with the Newport News Shipbuilding Company.

What the report writer first must realize is the separation between him and any of these three types of audiences. Few reports are written for horizontal audiences within the same unit, such as from one person in the Surface Ship Design Branch to another person or project group within the Surface Ship Design Branch itself. Instead, a report at least addresses horizontal audiences within a larger framework, such as from the Surface Ship Design Branch to the Systems Analysis Branch. Important reports usually have complex audiences, that is, vertical and horizontal, and sometimes external audiences as well.

An analysis of the problems generated by horizontal audiences—often assumed to pose few problems—illustrates the difficulties most writers face in all

---

[1]A reference in H. B. Benford and J. C. Mathes, *Your Future in Naval Architecture,* Richards Rosen, New York, 1968.

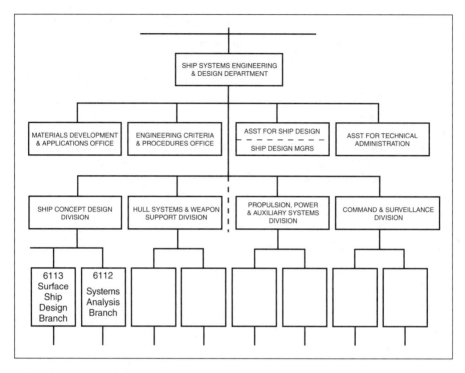

FIGURE I   Part of Organization Chart for Naval Ship Engineering Center

report writing situations. A systems engineer in the Systems Analysis Branch has little technical education in common with the naval architect in the Surface Ship Design Branch. In most colleges he takes only a few of the same mathematics and engineering science courses. The systems engineer would not know the wave resistance theory familiar to the naval architect, although he could use the results of his analysis. In turn, the naval architect would not know stochastics and probability theory, although he could understand systems models. But the differences between these audiences and writers go well beyond differences in training. In addition to having different educational backgrounds, the audiences will have different concerns, such as budget, production, or contract obligations. The audiences will also be separated from the writer by organizational politics and competition, as well as by personality differences among the people concerned.

When the writer addresses a horizontal audience in another organizational unit, he usually addresses a person in an organizational role. When addressed to the role rather than the person, the report is aimed at a department or a group. This means the report will have audiences in addition to the person addressed. It may be read primarily by staff personnel and subordinates. The addressee ultimately may act on the basis of the information reported, but at times he serves only to transfer the report to persons in his department who will use it. Furthermore, the report may have audiences in addition to those in the department addressed. It may be forwarded to other persons elsewhere, such as lawyers and comptrollers. The report travels routinely throughout organizational paths, and will have unknown or unanticipated audiences as well.

Consequently, even when on the same horizontal organizational level, the writer and his audience have little in common beyond the fact of working for the same organization, of having the same "rank" and perhaps of having the same educational level of attainment. Educational backgrounds can be entirely different; more important, needs, values, and uses are different. The report writer may recommend the choice of one switch over another on the basis of cost-efficiency analysis; his audiences may be concerned for business relationships, distribution patterns, client preferences, and budgets. Therefore, the writer should not assume that his audience has technical competence in the field, familiarity with the technical assignment, knowledge of him or of personnel in his group, similar value perspectives, or even complementary motives. The differences between writer and audience are distinctive, and may even be irreconcilable.

The differences are magnified when the writer addresses vertical audiences. Reports directed at vertical audiences, that is, between levels of an organization chart, invariably have horizontal audience components also. These complex report writing situations pose significant communication problems for the writer. Differences between writer and audience are fundamental. The primary audiences for the reports, especially informal reports, must act or make decisions on the basis of the reports. The reports thus have only instrumental value, that is, value insofar as they can be used effectively. The writer must design his report primarily according to how it will be used.

In addition to horizontal audiences and to vertical audiences, many reports are also directed to external audiences. External audiences, whether they consist of a few or many persons, have the distinctive, dissimilar features of the complex vertical audience. With external audiences these features invariably are exaggerated, especially those involving need and value. An additional complication is that the external audience can judge an entire organization on the basis of the writer's report. And sometimes most important of all, concerns for tact and business relationships override technical concerns.

In actual practice the writer often finds audiences in different divisions of his own company to be "external" audiences. One engineer encountered this problem in his first position after graduation. He was sent to investigate the inconsistent test data being sent to his group from a different division of the company in another city. He found that the test procedures being used in that division were faulty. However, at his supervisor's direction he had to write a report that would not "step on any toes." He had to write the report in such a manner as to have the other division correct its test procedures while not implying that the division was in any way at fault. An engineer who assumes that the purpose of his report is just to explain a technical investigation is poorly prepared for professional practice.

Most of the important communication situations for an engineer during his first five years out of college occur when he reports to his supervisor, department head, and beyond. In these situations, his audiences are action-oriented line management who are uninterested in the technical details and may even be unfamiliar with the assignment. In addition, his audiences become acquainted with him professionally through his reports; therefore, it is more directly the report than the investigation that is important to the writer's career.

Audience components and the significant design problems they pose are well illustrated by the various audiences for a formal report written by an engineer on the development of a process to make a high purity chemical, as listed in Figure 2, *Complex audience components for a formal report by a chemical engineer on a process to make a high purity chemical.* The purpose of the report was to explain the process; others would make a feasibility study of the process and evaluate it in comparison to other processes.

The various audiences for this report, as you can determine just by reading their titles, would have had quite different roles, backgrounds, interests, values, needs, and uses for the report. The writer's brief analysis of the audiences yielded the following:

> He could not determine the nature of many of his audiences, who they were, or what the specifics of their roles were.

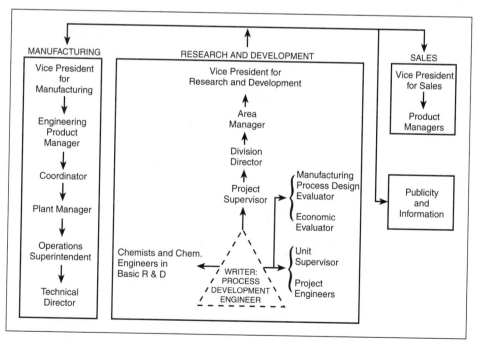

FIGURE 2   Complex Audience Components for a Formal Report by a Chemical Engineer on a Process to Make a High Purity Chemical

His audiences had little familiarity with his assignment.

His report would be used for information, for evaluation of the process, and for evaluation of the company's position in the field.

Some of his audiences would have from a minute to a half hour to glance at the report, some would take the report home to study it, and some would use it over extended periods of time for process analysis and for economic and manufacturing feasibility studies.

The useful lifetime of the report could be as long as twenty years.

The report would be used to evaluate the achievements of the writer's department.

The report would be used to evaluate the writing and technical proficiencies of the writer himself.

This report writer classified his audiences in terms of the conventional organization chart. Then to make them more than just names, titles, and roles he asked himself what they would know about his report and how they would use it. Even then he had only partially solved his audience problem and had just begun to clarify the design problems he faced. To do so he needed to analyze his audiences systematically.

A METHOD FOR SYSTEMATIC AUDIENCE ANALYSIS

To introduce the audience problem that report writers must face, we have used the conventional concept of the organization chart to classify audiences as *horizontal, vertical,* and *external.* However, when the writer comes to the task of performing an instrumentally useful audience analysis for a particular report, this concept of the organization and this classification system for report audiences are not very helpful.

First, the writer does not view from outside the total communication system modeled by the company organization chart. He is within the system himself, so his view is always relative. Second, the conventional outsider's view does not yield sufficiently detailed information about the report audiences. A single bloc on the organization chart looks just like any other bloc, but in fact each bloc represents one or several human beings with distinctive roles, backgrounds, and personal characteristics. Third, and most importantly, the outsider's view does not help much to clarify the specific routes of communication, as determined by audience needs, which an individual report will follow. The organization chart may describe the organization, but it does not describe how the organization functions. Thus many of the routes a report follows—and consequently the needs it addresses—will not be signaled by the company organization chart.

In short, the conventional concept of report audiences derived from organization charts is necessarily abstract and unspecific. For that reason a more effective method for audience analysis is needed. In the remaining portion of this . . . [selection], we will present a three-step procedure. The procedure calls for preparing an egocentric organization chart to identify individual report readers, characterizing these readers, and classifying them to establish priorities. Based upon an egocentric view of the organization and concerned primarily with what report readers need, this system should yield the information the writer must have if he is to design an individual report effectively.

Prepare an Egocentric Organization Chart

An egocentric organization chart differs from the conventional chart in two senses. First, it identifies specific individuals rather than complex organizational units. A bloc on the conventional chart may often represent a number of people, but insofar as possible the egocentric chart identifies particular individuals who are potential readers of reports a writer produces. Second, the egocentric chart categorizes people in terms of their proximity to the report writer rather than in terms of their hierarchical relationship to the report writer. Readers are not identified as organizationally superior, inferior, or equal to the writer but rather as near or distant from the writer. We find it effective to identify four different degrees of distance as is illustrated in Figure 3, *Egocentric organization chart*. In this figure, with the triangle representing the writer, each circle is an individual reader identified by his organizational title and by his primary operational concerns. The four degrees of distance are identified by the four concentric rings. The potential readers in the first ring are those people with whom the writer associates daily. They are typically those people in his same office or project group. The readers in the second ring are those people in other offices with whom the writer must normally interact in order to perform his job. Typically, these are persons in adjacent and management groups. The readers in the third ring are persons relatively more distant but still within the same organization. They are

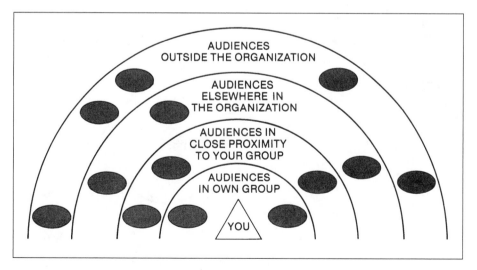

FIGURE 3   Egocentric Organization Chart

distant management, public relations, sales, legal department, production, purchasing, and so on. They are operationally dissimilar persons. The readers in the
fourth ring are persons beyond the organization. They may work for the same
company but in a division in another city. Or they may work for an entirely different organization.

Having prepared the egocentric organization chart, the report writer is able
to see himself and his potential audiences from a useful perspective. Rather than
seeing himself as an insignificantly small part of a complex structure—as he is apt
to do with the conventional organizational chart—the writer sees himself as a
center from which communication radiates throughout an organization. He sees
his readers as individuals rather than as faceless blocs. And he sees that what he
writes is addressed to people with varying and significant degrees of difference.

A good illustration of the perspective provided by the egocentric organization chart is the chart prepared by a chemical engineer working for a large corporation, Figure 4, *Actual egocentric organization chart of an engineer in a large
corporation*. It is important to notice how the operational concerns of the persons even in close proximity vary considerably from those of the development
engineer. What these people need from reports written by this engineer, then,
has little to do with the processes by which he defined his technical problems.

The chemical engineer himself is concerned with the research and development of production processes and has little interest in, or knowledge of, bud-
getary matters. Some of the audiences in his group are chemists concerned with

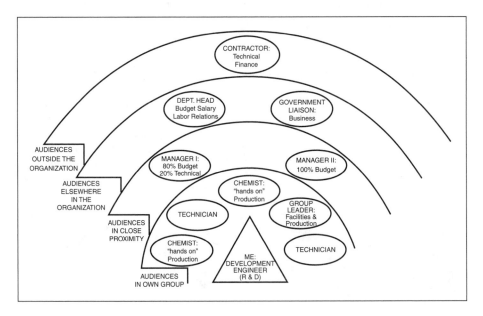

FIGURE 4 Actual Egocentric Organization Chart of an Engineer in a Large Corporation

production—not with research and development. Because of this they have, as he said, "lost familiarity with the technical background, and instead depend mostly on experience." Other audiences in his group are technicians concerned only with operations. With only two years of college, they have had no more than introductory chemistry courses and have had no engineering courses.

Still another audience in his group is his group leader. Rather than being concerned with development, this reader is concerned with facilities and production operations. Consequently, he too is "losing familiarity with the technical material." Particularly significant for the report writer is that his group leader in his professional capacity does not use his B.S.Ch.E. degree. His role is that of manager, so his needs have become administrative rather than technical.

The concerns of the chemical engineer/report writer's audiences in close proximity to his group change again. Instead of being concerned with development or production operations, these audiences are primarily concerned with the budget. They have little technical contact, and are described as "business oriented." Both Manager I and Manager II are older, and neither has a degree in engineering. One has a Ph.D. degree in chemistry, the other an M.S. degree in technology. Both have had technical experience in the lab, but neither can readily follow technical explanations. As the chemical engineer said, both would find it "difficult to return to the law."

The report writer's department head and other persons through whom the group communicates with audiences elsewhere in the organization, and beyond it, have additional concerns as well as different backgrounds. The department head is concerned with budget, personnel, and labor relations. The person in contact with outside funding units—in this case, a government agency—has business administration degrees and is entirely business oriented. The person in contact with subcontractors has both technical and financial concerns.

Notice that when this writer examined his audiences even in his own group as well as those in close proximity to him, he saw that the natures, the backgrounds, and especially the operational concerns of his audiences vary and differ considerably. As he widened the scope of his egocentric organization chart, he knew less and less about his audiences. However, he could assume they will vary even more than those of the audiences in close proximity.

Thus, in the process of examining the audience situation with an egocentric organization chart, a report writer can uncover not only the fact that audiences have functionally different interests, but also the nature of those functional differences. He can proceed to classify the audiences for each particular report in terms of audience needs.

Preparation of the egocentric organization chart is the first step of your procedure of systematic audience analysis. Notice that this step can be performed once to describe your typical report audience situation but must be particularized for each report to define the audiences for that report. Having prepared the egocentric chart once, the writer revises his chart for subsequent reports by adding or subtracting individual audiences.

Characterize the Individual Report Readers

In the process of preparing the egocentric organization chart, you immediately begin to think of your individual report readers in particular terms. In preparing the egocentric chart discussed above, the report writer mentioned such items as a reader's age, academic degrees, and background in the organization as well as his operational concerns. All of these particulars will come to mind when you think of your audiences as individuals. However, a systematic rather than piecemeal audience analysis will yield more useful information. The second step of audience analysis is, therefore, a systematic characterization of each person identified in the egocentric organization chart. A systematic characterization is made in terms of *operational, objective,* and *personal* characteristics.

The *operational characteristics* of your audiences are particularly important. As you identify the operational characteristics for a person affected by your report, try to identify significant differences between his or her role and yours. What are his professional values? How does he spend his time? That is, will his daily concerns and attitudes enable him to react to your report easily, or will they make it difficult for him to grasp what you are talking about? What does he know about your role, and in particular, what does he know or remember about your technical assignment and the organizational problem that occasioned your report to come to him? You should also consider carefully what he will need from your report. As you think over your entire technical investigation, ask yourself if that person will involve staff personnel in action on your report, of if he will in turn activate other persons elsewhere in the organization, when he receives the report. If he should, you must take their reactions into account when you write your report.

In addition, you should ask yourself, "How will my report affect his role?" A student engineer recently told us of an experience he had during summer employment when he was asked to evaluate the efficiency of the plant's waste treatment process. Armed with his fresh knowledge from advanced chemical engineering courses, to his surprise he found that, by making a simple change in the process, the company could save more than $200,000 a year. He fired off his report with great anticipation of glowing accolades—none came. How had his report affected the roles of some of his audiences? Although the writer had not considered the report's consequences when he wrote it, the supervisor, the manager, and related personnel now were faced with the problem of accounting for their waste of $200,000 a year. It should have been no surprise that they were less than elated over his discovery.

By *objective characteristics* we mean specific, relevant background data about the person. As you try to identify his or her educational background, you may note differences you might have otherwise neglected. Should his education seem to approximate yours, do not assume he knows what you know. Remember that the half-life of engineering education today is about five years. Thus, anyone five to ten years older than you, if you are recently out of college, probably will be only superficially familiar with the material and jargon of your advanced technical courses. If you can further identify his past professional experiences and roles,

you might be able to anticipate his first-hand knowledge of your role and technical activities as well as to clarify any residual organizational commitments and
value systems he might have. When you judge his knowledge of your technical
area, ask yourself, "Could he participate in a professional conference in my field
of specialization?"

For *personal characteristics,* when you identify a person by name, ask yourself
how often the name changes in this organizational role. When you note his or her
approximate age, remind yourself how differences in age can inhibit communication. Also note personal concerns that could influence his reactions to your report.

A convenient way to conduct the audience analysis we have been describing
and to store the information it yields is to use an analysis form similar to the one
in Figure 5, *Form for characterizing individual report readers* [p. 182]. It may be
a little time-consuming to do this the first time around, but you can establish a
file of audience characterizations. Then you can add to or subtract from this file
as an individual communication situation requires.

One final point: This form is a means to an end rather than an end in itself.
What is important for the report writer is that he thinks systematically about the
questions this form raises. The novice usually has to force himself to analyze his
audiences systematically. The experienced writer does this automatically.

Classify Audiences in Terms of How They Will Use Your Report

For each report you write, trace out the communication routes on your egocentric organization chart and add other routes not on the chart. Do not limit these
routes to those specifically identified by the assignment and the addresses of the
report. Rather, think through the total impacts of your report on the organization. That is, think in terms of the first, second, and even some third-order consequences of your report, and trace out the significant communication routes
involved. All of these consequences define your actual communication.

When you think in terms of consequences, primarily you think in terms of the
uses to which your report will be put. No longer are you concerned with your
technical investigation itself. In fact, when you consider how readers will use your
report, you realize that very few of your potential readers will have any real interest in the details of your technical investigation. Instead, they want to know the
answers to such questions as "Why was this investigation made? What is the significance of the problem it addresses? What am I supposed to do with the results of
this investigation? What will it cost? What are the implications—for sales, for production, for the unions? What happens next? Who does it? Who is responsible?"

It is precisely this audience concern for nontechnical questions that causes so
much trouble for young practicing engineers. Professionally, much of what the
engineer spends his time doing is, at most, of only marginal concern to many of
his audiences. His audiences ask questions about things which perhaps never
entered his thoughts during his own technical activities when he received the
assignment, defined the problem, and performed his investigation. These questions, however, must enter into his considerations when he writes his report.

NAME: TITLE:

A. OPERATIONAL CHARACTERISTICS:
   1. His role within the organization and consequent value system:

   2. His daily concerns and attitudes:

   3. His knowledge of your technical responsibilities and assignment:

   4. What he will need from your report:

   5. What staff and other persons will be activated by your report through him:

   6. How your report could affect his role:

B. OBJECTIVE CHARACTERISTICS:
   1. His education—levels, fields, and years:

   2. His past professional experiences and roles:

   3. His knowledge of your technical area:

C. PERSONAL CHARACTERISTICS:
   Personal characteristics that could influence his reactions—age, attitudes, pet concerns, etc.

FIGURE 5   Form for Characterizing Individual Report Readers

Having defined the communication routes for a report you now know what audiences you will have and what questions they will want answered. The final step in our method of audience analysis is to assign priorities to your audiences. Classify them in terms of how they will use your report. In order of their importance to you (not in terms of their proximity to you), classify your audiences by these three categories:

- *Primary audiences*—who make decisions or act on the basis of the information a report contains.
- *Secondary audiences*—who are affected by the decisions and actions.
- *Immediate audiences*—who route the report or transmit the information it contains.

The *primary audience* for a report consists of those persons who will make decisions or act on the basis of the information provided by the report. The report overall should be designed to meet the needs of these users. The primary audience can consist of one person who will act in an official capacity, or it can consist of several persons representing several offices using the report. The important point here is that the primary audience for a report can consist of persons from any ring on the egocentric organization chart. They may be distant or in close proximity to the writer. They may be his organizational superiors, inferiors, or equals. They are simply those readers for whom the report is primarily intended. They are the top priority users.

In theory at least, primary audiences act in terms of their organizational roles rather than as individuals with distinctive idiosyncracies, predilections, and values. Your audience analysis should indicate when these personal concerns are likely to override organizational concerns. A typical primary audience is the decision maker, but his actual decisions are often determined by the evaluations and recommendations of staff personnel. Thus the report whose primary audience is a decision maker with line responsibility actually has an audience of staff personnel. Another type of primary audience is the production superintendent, but again his actions are often contingent upon the reactions of others.

In addition, because the report enters into a system, in time both the line and staff personnel will change; roles rather than individuals provide continuity. For this reason, it is helpful to remember the words of one engineer when he said, "A complete change of personnel could occur over the lifetime of my report." The report remains in the file. The report writer must not assume that his primary audience will be familiar with the technical assignment. He must design the report so that it contains adequate information concerning the reasons for the assignment, details of the procedures used, the results of the investigation, and conclusions and recommendations. This information is needed so that any future component of his primary audience will be able to use the report confidently.

The *secondary audience* for a report consists of those persons other than primary decision makers or users who are affected by the information the report transmits into the system. These are the people whose activities are affected when a primary audience makes a decision, such as when production supervision has to adjust to management decisions. They must respond appropriately when a primary audience acts, such as when personnel and labor relations have to accommodate production line changes. The report writer must not neglect the needs of his secondary audiences. In tracing out his communication routes, he will identify several secondary audiences. Analysis of their needs will reveal what additional information the report should contain. This information is often omitted by writers who do not classify their audiences sufficiently.

The *immediate audience* for a report are those persons who route the report or transmit the information it contains. It is essential for the report writer to identify his immediate audiences and not to confuse them with his primary audiences. The immediate audience might be the report writer's supervisor or another middle management person. Yet usually his role will be to transmit information rather than to use the information directly. An information system has numerous persons who transmit reports but who may not act upon the information or who may not be affected by the information in ways of concern to the report writers. Often, a report is addressed to the writer's supervisor, but except for an incidental memo report, the supervisor serves only to transmit and expedite the information flow throughout the organizational system.

A word of caution: at times the immediate audience is also part of the primary audience; at other times the immediate audience is part of the secondary audience. For each report you write, you must distinguish those among your readers who will function as conduits to the primary audience.

As an example of these distinctions between categories of report audiences, consider how audiences identified on the egocentric organization chart, Figure 4 [p. 178], can be categorized. Assume that the chemical engineer writes a report on a particular process improvement he has designed. The immediate audience might be his Group Leader. Another would be Manager I, transmitting the report to Manager II. The primary audiences might be Manager II and the Department Head; they would ask a barrage of nontechnical questions similar to those we mentioned a moment ago. They will decide whether or not the organization will implement the improvement recommended by the writer. The Department Head also could be part of the secondary audience by asking questions relating to labor relations and union contracts. Other secondary audiences, each asking different questions of the report, could be:

> The person in contact with the funding agency, who will be concerned with budget and contract implications.
>
> The person in contact with subcontractors, determining how they are affected.
>
> The Group Leader, whose activities will be changed.
>
> The "hands on" chemist, whose production responsibilities will be affected.
>
> The technicians, whose job descriptions will change.

In addition to the secondary audiences on the egocentric organization chart, the report will have other secondary audiences throughout the organization— technical service and development, for example, or perhaps waste treatment.

At some length we have been discussing a fairly detailed method for systematic audience analysis. The method may have seemed more complicated than it actually is. Reduced to its basic ingredients, the method requires you, first, to identify all the individuals who will read the report, second, to characterize them, and third, to classify them. The *Matrix for audience analysis,* Figure 6, is a convenient device for characterizing and classifying your readers once you have identified them. At a glance, the matrix reveals what information you have and what information you still need to generate. Above all, the matrix forces you to think systematically. If you are able to fill in a good deal of specific information in each cell (particularly in the first six cells), you have gone a long way towards seeing how the needs of your audiences will determine the design of your report.

Character-istics / Types of audiences	Operational	Objective	Personal
Primary	①	④	⑦
Secondary	②	⑤	⑧
Immediate	③	⑥	⑨

FIGURE 6   Matrix for Audience Analysis

We have not introduced a systematic method for audience analysis with the expectation that it will make your communication task easy. We have introduced you to the problems you must account for when you design your reports—problems you otherwise might ignore. You should, at least, appreciate the complexity of a report audience. Thus, when you come to write a report, you are less likely to make false assumptions about your audience. To develop this attitude is perhaps as important as to acquire the specific information the analysis yields. On the basis of this attitude, you now are ready to determine the specific purpose of your report.

Richard W. Dodge

What to Report

At the time he wrote this article, Richard W. Dodge was the editor of Westinghouse Engineer.

Technical reports *can* be a useful tool for management—not only as a source of general information, but, more importantly, as a valuable aid to decision making. But to be effective for these purposes, a technical report must be geared to the needs of management.

Considerable effort is being spent today to upgrade the effectiveness of the technical report. Much of this is directed toward improving the writing abilities of engineers and scientists; or toward systems of organization, or format, for reports. This effort has had some rewarding effects in producing better written reports.

But one basic factor in achieving better reports seems to have received comparatively little attention. This is the question of audience needs. Or, expressed another way, *"What does management want in reports?"* This is an extremely basic question, and yet it seems to have had less attention than have the mechanics of putting words on paper.

A recent study conducted at Westinghouse sheds considerable light on this subject. While the results are for one company, probably most of them would apply equally to many other companies or organizations.

The study was made by an independent consultant with considerable experience in the field of technical report writing. It consisted of interviews with Westinghouse men at every level of management, carefully selected to present an accurate cross section. The list of questions asked is shown in Table 1. The results were compiled and analyzed, and from the report several conclusions are apparent. In addition, some suggestions for report writers follow as a natural consequence.

TABLE I   Questions Asked of Managers

1. What types of reports are submitted to you?
2. What do you look for *first* in the reports submitted to you?
3. What do you want from these reports?
4. To what depth do you want to follow any one particular idea?
5. At what level (how technical and how detailed) should the various reports be written?
6. What do you want emphasized in the reports submitted to you? (Facts, interpretations, recommendations, implications, etc.)
7. What types of decisions are you called upon to make or to participate in?
8. What type of information do you need in order to make these decisions?
9. What types of information do you receive that you don't want?
10. What types of information do you want but not receive?
11. How much of a typical or average report you receive is useful?
12. What types of reports do you write?
13. What do you think your boss wants in the reports you send him?
14. What percentage of the reports you receive do you think desirable or useful? (In kind or frequency.)
15. What percentage of the reports you write do you think desirable or useful? (In kind or frequency.)
16. What particular weaknesses have you found in reports?

WHAT MANAGEMENT LOOKS FOR IN ENGINEERING REPORTS

When a manager reads a report, he looks for pertinent facts and competent opinions that will aid him in decision making. He wants to know right away whether he should read the report, route it, or skip it.

To determine this, he wants answers fast to some or all of the following questions:

- What's the report about and who wrote it?
- What does it contribute?
- What are the conclusions and recommendations?
- What are their importance and significance?
- What's the implication to the Company?
- What actions are suggested? Short range? Long range?
- Why? By whom? When? How?

The manager wants this information in brief, concise, and meaningful terms. He wants it at the beginning of the report and all in one piece.

For example, if a summary is to convey information efficiently, it should contain three kinds of facts:

1. What the report is about;
2. The significance and implications of the work; and
3. The action called for.

To give an intelligent idea of what the report is about, first of all the problem must be defined, then the objectives of the project set forth. Next, the reasons for doing the work must be given. Following this should come the conclusions. And finally, the recommendations.

Such summaries are informative and useful, and should be placed at the beginning of the report.

The kind of information a manager wants in a report is determined by his management responsibilities, but how he wants this information presented is determined largely by his reading habits. This study indicates that management report reading habits are surprisingly similar. Every manager interviewed said he read the *summary* or abstract; a bare majority said they read the *introduction* and *background* sections as well as the *conclusions* and *recommendations;* only a few managers read the *body* of the report or the *appendix* material.

The managers who read the *background* section, or the conclusions and recommendations, said they did so ". . . to gain a better perspective of the material being reported and to find an answer to the all-important question: What do we do next?" Those who read the *body* of the report gave one of the following reasons:

1. Especially interested in subject
2. Deeply involved in the project
3. Urgency of problem requires it
4. Skeptical of conclusions drawn.

And those few managers who read the *appendix* material did so to evaluate further the work being reported. To the report writer, this can mean but one thing: If a report is to convey useful information efficiently, the structure must fit the manager's reading habits.

The frequency of reading chart in Figure 1 [p. 190] suggests how a report should be structured if it is to be useful to management readers.

SUBJECT MATTER INTEREST

In addition to what facts a manager looks for in a report and how he reads reports, the study indicated that he is interested in five broad technological areas. These are:

1. Technical problems
2. New projects and products
3. Experiments and tests
4. Materials and processes
5. Field troubles.

Managers want to know a number of things about each of these areas. These are listed in Table 2. Each of the sets of questions can serve as an effective checklist for report writers.

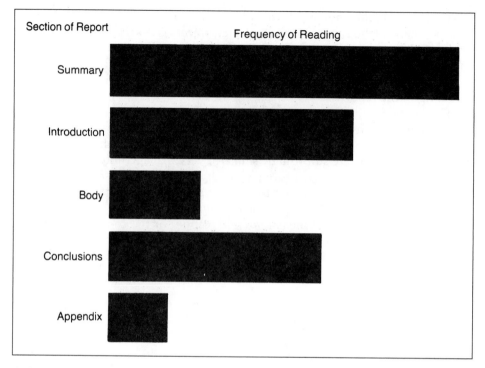

FIGURE I How Managers Read Reports

In addition to these subjects, a manager must also consider market factors and organization problems. Although these are not the primary concern of the engineer, he should furnish information to management whenever technical aspects provide special evidence or insight into the problem being considered. For example, here are some of the questions about marketing matters a manager will want answered:

- What are the chances for success?
- What are the possible rewards? Monetary? Technological?
- What are the possible risks? Monetary? Technological?
- Can we be competitive? Price? Delivery?
- Is there a market? Must one be developed?
- When will the product be available?

And, here are some of the questions about organization problems a manager must have answered before he can make a decision:

- Is it the type of work Westinghouse should do?
- What changes will be required? Organization? Manpower? Facilities? Equipment?
- Is it an expanding or contracting program?
- What suffers if we concentrate on this?

TABLE 2 What Managers Want to Know

Problems	Tests and Experiments
What is it?	What tested or investigated?
Why undertaken?	Why? How?
Magnitude and importance?	What did it show?
What is being done? By whom?	Better ways?
Approaches used?	Conclusions? Recommendations?
Thorough and complete?	Implications to Company?
Suggested solutions? Best? Consider others?	**Material and Processes**
What now?	Properties, characteristics, capabilities?
Who does it?	Limitations?
Time factors?	Use requirements and environment?
New Projects and Products	Areas and scope of application?
Potential?	Cost factors?
Risks?	Availability and sources?
Scope of application?	What else will do it?
Commercial implications?	Problems in using?
Competition?	Significance of application to Company?
Importance to Company?	**Field Troubles and Special**
More work to be done? Any problems?	**Design Problems**
Required manpower, facilities and equipment?	Specific equipment involved?
Relative importance to other projects and	What trouble developed? Any trouble history?
products?	How much involved?
Life of project or product line?	Responsibility? Others? Westinghouse?
Effect on Westinghouse technical position?	What is needed?
Priorities required?	Special requirements and environment?
Proposed schedule?	Who does it? Time factors?
Target date?	Most practical solution?
	Recommended action?
	Suggested product design changes?

These are the kinds of questions Westinghouse management wants answered about projects in these five broad technological areas. The report writer should answer them whenever possible.

LEVEL OF PRESENTATION

Trite as it may sound, the technical and detail level at which a report should be written depends upon the reader and his use of the material. Most readers— certainly this is true for management readers—are interested in the significant material and in the general concepts that grow out of detail. Consequently, there is seldom real justification for a highly technical and detailed presentation.

Usually the management reader has an educational and experience background different from that of the writer. *Never* does the management reader have the same knowledge of and familiarity with the specific problem being reported that the writer has.

Therefore, the writer of a report for management should write at a technical level suitable for a reader whose educational and experience background is in a field different from his own. For example, if the report writer is an electrical engineer, he should write his reports for a person educated and trained in a field such as chemical engineering, or mechanical engineering, or metallurgical engineering.

All parts of the report *should preferably* be written on this basis. The highly technical, mathematical, and detailed material—if necessary at all—can and should be placed in the appendix.

MANAGEMENT RESPONSIBILITIES

The information presented thus far is primarily of interest to the report writer. In addition, however, management itself has definite responsibilities in the reporting process. These can be summed up as follows:

1. Define the project and the required reports;
2. Provide proper perspective for the project and the required reporting;
3. See that effective reports are submitted on time; and
4. See that the reports are properly distributed.

An engineering report, like any engineered product, has to be designed to fill a particular need and to achieve a particular purpose within the specific situation. Making sure that the writer knows what his report is to do, how it is to be used, and who is going to use it—all these things are the responsibilities of management. Purpose, use, and reader are the design factors in communications, and unless the writer knows these things, he is in no position to design an effective instrument of communication—be it a report, a memorandum, or what have you.

Four conferences at selected times can help a manager control the writing of those he supervises and will help him get the kind of reports he wants, when he wants them.

Step 1—At the beginning of the project. The purpose of this conference is to define the project, make sure the engineer involved knows what it is he's supposed to do, and specify the required reporting that is going to be expected of him as the project continues. What kind of decisions, for example, hinge upon his report? What is the relation of his work to the decision making process of management? These are the kinds of questions to clear up at this conference.

If the project is an involved one that could easily be misunderstood, the manager may want to check the effectiveness of the conference by asking the engineer to write a memorandum stating in his own words his understanding of the project, how he plans to handle it, and the reporting requirements. This can ensure a mutual understanding of the project from the very outset.

Step 2—At the completion of the investigation. When the engineer has finished the project assignment—but before he has reported on it—the manager should have him come in and talk over the results of his work. What did he find out? What conclusions has he reached? What is the main supporting evidence for these conclusions? What recommendations does he make? Should any future action be suggested? What is the value of the work to the Company?

The broader perspective of the manager, plus his extensive knowledge of the Company and its activities, puts him in the position of being able to give the engineer a much better picture of the value and implications of the project.

The mechanism for getting into the report the kind of information needed for decision making is a relatively simple one. As the manager goes over the material with the engineer, he picks out points that need to be emphasized, and those that can be left out. This is a formative process that aids the engineer in the selection of material and evidence to support his material.

Knowing in advance that he has a review session with his supervisor, the chances are the engineer will do some thinking beforehand about the project and the results. Consequently, he will have formed some opinions about the significance of the work, and will, therefore, make a more coherent and intelligent presentation of the project and the results of his investigation.

This review will do something for the manager, too. The material will give him an insight into the value of the work that will enable him to converse intelligently and convincingly about the project to others. Such a preview may, therefore, expedite decisions influencing the project in one way or another.

Step 3—After the report is outlined. The manager should schedule a third conference after the report is outlined. At this session, the manager and the author should review the report outline step-by-step. If the manager is satisfied with the outline, he should tell the author so and tell him to proceed with the report.

If, however, the manager is not satisfied with the outline and believes it will have to be reorganized before the kind of report wanted can be written, he must make this fact known to the author. One way he can do this is to have the author tell him why the outline is structured the way it is. This usually discloses the organizational weakness to the author and consequently he will be the one to suggest a change. This, of course, is the ideal situation. However, if the indirect approach doesn't work, the more direct approach must be used.

Regardless of the method used to develop a satisfactory outline, the thing the manager must keep in mind is this: It's much easier to win the author's consent to structural changes at the outline stage, i.e., *before* the report is written than afterward. Writing is a personal thing; therefore, when changes are suggested in organization or approach, these are all too frequently considered personal attacks and strained relations result.

Step 4— After the report is written. The fourth interview calls for a review and approval by the manager of the finished report and the preparation of a distribution list. During this review, the manager may find some sections of the report that need changing. While this is to be expected, he should limit the extent of these changes. The true test of any piece of writing is the clarity of the statement. If it's clear and does the job, the manager should leave it alone.

This four-step conference mechanism will save the manager valuable time, and it will save the engineer valuable time. Also, it will ensure meaningful and useful project reports—not an insignificant accomplishment itself. In addition, the process is an educative one. It places in the manager's hands another tool he can use to develop and broaden the viewpoint of the engineer. By eliminating misunderstanding and wasted effort, the review process creates a more helpful and effective working atmosphere. It acknowledges the professional status of the engineer and recognizes his importance as a member of the engineering department.

Christian K. Arnold

The Writing of Abstracts

The late Christian K. Arnold was an instructor in English and a technical writer and editor before becoming Associate Executive Secretary of the Association of State Universities and Land-Grant Colleges.

The most important section of your technical report or paper is the abstract. Some people will read your report from cover to cover; others will skim many parts, reading carefully only those parts that interest them for one reason or another; some will read only the introduction, results, and conclusions; but everyone who picks it up will read the abstract. In fact, the percentage of those who read beyond the abstract is probably related directly to the skill with which the abstract is written. The first significant impression of your report is formed in the reader's mind by the abstract; and the sympathy with which it is read, if it is read at all, is often determined by this first impression. Further, the people your organization wants most to impress with your report are the very people who will probably read no more than the abstract and certainly no more than the abstract, introduction, conclusions, and recommendations. And the people you should want most to read your paper are the ones for whose free time you have the most competition.

Despite its importance, you are apt to throw your abstract together as fast as possible. Its construction is the last step of an arduous job that you would rather have avoided in the first place. It's a real relief to be rid of the thing, and almost anything will satisfy you. But a little time spent in learning the "rules" that govern the construction of good abstracts and in practicing how to apply them will pay material dividends to both you and your organization.

The abstract—or summary, foreword, or whatever you call the initial thumbnail sketch of your report or paper—has two purposes: (1) it provides the specialist in the field with enough information about the report to permit him to decide whether he could read it with profit, and (2) it provides the administrator or executive with enough knowledge about what has been done in the study or project and with what results to satisfy most of his administrative needs.

It might seem that the design specifications would depend upon the purpose for which the abstract is written. To satisfy the first purpose, for instance, the abstract needs only to give an accurate indication of the subject matter and scope of the study; but, to satisfy the second, the abstract must summarize the results and conclusions and give enough background information to make the results understandable. The abstract designed for the first purpose can tolerate any technical language or symbolic shortcuts understood at large by the subject-matter group; the abstract designed for the second purpose should contain no terms not generally understood in a semitechnical community. The abstract for the first purpose is called a *descriptive abstract;* that for the second, an *informative abstract.*

The following abstract, prepared by a professional technical abstracter in the Library of Congress, clearly gives the subject-matter specialist all the help he needs to decide whether he should read the article it describes:

> Results are presented of a series of cold-room tests on a Dodge diesel engine to determine the effects on starting time of (1) fuel quantity delivered at cranking speed and (2) type of fuel-injection pump used. The tests were made at a temperature of −10°F with engine and accessories chilled at −10°F at least 8 hours before starting.

Regardless of however useful this abstract might be on a library card or in an index or an annotated bibliography, it does not give an executive enough information. Nor does it encourage everyone to read the article. In fact, this abstract is useless to everyone except the specialist looking for sources of information. The descriptive abstract, in other words, cannot satisfy the requirement of the informative abstract.

But is the reverse also true? Let's have a look at an informative abstract written for the same article:

> A series of tests was made to determine the effect on diesel-engine starting characteristics at low temperatures of (1) the amount of fuel injected and (2) the type of injection pump used. All tests were conducted in a cold room maintained at −10°F on a commercial Dodge engine. The engine and all accessories were "cold-soaked" in the test chamber for at least 8 hours before each test. Best starting was obtained with 116 cu mm of fuel, 85 per cent more than that required for maximum power. Very poor starting was obtained with the lean setting of 34.7 cu mm. Tests with two different pumps indicated that, for best starting characteristics, the pump must deliver fuel evenly to all cylinders even at low cranking speeds so that each cylinder contributes its maximum potential power.

This abstract is not perfect. With just a few more words, for instance, the abstracter could have clarified the data about the amount of fuel delivered: do the figures give flow rates (what is the unit of time?) or total amount of fuel injected (over how long a period?)? He could easily have defined "best" starting. He could have been more specific about at least the more satisfactory type of pump: what is the type that delivers the fuel more evenly? Clarification of these points would not have increased the length of the abstract significantly.

The important point, however, is not the deficiencies of the illustration. In fact, it is almost impossible to find a perfect, or even near perfect, abstract, quite possibly because the abstract is the most difficult part of the report to write. This difficulty stems from the severe limitations imposed on its length, its importance to the over-all acceptance of the report or paper, and, with informative abstracts, the requirement for simplicity and general understandability.

The important point, rather, is that the informative abstract gives everything that is included in the descriptive one. The informative abstract, that is, satisfies not only its own purpose but also that of the descriptive abstract. Since values are obtained from the informative abstract that are not obtained from the descriptive, it is almost always worthwhile to take the extra time and effort necessary to produce a good informative abstract for your report or memo. Viewed from the standpoint of either the total time and effort expended on the writing job as a whole or the extra benefits that accrue to you and your organization, the additional effort is inconsequential.

It is impossible to lay down guidelines that will lead always to the construction of an effective abstract, simply because each reporting job, and consequently each abstract, is unique. However, general "rules" can be established that, if practiced conscientiously and applied intelligently, will eliminate most of the bugs from your abstracts.

1. *Your abstract must include enough specific information about the project or study to satisfy most of the administrative needs of a busy executive.* This means that the more important results, conclusions, and recommendations, together with enough additional information to make them understandable, must be included. This additional information will most certainly include an accurate statement of the problem and the limitations placed on it. It will probably include an interpretation of the results and the principal facts upon which the analysis was made, along with an indication of how they were obtained. Again, *specific* information must be given. One of the most common faults of abstracts in technical reports is that the information given is too general to be useful.

2. *Your abstract must be a self-contained unit, a complete report-in-miniature.* Sooner or later, most abstracts are separated from the parent report, and the abstract that cannot stand on its own feet independently must then either be rewritten or will fail to perform its job. And the rewriting, if it is done, will be done by someone not nearly as sympathetic with your study as you are. Even if it is not separated from the report, the abstract must be written as a complete, independent unit if it is to be of the most help possible to the executive. This rule automatically eliminates the common deadwood phrases like "this report contains . . ." or "this report is a report on . . ." that clutters up many abstracts. It also eliminates all references to sections, figures, tables, or anything else contained in the report paper.

3.   *Your abstract must be short.* Length in an abstract defeats every purpose for which it is written. However, no one can tell you just how short it must be. Some authorities have attempted to establish arbitrary lengths, usually in terms of a certain percentage of the report, the figure given normally falling between three and ten per cent. Such artificial guides are unrealistic. The abstract for a 30-page report must necessarily be longer, percentagewise, than the abstract for a 300-page report, since there is certain basic information that must be given regardless of the length of the report. In addition, the information given in some reports can be summarized much more briefly than can that given in other reports of the same over-all dimensions. Definite advantages, psychological as well as material, are obtained if the abstract is short enough to be printed entirely on one page so that the reader doesn't even have to turn a page to get the total picture that the abstract provides. Certainly, it should be longer than the interest span of an only mildly interested and very busy executive. About the best practical advice that can be given in a vacuum is to make your abstract as short as possible without cutting out essential information or doing violence to its accuracy. With practice, you might be surprised to learn how much information you can crowd into a few words. It helps, too, to learn to blue-pencil unessential information. It is perhaps important to document that "a meeting was held at the Bureau of Ordnance on Tuesday, October 3, 1961, at 2:30 P.M." somewhere, but such information is just excess baggage in your abstract: it helps neither the research worker looking for source material nor the administrator looking for a status or information summary. Someone is supposed to have once said, "I would have written a shorter letter if I had had more time." Take the time to make your abstracts shorter; the results are worth it. But be careful not to distort the facts in the condensing.

4.   *Your abstract must be written in fluent, easy-to-read prose.* The odds are heavily against your reader's being an expert in the subject covered by your report or paper. In fact, the odds that he is an expert in your field are probably no greater than the odds that he has only a smattering of training in any technical or scientific discipline. And even if he were perfectly capable of following the most obscure, tortured technical jargon, he will appreciate your sparing him the necessity for doing it. T. O. Richards, head of the Laboratory Control Department, and R.A. Richardson, head of the Technical Data Department, both of the General Motors Corporation, have written that their experience shows the abstract cannot be made too elementary. "We never had [an abstract] . . . in which the explanations and terms were too simple." This requirement immediately eliminated the "telegraphic" writing often found in abstracts. Save footage by sound practices of economy and not by cutting out the articles and the transitional devices needed for smoothness and fluency. It also eliminates those obscure terms that you defend on the basis of "that's the way it's always said."

5.   *Your abstract must be consistent in tone and emphases with the report paper, but it does not need to follow the arrangement, wording, or proportion of the original.* Data, information, and ideas introduced into the abstract must also appear in the report or paper. And they must appear with the same emphases. A conclusion or recommendation that is qualified in the report paper must not turn up without the qualification in the abstract. After all, someone might read both the abstract and the report. If this reader spots an inconsistency or is confused, you've lost a reader.

6.   *Your abstract should make the widest possible use of abbreviations and numerals, but it must not contain any tables or illustrations.* Because of the space limitations imposed on abstracts, the rules governing the use of abbreviations and numerals are relaxed for it. In fact, all figures except those standing at the beginning of sentences should be written as numerals, and all abbreviations generally accepted by such standard sources as the American Standards Association and "Webster's Dictionary" should be used.

By now you must surely see why the abstract is the toughest part of your report to write. A good abstract is well worth the time and effort necessary to write it and is one of the most important parts of your report. And abstract writing probably contributes more to the acquisition of sound expository skills than does any other prose discipline.

Vincent Vinci

Ten Report Writing Pitfalls: How to Avoid Them

Vincent Vinci was Director of Public Relations for Lockheed Electronics when he wrote this article.

The advancement of science moves on a pavement of communications. Chemists, electrical engineers, botanists, geologists, atomic physicists, and other scientists are not only practitioners but interpreters of science. As such, the justification, the recognition and the rewards within their fields result from their published materials.

Included in the vast field of communications is the report, a frequently used medium for paving the way to understanding and action. The engineering manager whose function is the direction of people and programs receives and writes many reports in his career. And therefore the need for technical reports that communicate effectively has been internationally recognized.

Since scientific writing is complicated by specialized terminology, a need for precision and the field's leaping advancement, the author of an engineering report can be overwhelmed by its contents. The proper handling of contents and communication of a report's purpose can be enhanced if the writer can avoid the following 10 pitfalls.

PITFALL 1: IGNORING YOUR AUDIENCE

In all the forms of communications, ignoring your audience in the preparation of a report is perhaps the greatest transgression. Why? All other forms of communication, such as instruction manuals, speeches, books and brochures, are directed to an indefinable or only partially definable audience. The report, on the other hand, is usually directed to a specific person or group and has a specific purpose.

So, it would certainly seem that if one knows both the "who" and the "why," then a report writer should not be trapped by this pitfall.

But it is not enough to know the who and why, you need to know "how." To get to the how, let's assume that the reader is your boss and has asked you to write a trip report. You are to visit several plants and report on capital equipment requirements. Before you write the first word, you will have to find out what your boss already knows about these requirements. It is obvious that he wants a new assessment of the facilities' needs. But, was he unsatisfied with a recent assessment and wants another point of view, or is a new analysis required because the previous report is outdated—or does he feel that now is the time to make the investment in facilities so that production can be increased over the next five years? That's a lot of questions, but they define both the who and why of your trip and, more importantly, your report.

By this time you may get the feeling that I am suggesting you give him exactly what he wants to read. The answer is yes and no. No, I don't mean play up to your boss's likes and dislikes. I do mean, however, that you give him all the information he needs to make a decision—the pros and the cons.

I mean also that the information be presented in a way that he is acclimated to in making judgments. For example, usually a production-oriented manager or executive (even the chief executive) will think in terms of his specialty. The president of a company who climbed the marketing ladder selling solvents will think better in marketing terms. Therefore, perhaps the marketing aspects of additional equipment and facilities should be stressed. You should also be aware that if you happen to be the finance director, your boss will expect to see cost/investment factors too.

A simple method for remembering, rather than ignoring, your audience is to place a sheet of paper in front of you when you start to write your report. On the paper have written in bold letters **WHO, WHY** and **HOW,** with the answers clearly and cogently defined. Keep it in front of you throughout the preparation of your report.

PITFALL 2: WRITING TO IMPRESS

Nothing turns a reader off faster than writing to impress. Very often reports written to leave a lasting scholarly impression on top management actually hinder communication.

Generally, when a word is used to impress, the report writer assumes that the reader either knows its meaning or will take the trouble to look it up. Don't assume that a word familiar to you is easily recognized by your reader. I recall a few years ago, there was a word "serendipity" which became a fashionable word to impress your reader with. And there was "fulsome" and "pejorative," and more. All are good words, but they're often misused or misapplied. They were shoved into reports to impress, completely disregarding the reader. Your objective is that your reader comprehend your thoughts, and there should be a minimum of impediments to understanding—understanding with first reading, and no deciphering.

Unfortunately, writing to impress is not merely restricted to use of obscure words but also includes unnecessary detail and technical trivia. Perhaps the scientist, chemist, chemical engineer and others become so intrigued with technical fine points that the meaningful (to your audience) elements of a report are buried. And quite often the fault is not so much a lack of removing the chaff from the grain but an attempt to technically impress the reader. Of course there exist reports that are full of technical detail because the nature of the communication is to impart a new chemical process, compound or technique. Even when writing this kind of report, you should eliminate any esoteric technical facts that do not contribute to communication, even though you may be tempted to include them to exhibit your degree of knowledge in the field.

PITFALL 3: HAVING MORE THAN ONE AIM

A report is a missile targeted to hit a point or achieve a mission. It is not a barrage of shotgun pellets that scatter across a target indiscriminately.

Have you ever, while reading a report, wondered where or what it was leading to—and even when you are finished you weren't quite sure? The writer probably had more than one aim, thereby preventing you from knowing where the report was heading.

Having more than one aim is usually the sign of a novice writer, but the pitfall can also trip up an experienced engineer if he does not organize the report toward one objective.

It is too easy to say that your report is being written to communicate, to a specific audience, information about your research, tests, visit, meeting, conference, field trip, progress or any other one of a range of activities that may be the subject. If you look at the first part of the sentence, you will see that "specific audience" and "information" are the key words that have to be modified to arrive at the goal of your report. For instance, you must define the specific audience such as the "members of the research council," "the finance committee" or "the chief process engineer and his staff."

Secondly, you need to characterize the information, such as "analysis of a new catalytic process," "new methods of atomic absorption testing" or "progress on waste treatment programs." You should be able to state the specific purpose of your report in one sentence: e.g., "The use of fibrous material improves scrubber efficiency and life—a report to the product improvement committee."

When you have arrived at such a definition of your purpose and audience, you can then focus both the test results and analysis toward that purpose, tempered with your readers in mind.

The usual error made in writing reports is to follow the chronology of the research in the body of the report with a summary of a set of conclusions and recommendations attached. The proper procedure to follow is to write (while focusing on your report goal) the analysis first (supported by test essentials or any other details), then your introduction or summary—sort of reverse chronology. But be sure that your goal and audience are clearly known because they become the basis of organizing your report.

PITFALL 4: BEING INCONSISTENT

If you work for an international chemical firm, you may be well aware of problems in communicating with plant managers and engineers of foreign installations or branches. And I'm not referring to language barriers, because for the most part these hurdles are immediately recognized and taken care of. What is more significant is units of measure. This problem is becoming more apparent as the United States slowly decides whether or not to adopt the metric system. Until it is adopted your best bet is to stick to one measurement system throughout the report. Preferably, the system chosen should be that familiar to your audience. If the audience is mixed, you should use both systems with one (always the same one) in parentheses. Obviously, don't mix units of measure because you will confuse or annoy your readers.

Consistency is not limited to measurements but encompasses terms, equations, derivations, numbers, symbols, abbreviations, acronyms, hyphenation, capitalization and punctuation. In other words, consistency in the mechanics of style will avoid work for your reader and smooth his path toward understanding and appreciating the content of the report.

If your company neither has a style guide nor follows the general trends of good editorial practice, perhaps you could suggest instituting a guide. In addition to the U.S. Government Printing Office Style Manual, many scientific and engineering societies have set up guides which could be used.

PITFALL 5: OVERQUALIFYING

Chemical engineers, astronomers, geologists, electrical engineers, and scientists of any other discipline have been educated and trained to be precise. As a result, they strive for precision, accuracy, and detail. That tends to work against the scientist when it comes to writing. Add to that the limited training received in the arts, and you realize why written expression does not come easily.

Most reports, therefore, have too many modifiers—adjectives, clauses, phrases, adverbs and other qualifiers. Consider some examples: the single-stage, isolated double-cooled refractory process breakdown, or the angle of the single-rotor dc hysteresis motor rotor winding. To avoid such difficult-to-comprehend phrases, you could in the first example write "the breakdown of the process in single-stage, isolated double-cooled refractories," and in the second, "the angle of the rotor winding in single-rotor dc hysteresis motors can cause . . . ," and so on. This eliminates the string of modifiers and makes the phrase easier to understand.

Better still, if your report allows you to say at the beginning that the following descriptions are only related to "single-stage, isolated double-cooled refractories" or "single-rotor dc hysteresis motors," you can remove the cumbersome nomenclature entirely.

In short, to avoid obscuring facts and ideas, eliminate excessive modifiers. Try to state your idea or main point first and follow with your qualifying phrases.

PITFALL 6: NOT DEFINING

Dwell, lake, and barn, all are common words. Right? Right and wrong. Yes, they are common to the nonscientist. To the mechanical engineer, dwell is the period a cam follower stays at maximum lift; to a chemical engineer, lake is a dye compound; and to an atomic physicist, a barn is an atomic cross-sectional area (10^{-24} cm^2).

These three words indicate two points: first, common words are used in science with other than their common meanings; and second, terms need to be defined.

In defining terms you use in a report, you must consider what to define and how to define. Of the two, I consider what to define a more difficult task and suggest that you review carefully just which terms you need defined. If you analyze the purpose, the scope, the direction and your audience (reader/user), you will probably get a good handle on such terms.

"How to define" ranges from the simple substitution of a common term for an uncommon one, to an extended or amplified explanation. But whatever the term, or method of definition, you need to slant it both to the reader and to the report purpose.

PITFALL 7: MISINTRODUCING

Introductions, summaries, abstracts and forewords—whatever you use to lead your reader into your report, it should not read like an exposition of a table of contents. If it does, you might as well let your audience read the table of contents.

The introduction, which should be written after the body of the report, should state the subject, purpose, scope, and the plan of the report. In many cases, an introduction will include a summary of the findings or conclusions. If a report is a progress report, the introduction should relate the current report to previous reports. Introductions, then, not only tell the sequence or plan of the report, but tell the what, how and why of the subject as well.

PITFALL 8: DAZZLING WITH DATA

Someone once said that a good painter not only knows what to put in a painting, but more importantly he knows what to leave out. It's much the same with report writing. If you dazzle your reader with tons of data, he may be moved by the weight of the report but may get no more out of it than that.

The usual error occurs in supportive material that many engineers and scientists feel is unnecessary to give a report scientific importance. The truth is that successful scientific writing (which includes reports) is heavily grounded in reality, simplicity, and understanding—not quantity.

The simplest way to evaluate the relevancy of information is to ask yourself after writing a paragraph, "What can I remove from this paragraph without destroying its meaning and its relationship to what precedes and what will follow?" Then, ask another question, "Does my reader require all that data to comprehend, evaluate or make a decision with?" If you find you can do without excess words, excess description and excessive supportive data, you will end up with a tighter, better and more informative report.

These principles should also be used to evaluate graphs, photographs, diagrams and other illustrations. Remember, illustrations should support or aid comprehension rather than being a crutch on which your report leans. The same should be kept in mind when determining just how much you should append to your report. There is no need to copy all your lab notes to show that detailed experimentation was performed to substantiate the results. A statement that the notes exist and are available will suffice.

PITFALL 9: NOT HIGHLIGHTING

Again, I believe the analogy of the painter applies. A good painter also knows what to highlight and what to subdue in a portrait or scene.

If you don't accent the significant elements, findings, illustrations, data, tests, facts, trends, procedures, precedents, or experiments pertinent to the subject and object of your report, you place the burden of doing so on your reader. As a result, he may consider the report a failure, draw his own conclusions, or hit upon the significant elements by chance. In any event, don't leave it up to your reader to search out the major points of your report.

Highlighting is one step past knowing what goes into your report and what to leave out (see Pitfall 8 above). All the key points of your report should define and focus on the purpose of your report. They must be included in your summary or conclusions, but these sections are not the only places to highlight. Attention should be called to key elements needed for the understanding of your material throughout the body of the report. Several methods may be used: you can underline an important statement or conclusion, you can simply point out that a particular illustration is the proof of the results of an experiment, or, as most professional writers do, you can make the key sentence the first or last sentence of a paragraph.

PITFALL 10: NOT REWRITING

Did you ever hear of an actor who hadn't rehearsed his lines before stepping before an audience? An actor wouldn't chance it—his reputation and his next role depend on his performance. The engineer shouldn't chance it either. Don't expect the draft of your report to be ready for final typing and reproduction without rewriting.

Once you have judged what your report will contain and how it will be organized, just charge ahead and write the first draft. Don't worry about choosing the precise word, turning that meaningful phrase, or covering all the facts in one paragraph or section. Once you have written your first draft (and the quicker you accomplish this the more time you will have to perfect the text), you are in a better position to analyze, tailor, and refine the report as a whole. Now you are also able to focus all the elements toward your purpose and your audience.

As you begin the rewriting process simply pick up each page of your draft, scan it, and ask yourself what role the material on that page plays in the fulfillment of the report's objective and understanding. You will find that this will enable you to delete, add, change and rearrange your material very quickly.

After you have completed this process, then rewrite paragraph by paragraph, sentence by sentence, and word by word. Your final step is to repeat the procedure of examining each page's contents. When you are satisfied with its flow and cohesion, then you will have a good report, one you know will be well received and acted upon.

Walter E. Oliu, Charles T. Brusaw, and Gerald J. Alred

Creating Tables and Illustrations

Walter E. Oliu is a technical writer with the U.S. Nuclear Regulatory Commission. Charles T. Brusaw is retired from NCR Corporation where he was a senior program instructor. Gerald J. Alred is professor of English at the University of Wisconsin–Milwaukee.

DESIGNING AND INTEGRATING VISUALS WITH TEXT

To make the most effective use of visuals and to integrate them smoothly with the text of your document, consider your graphics requirements even before you begin to write. Plan your visuals—tables, graphs, drawings, charts, maps, or photographs—when you're planning the scope and organization of your final work, whether it's a report, newsletter, brochure, or Web site. Make graphics an integral part of your outline, noting approximately where each should appear throughout the text. At each place where you plan to include a visual, either make a rough sketch of the visual or write "illustration of . . ." and enclose each suggestion in a box in your outline. You can indicate the placement of visuals in your outline by hand or on the computer. You can also copy and paste graphics directly into your outline at the appropriate places using your computer's clipboard feature. . . . Outlines are a means to an end, not an end in themselves, so like other information in an outline, these boxes and sketches can be moved, revised, or deleted as required. Planning your graphics requirements from the beginning stages of your outline preparation ensures their appropriate integration throughout all versions of the draft to the finished work. The following guidelines apply to most visual materials you might use to supplement or clarify the information in your text. . . . These tips will help you create and present your visual materials effectively.

1. Make clear in the text why the illustration is included. The amount of description each illustration requires will vary with its importance. An illustration showing an important feature or system may be central to an entire discussion. The complexity of the illustration will also affect the discussion, as will the background your readers bring to the information. Nonexperts require lengthier explanations than experts do, as a rule.
2. Keep the illustration to the point and uncluttered by including only information necessary to the discussion in the text and by eliminating unnecessary labels, arrows, boxes, and lines.
3. Keep terminology consistent. Do not refer to something as a "proportion" in the text and as a "percentage" in the illustration. Define all acronyms in the text, figure, or table. If any symbols are not self-explanatory, include a key that defines them.
4. Specify the units of measurement used or include a scale of relative distances, when appropriate. Ensure that relative sizes are clear or indicate distance by a scale, as on a map.
5. Position the lettering of any explanatory text or labels horizontally for ease of reading, if possible.
6. Give each illustration a concise title that clearly describes its contents.
7. Assign a figure or table number, if your final product is a document containing five or more illustrations. The figure or table number precedes the title:

   ■ Figure 1. Projected sales for 1998–2001

   Note that graphics (photographs, drawings, maps, etc.) are generally labeled "figures," while tables are labeled "tables."
8. In documents with more than five illustrations, list the illustrations by title, together with figure and page numbers, or table and page numbers, following the table of contents. The figures so listed should be titled "List of Figures." The tables so listed should be titled "List of Tables."
9. Refer to illustrations in the text of documents by their figure or table number.
10. If an illustration is central to a discussion, illuminating or strongly reinforcing it, place it as close as possible to the text where it is discussed. However, no illustration should precede its first text mention. Its appearance without an introduction in the text will confuse readers. If the illustration is lengthy and detailed, place it in an appendix, and be sure to refer to the illustration in the text of your document.
11. Allow adequate white space on the page around and within the illustration. This principle is true for Web sites as well as for documents.
12. If you wish to use an illustration that is copyrighted, first obtain a written permission from the copyright holder. Acknowledge such borrowings in a source or credit line below the caption for a figure and below any footnotes at the bottom of a table. Information that is in the public domain can be used without obtaining written permission to reproduce it. Publications of the federal government, for example, are not copyrighted, but you should acknowledge their source in a credit line. (Such a credit line appears in Figure 1.)

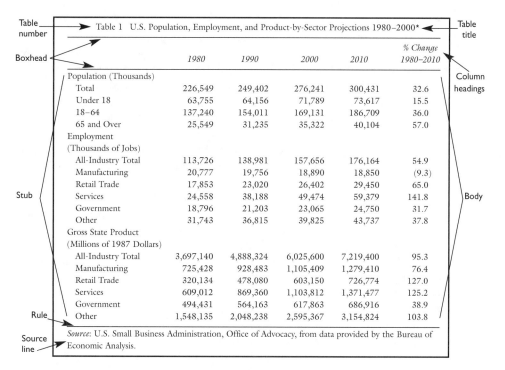

Table number

Table title

Boxhead

Column headings

Stub

Body

Rule

Source line

Table 1   U.S. Population, Employment, and Product-by-Sector Projections 1980–2000*

	1980	1990	2000	2010	% Change 1980–2010
Population (Thousands)					
Total	226,549	249,402	276,241	300,431	32.6
Under 18	63,755	64,156	71,789	73,617	15.5
18–64	137,240	154,011	169,131	186,709	36.0
65 and Over	25,549	31,235	35,322	40,104	57.0
Employment (Thousands of Jobs)					
All-Industry Total	113,726	138,981	157,656	176,164	54.9
Manufacturing	20,777	19,756	18,890	18,850	(9.3)
Retail Trade	17,853	23,020	26,402	29,450	65.0
Services	24,558	38,188	49,474	59,379	141.8
Government	18,796	21,203	23,065	24,750	31.7
Other	31,743	36,815	39,825	43,737	37.8
Gross State Product (Millions of 1987 Dollars)					
All-Industry Total	3,697,140	4,888,324	6,025,600	7,219,400	95.3
Manufacturing	725,428	928,483	1,105,409	1,279,410	76.4
Retail Trade	320,134	478,080	603,150	726,774	127.0
Services	609,012	869,360	1,103,812	1,371,477	125.2
Government	494,431	564,163	617,863	686,916	38.9
Other	1,548,135	2,048,238	2,595,367	3,154,824	103.8

Source: U.S. Small Business Administration, Office of Advocacy, from data provided by the Bureau of Economic Analysis.

FIGURE I   Sample Table

A discussion of visuals commonly used in on-the-job writing follows. Your topic will ordinarily determine the best visual material to use.

Tables

A table is useful for showing large numbers of specific, related data in a brief space. Because a table displays its information in rows and columns, the reader can easily compare data in one column with data in another. If such data were presented in the text, the reader would read through groups of numbers and possibly not recognize their significance. Tables typically include the following elements (Figure 1):

- *Table number.* Table numerals are usually Arabic and should be assigned sequentially to the tables throughout the text.
- *Table title.* The title, which is placed just above the table, should describe concisely what the table represents.
- *Boxhead.* The boxhead contains the column headings, which should be brief but descriptive. Units of measurement, where necessary, should either be specified as part of the heading or enclosed in parentheses beneath the heading. Standard abbreviations and symbols are acceptable. Avoid vertical lettering whenever possible.

- *Stub.* The lefthand vertical column of a table, called the stub, lists the items about which information is given in the body of the table.
- *Body.* The body comprises the data below the column headings and to the right of the stub. Within the body, columns should be arranged so that the terms to be compared appear in adjacent rows and columns. Where no information exists for a specific item, substitute a row of dots or a dash to acknowledge the gap.
- *Rules.* Rules are the lines that separate the table into its various parts. Horizontal rules are placed below the title, below the body of the table, and between the column headings and the body of the table. Tables should not be closed at the sides. The columns within the table may be separated by vertical rules if such lines aid clarity.
- *Source line.* The source line, which identifies where the data were obtained, appears below the table (when a source line is appropriate). Many organizations place the source line below the footnotes.
- *Footnotes.* Footnotes are used for explanations of individual items in the table. Symbols (*, †) or lower case letters (sometimes in parentheses) rather than numbers are ordinarily used to key table footnotes because numbers might be mistaken for numerical data within the table and they could confuse the numbering system for text footnotes.
- *Continuing tables.* When a table must be divided so that it can be continued on another page, repeat the column headings and give the table number at the head of each new page with a "continued" label ("Table 3, continued").

To list relatively few items that would be easier for the reader to grasp in tabular form, use an informal table.

■ The sound-intensity levels (decibels) for the three frequency bands (in hertz) were determined to be

Frequency Band (Hz)	Decibels
600–1200	68
1200–2400	62
2400–4800	53

Although informal tables do not need titles or table numbers to identify them, they do require column headings that accurately describe the information listed.

Graphs

Graphs, like tables, present numerical data in visual form. Graphs (also called charts) have several advantages over tables. Trends, movements, distributions, and cycles are more readily apparent in graphs than they are in tables. Further, by providing a means for ready comparisons, a graph often shows a significance in the data not otherwise immediately evident. Be awares, however, that although graphs present statistics in a more interesting and comprehensible form than tables do, they are less accurate. For this reason, they are often accompanied by tables that give exact figures. (Note the difference between the graph and table showing the same data in Figure 2.) If the graph remains uncluttered, the exact data can be

Table 3: U.S. Goods Imports

Imports	1997	1998	1999	98–99	94–99	92–99
	Billions of Dollars			Percent Change		
Total (BOP Basis)*	$876.4	$917.2	$1,030.2	12.3%	54.1%	92.0%
Food, feeds, and beverages	$39.7	$41.2	$43.6	5.7%	40.6%	57.9%
Industrial supplies and materials	$213.8	$200.1	$222.6	10.7%	36.7%	59.9%
Capital goods, except autos	$253.3	$269.6	$296.9	10.1%	61.0%	121.0%
Autos and auto parts	$139.8	$149.1	$179.5	20.4%	51.7%	95.6%
Consumer goods	$193.8	$216.5	$239.6	10.7%	63.8%	95.3%
Other	$29.3	$35.4	$43.9	24.1%	106.1%	148.0%

Source: U.S. Department of Commerce, Balance of Payment Basis for Total.
*Census Basis for Sectors.

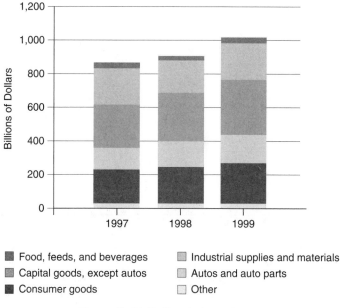

■ Food, feeds, and beverages   ▨ Industrial supplies and materials
▨ Capital goods, except autos   ▨ Autos and auto parts
■ Consumer goods   ▫ Other

Figure 3: U.S. Goods Imports

Source: U.S. Department of Commerce, Census Basis.

IGURE 2   Figure and Table Showing the Same Data

added to each column, thereby giving the reader both a quick overview of the data and accurate figures. (See Figures 8 and 9 on pages 216 and 217.) There are many different kinds of graphs, most notably line graphs, bar graphs, pie graphs, and picture graphs. All kinds of graphs can now be easily rendered on a computer after the data has been entered into a spreadsheet or database application.

Line Graphs. The line graph, which is the most widely used of all graphs, shows the relationship between two sets of figures. The graph is composed of a vertical axis and a horizontal axis that intersect at right angles. Each axis represents one set of data. The relationship between the two sets is indicated by points plotted along appropriate intersections of the two axes. Once plotted, the points are connected to form a continuous line, and the relationship between the two sets of data becomes readily apparent.

The line graph's vertical axis usually represents amounts (the vertical axis in Figure 3 represents the percentage of change for a 50-year period) and its horizontal axis usually represents increments of time (the horizontal axis in Figure 3 represents 10-year periods).

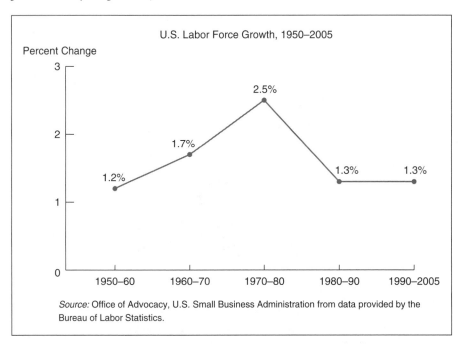

FIGURE 3   Single-Line Graph

Line graphs with more than one plotted line allow for comparisons between two sets of statistics. In creating such graphs, be certain to identify each plotted line with a label or a legend, as shown in Figure 4. You can emphasize the difference between the two lines by shading the space that separates them. Line graphs can also be shaded to emphasize the quantity or magnitude of the data depicted, as in Figures 4 and 5. The following guidelines apply to most line graphs:

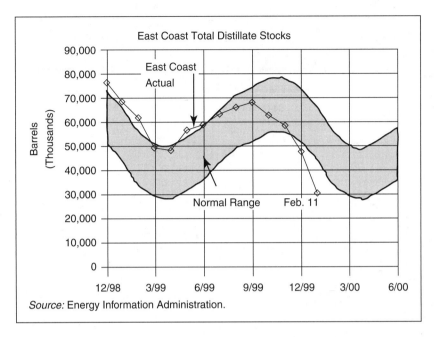

FIGURE 4   Double-Line Graph with Shading

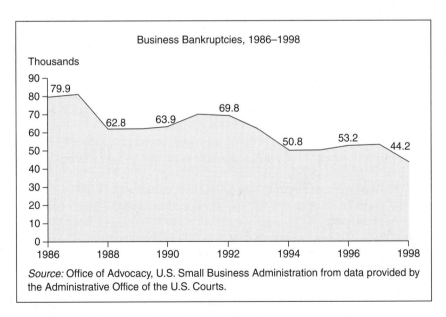

FIGURE 5   Single-Line Graph with Shading

1. Give the graph a title that describes the data clearly and concisely.
2. Indicate the zero point of the graph (the point where the two axes meet). If the range of data shown makes it inconvenient to begin at zero, insert a break in the scale, as in Figure 6; otherwise, the graph would show a large area with no data.

FIGURE 6   Line Graph with Vertical Axis Broken

3. Divide the vertical axis into equal portions, from the least amount at the bottom to the greatest amount at the top. The caption for this scale may be placed at the upper left (as in Figure 3), or, as is more often the case, vertically along the vertical axis, (as in Figure 4).
4. Divide the horizontal axis into equal units from left to right, and label them to show what values each represents.
5. The angle at which the curved line rises and falls is determined by the number of points plotted. Attempting to show trends with too few data points will distort the depiction of the trends. (See Figure 7.)
6. Keep grid lines to a minimum so that the curved lines stand out. Because precise values are usually shown in a table of data accompanying a graph, detailed grid lines are unnecessary.
7. Include a label or a key when necessary.
8. If the information comes from another source, include a source line under the graph at the lower left, as in Figures 1 through 5.
9. Place explanatory footnotes directly below the figure caption. (See Figure 11.)
10. Make all lettering read horizontally if possible, except the caption for the vertical axis, which is usually positioned vertically.

Bar Graphs. Bar graphs consist of horizontal or vertical bars of equal width but scaled in length or height to represent some quantity. They are commonly used to show the following proportional relations:

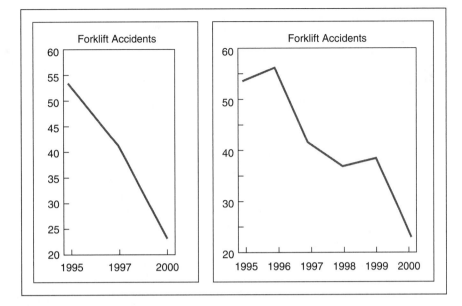

FIGURE 7   Two Graphs: Distorted Expression of Data (left) and Distortion-Free Expression of Data

- Different types of information during different periods of time (Figure 8).
- Quantities of the same kind of information at different points in time (Figure 9).
- Quantities of different information during a fixed period of time (Figure 10).
- Quantities of the different parts that make up a whole (Figure 11).

Note that in Figure 14, showing Christmas tree sales, the exact sales quantities appear at the end of each bar. This eliminates the need to have an accompanying table giving the price data. If the bars are not labeled, as in Figures 9 and 12, the different portions must be clearly indicated by shading, crosshatching, or other devices. Include a key that represents the various subdivisions.

Bar graphs can also indicate what proportion of a whole the various component parts represent. In such a graph, the bar, which is theoretically equivalent to 100 percent, is divided according to the proportion of the whole that each item sampled represents. (Compare the displays of the same data in Figures 11 and 13.) In some bar graphs, the completed bar does not represent 100 percent, because not all the parts of the whole have been included in the sample. (See Figure 12.)

Pie Graphs. A pie graph presents data as wedge-shaped sections of a circle. The circle equals 100 percent, or the whole, of some quantity (a tax dollar, a bus fare, the hours of a working day), with the wedges representing the various parts into which the whole is divided. In Figure 13, for example, the circle stands for a city tax dollar and is divided into units equivalent to the percentages of the tax dollar spent on various city services. Note that the slice representing salaries is slightly

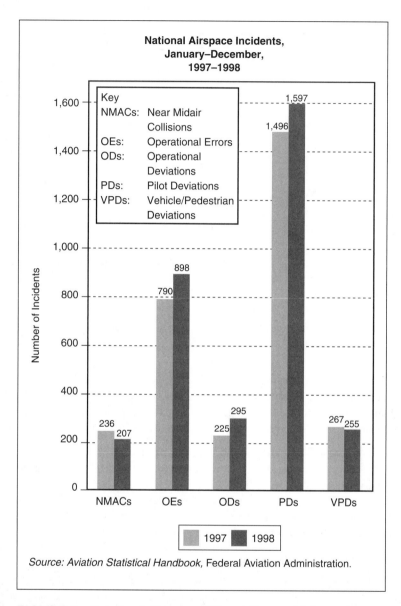

FIGURE 8   Bar Graph Showing Different Types of Information During Different Periods of Time

offset (exploded) from the others to emphasize that data. This feature is commonly available on computer software that produces pie graphs.

The relationships among the various statistics presented in a pie graph are easy to grasp, but the information is often rather general. For this reason, a pie graph is often accompanied by a table that presents the actual figures on which the percentages in the graph are based.

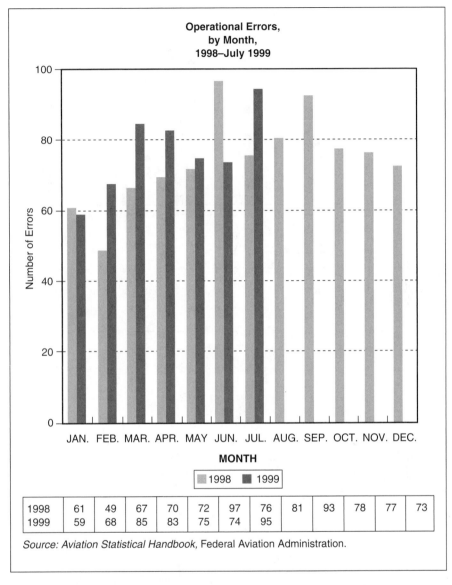

FIGURE 9   Bar Graph Showing Quantities of the Same Kind of Information
at Different Periods of Time

When you construct a pie graph, keep the following points in mind:

1. The complete 360° circle is equivalent to 100 percent.
2. When possible, begin at the 12 o'clock position and sequence the wedges clock-
   wise, from largest to smallest. (This is not always possible because the default setting for some charting software sequences the data counterclockwise.)

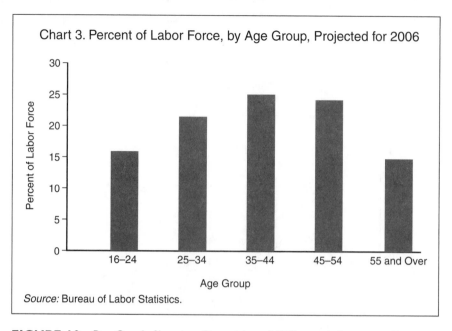

FIGURE 10 Bar Graph Showing Quantities of Different Information During a Fixed Period of Time

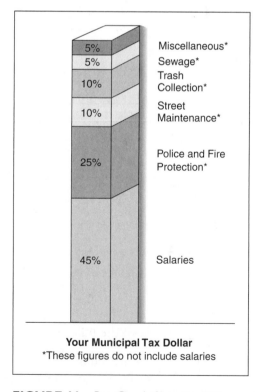

FIGURE 11 Bar Graph Showing Different Quantities of Different Parts of a Whole

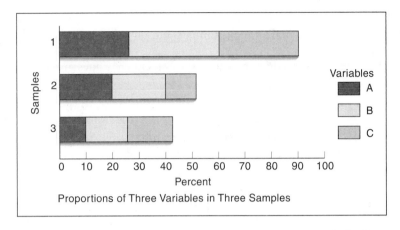

FIGURE 12   Bar Graph in Which Not All Parts of a Whole Have Been Included

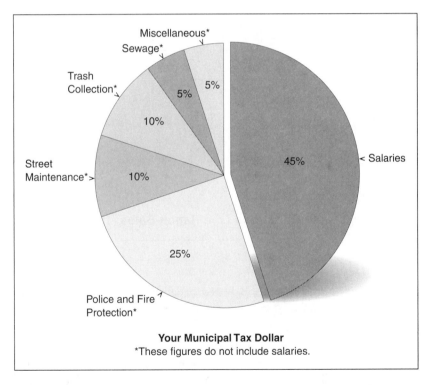

FIGURE 13   Exploded Pie Graph (with Same Data as in Figure 11)

3. Each wedge appears with a distinctive pattern or is shown in various shades of gray.
4. Keep all labels horizontal and, most important, give the percentage value of each wedge.
5. Check to see that all wedges, as well as the percentage values given for them, add up to 100 percent.

Although pie graphs have a strong visual impact, they also have drawbacks. If more than five or six items of information are presented, the graph looks cluttered, and unless percentages are labeled on each section, the reader cannot compare the values of the sections as accurately as on a bar graph.

Picture and Dimensional-Column Graphs. Picture graphs are modified bar graphs that use picture symbols to represent the item for which data are presented. Each symbol corresponds to a specified quantity of the item, as shown in Figure 14.

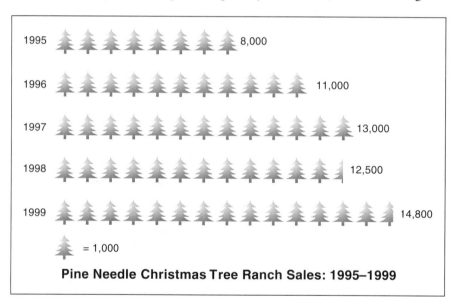

FIGURE 14   Picture Graph

Note that precise figures are also included because the picture symbol can indicate only approximate figures. Here are some tips on preparing picture graphs:

1. Make the symbol self-explanatory.
2. Have each symbol represent a specific number of units.
3. Show larger quantities by increasing the number of symbols rather than by creating a larger symbol because if the latter is done it is difficult to judge relative size accurately.

Consider a common on-the-job reporting requirement—tracking a series of expenses over a given period of time. Assume that you wish to show your company's expenses over a three-month period for security, courier, mail, and custodial services. Once you enter the data for these services into a spreadsheet program, you can display them in a variety of graph styles. As you select from among the options available, keep in mind your reader's need to interpret the data accurately and quickly and keep the graph style as simple as possible for the information shown.

Graphs that depict columns as three-dimensional pillars are popular—they give the data a solid, almost building-block appearance. They can, however, obscure rather than clarify the information, depending on how they are displayed. Consider the graph in Figure 15. Although the data are accurate, they cannot be interpreted as shown. The axis showing expenditures cannot be correlated with most columns representing the various services. The graph also obscures the columns for courier and guard-force services. Finally, this graph style does not allow readers to spot trends for expenditures over the three-month period, which is key information to decision-makers.

The graph in Figure 16 presents more clearly the same data as that shown in Figure 15. The trends of expenditures are easy to spot and all the data are at least visible. Yet this graph is not ideal. To interpret the information, the reader would need to put a ruler on the page and align the tops of the columns with the axis showing expenditures.

The three-dimensional appearance can also cause confusion: Is the front or back of each column the correct data point? Also somewhat confusing is that, at first glance, the reader is tricked into interpreting the spaces between the column clusters as columns.

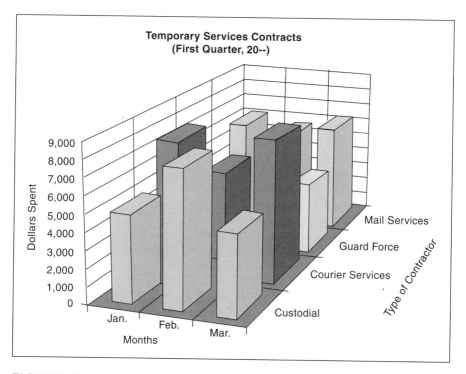

FIGURE 15   Three-Dimensional Column Graph

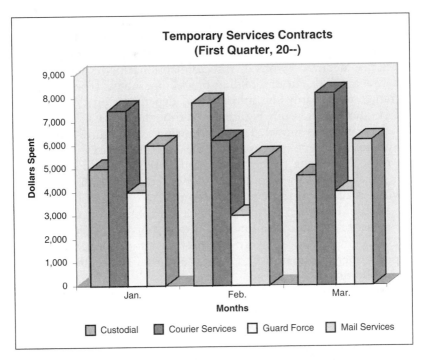

FIGURE 16 Three-Dimensional Column Graph (Showing Same Data as in Figure 15)

The graphs in Figures 17 and 18 would best represent the data, depending on your intent. Figure 17 avoids the ambiguity of the graph in Figure 16 by showing the data in two dimensions. It also displays the horizontal lines for expenditures on the vertical axis, thus making the data for each column easier to interpret. If you wished to show relative expenses among the four variables for a given quarter, this graph would be ideal. However, if you wished to show trends for the entire three-month period in contract expenditures at a glance, the graph in Figure 18 is preferable.

When precise dollar amounts for each service are equally important, you can provide a table showing that information.

As Figures 15 through 18 show, the more complicated a graph looks, the harder it is to interpret. On balance, simple is better for the reader. Use this principle when you review your computer graphics on screen in several styles and consider your reader's needs before deciding which style to use.

Drawings

A drawing is useful when your reader needs an impression of an object's general appearance or an overview of a series of steps or directions. . . . Drawings are the graphic of choice when it is necessary to focus on details or relationships that a photograph cannot capture. A drawing can emphasize the significant piece of a

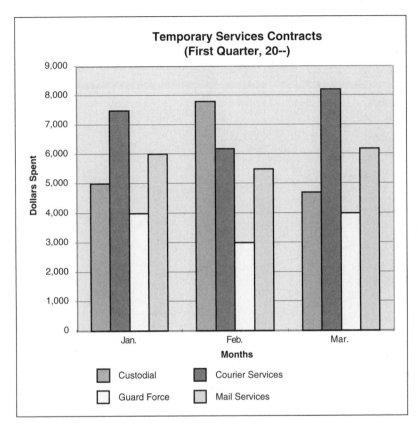

FIGURE 17   Two-Dimensional Column Graph (Showing Same Data as in Figures 15 and 16)

mechanism, or its function, and omit what is not significant. A cutaway drawing can show the internal parts of a piece of equipment in such a way that their relationship to the overall equipment is clear (Figure 19).

Drawings are also an ideal option when you must illustrate something but find it either impractical or too expensive to reproduce a photograph (Figure 20). However, if the actual appearance of an object (a dented fender) or a phenomenon (a wind-tunnel experiment) is necessary to your document, a photograph is essential. An exploded-view drawing can show the proper sequence in which parts fit together or the details of each individual part (Figure 21).

Drawings requiring a high degree of accuracy and precision are generally rendered by graphics specialists. General-interest images needed to illustrate newsletters and brochures can be found in the clip-art libraries that come with desktop computer-graphics programs and even with word-processing programs. They contain thousands of noncopyrighted symbols, shapes, and images of people, equipment, furniture, buildings, and the like. . . .

Many organizations have their own format specifications for drawings. In the absence of such specifications, the following tips should be helpful:

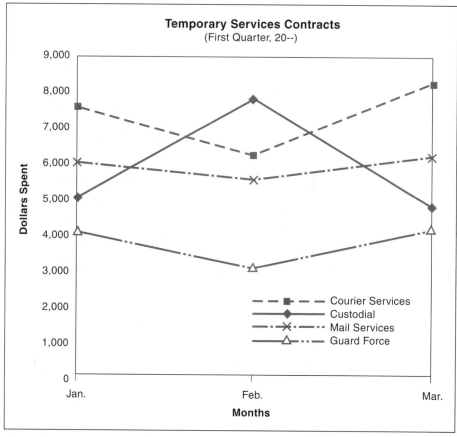

FIGURE 18 Line Graph (Showing Same Data as in Figures 15 through 17)

1. Show the equipment from the point of view of the person who will use it.
2. When illustrating a subsystem, show its relationship to the larger system of which it is a part.
3. Draw the different parts of an object in proportion to one another, unless you indicate that certain parts are enlarged.
4. Where a sequence of drawings is used to illustrate a process, arrange them from left to right and from top to bottom.
5. Label parts in the drawing so that text references to them are clear and consistent.
6. Depending on the complexity of what is shown, label the parts themselves, or use a letter or number key. . . .

Flowcharts

A flowchart is a diagram of a process that involves stages, with the sequence of stages shown from beginning to end The flowchart presents an overview of the process that allows the reader to grasp the essential steps quickly and easily. The

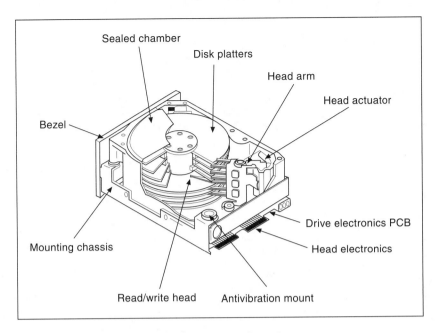

FIGURE 19   Cutaway Drawing of a Hard Disk Drive

Exercises You Can Do

Before beginning keying and during breaks throughout the day, take time to do the following stretching exercises:

Gently press the hand against a firm flat surface, stretching the fingers and wrist. Hold for five seconds.

Rest the forearm on the edge of a table. Grasp the fingers of one hand and gently bend back the wrist, stretching the hands and wrist. Hold for five seconds.

FIGURE 20   Drawing of Stretching Exercises to Prevent Repetitive-Stress Injuries

FIGURE 21   Exploded-View Drawing of a Power-Steering Valve
Source: Courtesy of the Harnischfeger Corporation.

process being illustrated could range from the stages by which bauxite ore is refined into aluminum ingots for fabrication to the steps involved in preparing a manuscript for publication.

Flowcharts can take several forms to represent the steps in a process. They can consist of labeled blocks . . . , pictorial representations . . . , or standardized symbols. . . . The items in any flowchart are always connected according to the sequence in which the steps occur. The normal direction of flow in a chart is left to right or top to bottom. When the flow is otherwise, be sure to indicate it with arrows.

Flowcharts that document computer programs and other information-processing procedures use standardized symbols. The standards are set forth in

U.S.A. Standard Flowchart Symbols and Their Usage in Information Processing, published by the American National Standards Institute, publication X3.5. When creating a flowchart, follow . . . [these] guidelines . . . :

1. With labeled blocks and standardized symbols, use arrows to show the direction of flow, especially if the flow is opposite to the normal direction. With pictorial representations, use arrows to show the direction of all flow.
2. Label each step in the process, or identify it with a conventional symbol. Steps can also be represented pictorially or by captioned blocks.
3. Include a key if the flowchart contains symbols that your reader may not understand.
4. Leave adequate white space on the page. Do not crowd your steps and directional arrows too close together.

Organizational Charts

An organizational chart shows how the various components of an organization are related to one another. Such an illustration is useful when you want to give readers an overview of an organization or indicate the lines of authority within the organization (Figure 22).

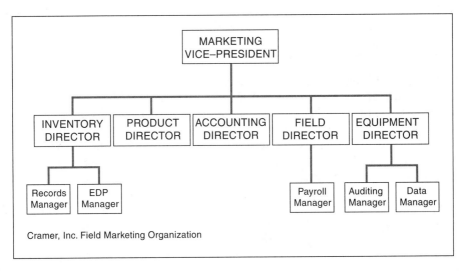

FIGURE 22   Organizational Chart

The title of each organizational component (office, section, division) is placed in a separate box. These boxes are then linked to a central authority. If your readers need the information, include the name of the person occupying the position identified in each box. As with all illustrations, place the organizational chart as close as possible to the text that refers to it.

Maps

Maps can be used to show the specific geographic features of an area (roads, mountains, rivers) or to show information according to geographic distribution (population, housing, manufacturing centers, and so forth).

Bear these points in mind as you create maps for use with your text. . . .

1. Make sure all boundaries within the map are clearly identified. Eliminate unnecessary boundaries.
2. Eliminate unnecessary information from your map. For example, if population is the focal point, do not include mountains, roads, rivers, and so on.
3. Include a scale of miles or feet, or of kilometers or meters, to give your reader an indication of the map's proportions.
4. Indicate which direction is north.
5. Emphasize key features by using shading, dots, crosshatching, or appropriate symbols, and include a key telling what the different colors, shadings, or symbols represent. . . .

Photographs

Photographs are vital to many publications. They are often to the best way to show the surface appearance of an object or to record an event or the development of a phenomenon over a period of time. Not all representations, however, call for photographs. They cannot depict the internal workings of a mechanism or below-the-surface details of objects or structures. Such details are better shown in drawings or diagrams.

Highlighting Photographic Objects. If you are taking the photo, stand close enough to the object so that it fills your picture frame. To get precise and clear photographs, choose camera angles carefully. A camera will photograph only what it is aimed at; accordingly, select important details and the camera angles that will record these details. To show the relative size of an unfamiliar object, place a familiar object—such as a ruler, a book, a tool, or a person—near the object that is to be photographed. . . .

Ask the printer reproducing your publication for special handling requirements if you use glossy photographs. If you use digitized photos, ask about the preferred resolution of the images. The higher the resolution, the better the quality. You can create digitized photos using scanners and digital cameras, so ensure that the equipment has the necessary memory for the resolution required by the printer.

Using Color. The advantages of using color photographs are obvious: Color is in many cases the only way to communicate crucial information. In medical, chemical, geological, and botanical publications, for example, readers often need to know exactly what an object or phenomenon looks like to accurately interpret it. In these circumstances, color reproduction is the only legitimate option available.

Posting color images at a Web site is no more complicated or expensive than posting black-and-white images. For publications, however, the preparation and printing of color photographs are complex technical tasks performed by graphics and printing professionals. If you are planning to use color photographs in your publication, discuss with these professionals the type, quality, and number of photographs required. Generally, they prefer that you provide color transparencies (slides) and color negatives (negatives of color prints) for reproduction.

Be mindful, too, that color reproduction is significantly more expensive than black-and-white reproduction. Color can also be tricky to reproduce accurately without losing contrast and vividness. For this reason, the original photographs must be sharply focused and rich in contrast.

Digital Shortcuts: Creating Computer Graphics

Computer software permits you to create a large variety of visual images, to access and modify clip-art images, and to display numerical information in graphs and charts. The computer visuals you create afford numerous advantages over pen-and-paper renderings in that they can be

- Saved for future use and updated as necessary
- Printed as is on paper or used in reports, other documents and Web sites
- Sent to presentation software programs to produce transparencies and 35-mm slides for meetings
- Communicated to others electronically via email

Your computer may also allow you to share data among documents created in different applications. Data sharing automates redundant tasks. For example, the data entered into spreadsheet software can be reproduced in graphical form (usually as pie charts and bar graphs), which can be embedded into documents (in word-processing software) and presentations (in presentation software). When the spreadsheet data are updated, the corresponding graphic in the document or presentation is automatically redrawn to reflect the changes, thus saving you the time and effort of manually revising the graphic.

USING GRAPHICS TO COMMUNICATE INTERNATIONALLY

Business and technical communicators increasingly will use graphics in the years to come to communicate with international audiences. This trend will grow because of the globalization of markets: Companies want to expand their markets in Saudi Arabia, China, South America, the countries of the former Soviet Union, and dozens of other places. Global markets mean increased global trade because of the prevalence of multinational corporations, the international subsidiaries of many companies, multination trade agreements, the increasing diversity of the U.S. workforce, and even increases in immigration. The audiences for these communications will include clients, business partners, colleagues, and current and potential employees and customers. Even though English is rapidly becoming the global language of business and science, many people speak it as a second or third language. For this reason, graphics offer distinct advantages for communicating in a global business climate.

- Graphics can communicate a message more effectively than text, particularly in the context of safety warnings or cautions.
- Graphics can sometimes replace technical terms that are difficult to translate.

Despite their unquestionable value in communicating with international audiences, symbols, images, and even colors are not free from cultural associations: They are context dependent, and context is culturally determined. Thus, there are no universally accepted graphics standards, with the exception of the symbols used in mathematics and certain scientific and engineering disciplines (e.g., the voltage symbol used in electrical engineering).

Writers who create documents for international readers can avoid confusing and possibly offending a global audience by understanding the following cultural differences in connotation.

Punctuation Marks

Punctuation marks, like words, are language specific. For example, in North America the question mark generally represents the need for information or the Help function in a computer manual or program. In many countries, this symbol is not understood at all.

Religious Symbols

Religious symbols that carry simultaneous religious and nonreligious meanings have long been used in North America, where it is common to use the cross as a symbol for first aid or hospital. In Muslim countries, a cross represents Christianity; a crescent (usually green) is a symbol for first aid.

Colors

The use of a particular color can distort or even change the meaning of graphics symbols. Red commonly indicates warning or danger in North America, Europe, and Japan. In China, however, red symbolizes joy. In Europe and North America, blue generally has a positive connotation; in Japan, the color represents villainy. In Europe and North America, yellow represents caution or cowardice; in Arab countries, yellow generally means fertility or strength.

People, Parts of the Body, and Gestures

Depicting people and parts of the body in graphics can be problematic (see Table 1). If your graphics will reach an international audience, it is better to avoid depictions of people eating or representations of bare arms and feet. Nudity in advertising, for example, generally is acceptable in Europe but not in North America. Even showing isolated body parts could lead to communications

TABLE I   International Implications of Gestures and Body Language

Body part	Gesture	Country	Interpretation
Head nodding	Up and down	Bulgaria	No
Left hand	Showing palm	Muslim countries	Dirty, unclean
Index finger	Pointing to others	Venezuela, Sri Lanka	Rude
Index-to-thumb circle	Circular "OK"	Germany, Netherlands	Rude
Ankle and leg	Crossing over knee	Indonesia, Syria	Rude
Eye	Touching finger below eye	Honduras	Caution

difficulties. For example, some Middle Eastern cultures regard the display of the soles of one's shoes to be disrespectful and offensive. Therefore, a technical manual that attempts to demonstrate the ease of running a software program by showing a user with his feet up on a desk would be considered offensive in the Middle East.

Communicators producing instructions often use hand gestures such as the victory sign ( ) or the "OK" sign ( ) as positive motivators. However, the meaning of each of these gestures varies by culture. In Australia, for example, the victory sign conveys the same meaning as holding up the middle finger in North America. Similarly, the gesture that means "OK" in North America can mean "worthless" in France, can mean "money" in Japan, and is a sexual insult in many other parts of the world. Even a smile may have different connotations. In Japan, smiling can be a sign of joy or can be used to hide displeasure; in some Asian cultures, smiling may be considered a sign of weakness.

Further, a manual that contains a pointed finger to indicate "turn the page" might offend someone in Venezuela. A writer preparing a manual for export to Honduras could indicate "caution" by using a picture of a person touching a finger below the eye.

Cultural Symbols

Signs and symbols are so culturally rooted that we often lose sight of the fact that they may be understood only in our culture. A Michigan manufacturer of fans wants to use easily understood symbols to represent the two speeds of its product: fast and slow. The technical communicators selected a rabbit and a turtle. However, recognition of these animals as symbols of speed and slowness requires familiarity with Aesop's fables—a Western tradition.

Technology Symbols

The symbols or icons we create to represent technology are laden with cultural assumptions. If users do not have regular contact with fax machines, photocopiers, computers, or cellular phones, technical and business communicators cannot predict how users will interpret representations of these devices.

Reading Practices

Whether text is read right to left or left to right influences how graphics are sequenced. In Israel, for example, text is read from right to left. For this audience, you will need to alter the design and sequencing of text and graphics.

Directional Signs

The signs we use to represent direction or time are open to misinterpretation. For example, the arrow sign on shipping cartons can be interpreted to mean either that the carton should be placed with the arrow pointing up to the top of the carton, or pointing down to the carton's most stable position. Western cultures tend to indicated the future (or something positive) by pointing to the right ($\rightarrow$) and the past (or something negative) by pointing to the left ($\leftarrow$); in the Chinese culture, left represents honor and right self-destruction.

Careful attention to the different connotations visual elements may have for an international audience makes translations easier, saves a company from potential embarrassment, and, over time, earns respect for the company and its products and services.

David W. Ewing

Strategies of Persuasion

David W. Ewing was Executive Editor-Planning of the Harvard Business Review *and a member of the faculty of the Harvard Business School.*

When we review reports, letters, and memoranda that get the intended results, we find a fascinating diversity of approaches. Some are gentle in approach, taking readers by the hand and leading them to a certain finding or recommendation. Others are brisk and abrupt. Some are objective in approach, carefully examining both sides of an idea, like a judge writing a difficult decision—at least until the end. Others burst with impatience to explain one side, and only one side, of a proposal or argument. Some flow swimmingly; others erupt like Mount Vesuvius.

How is it possible that communications taking such different approaches can all be effective? It is not sufficient to answer glibly, "It depends on the situation," or "Communications should mirror the personal style of the communicator." How does a written communication depend on the situation? Which style of the communicator should be reflected in a given communication? Examine the writing of most business executives, professionals, and public leaders, and you will find not one but *many* styles of exposition.

RULES EVERY PERSUADER SHOULD KNOW

The explanation lies in a set of relationships among the communicator, the reader, the message, and the time-space environment. These relationships work in predictable ways and are an important part of the knowledge of every good business and professional writer. They come into play in the planning stages of writing, when the writer is considering how he or she will proceed, and in the main body of the presentation, to accomplish what he or she has promised in the

opening paragraphs. The relationships affect some of the most important decisions a writer makes—the choice of ideas to use, the comparative emphasis to be
given various arguments pro and con, the types of reasons and supporting material used, the establishment of credibility, and other matters.

It is convenient but self-defeating to follow fixed prescriptions for persuasion, such as to put your strongest arguments first or last or to identify with the
readers. Such nostrums were fine for the age of patent medicines and snake-oil
peddlers, but not for the age of diagnostic medicine. What is more, they are
belittling. They assume that writers are witless. Good writers vary their
approaches in response to their readings of different situations. Just as a good
golfer plays an approach to a green differently depending on the wind, so a
good writer uses different strategies depending on the crosscurrents of mood
and feeling.

How should you choose your approach to a group of readers? What elements
of the approach should be tailored to the situation? These are the topics of
this . . . [selection]. Let us assume that the substance of the intended message
is clear in your mind. . . .

1. Consider Whether Your Views Will Make Problems for Readers.

J. C. Mathes and Dwight W. Stevenson tell of the student engineer who was
asked to evaluate the efficiency of the employer plant's waste-treatment process.[1]
He found that, by making a simple change, the company could save more than
$200,000 a year. Anticipating an enthusiastic response, he wrote up and deliv-
ered his report. Although he waited with great expectations, no accolades came.
Why? What he hadn't counted on was that now his supervisors would have to
explain to their bosses why they had allowed a waste of $200,000 a year. They
were far from elated to read his report.

"If you want to make a man your enemy," Henry C. Link once said, "tell
him simply, 'You are wrong.' This method works every time." Under the illusion
that their sciences are "hard," physicists, biologists, and others may assume that
it is necessary only to worry about setting the facts forth accurately. From pos-
terity's standpoint, perhaps yes—but not from the standpoint of writing for current results. As Kenneth Boulding, head of the American Association for the
Advancement of Science, has pointed out, the so-called "hard" sciences are in
many ways "soft," and vice versa. "You can knock forever on a deaf man's door,"
said Zorba the Greek, and the deaf man can be a physicist as well as a marketing
manager or public official.

If your views are bad news for readers, you proceed to report them, but with
empathy and tact and an effort to put yourself in the readers' shoes. You work as
carefully as if you were licking honey off a thorn.

---

[1] J. C. Mathes and Dwight W. Stevenson, *Designing Technical Reports* (Indianapolis: Bobbs-Merrill, 1976), pp. 18–19.

2. Don't Offer New Ideas, Directives, or Recommendations for Change until Your Readers Are Prepared for Them.

"Should I state my surprising findings at the very beginning of my memorandum?" a writer asks. "Should I go slow with my heretical proposal and hold the reader's hand?" asks another.

Generally speaking, the answer to all such questions depends on the extent of your audience's resistance to change, the amount of change you are asking for, the uncertainty in readers' minds as to your understanding of their situation, and what psychologists call the "perceived threat" of your communication, that is, how much it seems (to readers) to upset their values and interests. The more change, uncertainty, and/or threat, the slower you should proceed, the more carefully you should prepare your readers.

For instance, if your boss is enthusiastic about a new promotion scheme that he (or she) has paid a consultant $25,000 to devise, naturally you will want to go slow in shooting it down (at least if you want to stay in his good graces). In fact, any written criticism of the scheme is probably out of order until you have had a chance to talk with him and get a feel for the proper timing of any forthcoming criticism. When you do commit yourself to writing, you should probably review the arguments for the new scheme as fairly as possible, making it crystal clear to him that you understand them. Only then does it become timely to turn to the facts or conditions that, in your opinion, raise serious questions about the plan.

On the other hand, suppose the faulty promotion scheme is of little personal interest to your boss—it is not his (or her) "baby." Now the situation is different. You can launch right into the shortcomings, throwing your heaviest objections first. The fact is that the boss may not even want to know about the lesser objections, much less the supporting arguments once advanced for the plan; the main things he should know are that (a) the plan is in trouble and (b) the major, most compelling reasons why.

Clearly, this strategy is plain, everyday common sense—what you would normally do in communicating orally instead of in writing. Only in writing you must be more explicit and thorough, because a document lacks the expressiveness and visual advantages of a spoken dialogue.

Now consider another type of situation. Suppose it is your unhappy task to write a department manager that the extra appropriations he (or she) was promised have been cancelled. First of all, how would you handle it if you saw him often and could talk with him personally? . . . You would not (we hope) pussyfoot around trying to withhold the bad news from him. Also, once you had indicated the main message, you would probably backtrack a little and make it clear that the step is being taken with reluctance.

"Joe, it looks as if we're not going to be able to give you the extra budget we promised," you might say, getting down to business. That is the message in a nutshell—now for the review of common ground. "We know how much you have counted on getting those people and funds. There's no doubt you could manage them well and put them to good use. And we know that the morale of your people is involved in this, too. But the fact is that the sales we counted on are not coming in. We've got to cut somewhere, and, frankly, we feel it's got to be your department because. . . ."

If you are communicating by letter or memorandum, the strategy is exactly the same (only "Joe" may now be "Mr. Wyncoop"). After your lead, you review the main needs as he and you understand them, perhaps spelling them out more than you would have in a face-to-face meeting but choosing the same ones. Only then do you turn to the new conditions that make it necessary to do an about-face.

3. Your Credibitility with Readers Affects Your Strategy.

In general, communication research indicates that the chances of opinion change vary with the communicator's authority with his or her readers. In their succinct summary of the field, *Persuasion,* social scientists Marvin Karlins and Herbert I. Abelson point out that credibility itself is a variable; that is, it can be influenced by the words of the communicator.[2] Above all, as psychologists repeatedly emphasize, credibility lies in the eye of the beholder. . . . [3]

In written communications there are two types of credibility. It may be given or it may be acquired. Between the two lies a world of practical difference.

Given credibility may result from your position in an organization. If, let us say, you are the boss writing directions to a subordinate, your credibility is likely to be high. Given credibility also may result from reputation—a well-known chemist has more credibility in communications about polymers than a good industrial engineer has, but the latter would possess more credibility in communications about time-and-motion studies. It may result from the individuals and groups the writer is associated with—if he or she is a member of the same trade union the reader belongs to, a union held in high esteem by both, he or she has more credibility in a memorandum on grievance procedures than a member of the board of directors would have.

Though you may be high in given credibility, you may yet need to remind some readers of the fact. In the case of an obvious credential, such as a position in an important organization, a letterhead may be enough to do the trick. Another device is to insert a few lines of biographical data at the head of a report or brochure. If you have had experience or associations that carry weight with the reader, perhaps you can interject them early in the message. "During a visit I had last week with Zach Jarvis," you might say, knowing that Dr. Zachary P. Jarvis is a magic name with your reader, or "The Executive Committee of the Aberjona Basin Association asked me to join their meeting on Monday . . ." knowing that group carries a triple-A rating in the mind of the reader.

---

[2]Marvin Karlins and Herbert I. Abelson. *Persuasion* (New York: Springer, 1970); see pp.107–132.
[3]See, for example, Ralph L. Rosnow and Edward J. Robinson, Eds. *Experiments in Persuasion* (New York: Academic, 1967).

Of course, you do not want to overplay your hand at such name dropping. A report to a fairly diverse audience by a famous black organization began simply:

> The National Association for the Advancement of Colored People has for many years been dedicated to the task of defending the economic, social and political rights and interests of black Americans. The growing national debate about energy has led us to examine the question to ascertain the implications for black Americans.[4]

Again, an attorney of the American Civil Liberties Union, in a letter to members of the organization soliciting donations, began:

> My dear friend:

> I am the ACLU lawyer who went into court last April to defend freedom of speech in Skokie, Illinois, for a handful of people calling themselves "nazis."

> The case has had an enormous impact on my life.

> It has also gravely injured the ACLU financially. . . .[5]

Acquired credibility, on the other hand, is earned by thoughts and facts in the written message. I may not know you from Adam. Yet if you send me a letter or report that carefully, helpfully describes something I am interested in, you gain credibility in my estimation.

Some studies suggest that, if you are low in given credibility and seek to acquire it with an audience, a useful technique is to cite ideas or evidence that support the reader's existing views.[6] As Disraeli once said, "My idea of an agreeable person is a person who agrees with me." The very fact that you feel confident and knowledgeable enough to articulate these views is likely to lift you several notches in the reader's estimation.

Still another approach is that old standby of persuaders—identifying yourself, in an early section, with the goals and interests of the audience. Possibly the most famous example of this strategy is the opening of Marc Antony's funeral oration, in Shakespeare's *Julius Caesar:* "I come to bury Caesar, not to praise him. . . ."

Finally, you can acquire credibility by citing authorities who rate highly with your intended audience, or by exhibiting documentary evidence that, because of its source, lends prestige and authority to your proposals or ideas.

"For success in negotiation," says C. Northcote Parkinson and Nigel Rowe, "it is vitally important that people will believe what you say and assume that any promise you make will be kept. But it is no good saying: 'Trust me. Rely on my word.' Only politicians say that."[7]

---

[4] *The Wall Street Journal,* January 12, 1978.
[5] David Goldberger, letter dated March 20, 1978.
[6] Karlins and Abelson, op. cit., pp. 115–119.
[7] C. Northcote Parkinson and Nigel Rowe, "Better Communcation: Business's Best Defense," *The McKinsey Quarterly,* Winter 1978, p. 26.

Even if you have prestigious credentials, you cannot take too much for granted. In an age of television sets, radios, cassettes, and record players in every home, credibility—at least with the public—may come quicker for the singer or comedian than for the judge, business executive, or medical researcher. In fact, because of an association with an organization or profession, you may be stereotyped as a member of "them" or "the establishment." It may behoove you to establish that you are a person with a name, a personality, certain interests, certain experiences—not just a nameless representative.

4. If Your Audience Disagrees with Your Ideas or is Uncertain about Them, Present Both Sides of the Argument.

Behavioral scientists generally find that, if an audience is friendly to a persuader, or has no contrary views on the topic and will get none in the future, a one-sided presentation of a controversial question is most effective.[8] For instance, if your point is that sales of product X in the St. Louis territory could be doubled and you are writing to enthusiastic salespeople of product X, your best course is to concentrate on facts and examples showing the enormous potential of product X. There is no shortage of evidence showing that people generally prefer reading material that confirms their beliefs, and that they develop resistance to material that repudiates their beliefs. (As we shall see presently, however, this does not mean you cannot change their minds.)

But suppose your audience has not made up its mind, so far as you know? In this case you would do well to deal with *both* sides of the argument (or all sides, if there are more than two). Follow the same approach if the reader disagrees with you at the outset. For one thing, a two-sided presentation suggests to an uncertain or hostile audience that you possess objectivity. For another, it helps the reader remember your view by putting the pros and cons in relationship to one another. Also, it meets the reader's need to be treated as a mature, informed individual. As Karlins and Abelson point out:

> Conspicuously underlying your presentation is the assumption that the audience would be on your side if they only knew the truth. The other points of view should be presented with the attitude "it would be natural for you to have this idea if you don't know all the facts, but when you know all the facts, you will be convinced."[9]

Karlins and Abelson tested the reactions of audiences in postwar Germany to Voice of America broadcasts. They found that the most persuasive programs were those that included admissions of shortcomings in United States living conditions.[10]

---

[8]Experiments supporting this conclusion are reported by Carl I. Hovland, Arthur A. Lumsdaine, and F. Sheffield in *Experiments on Mass Communication* (Princeton: Princeton University Press, 1949). Cited by Karlins and Abelson, op. cit., p. 22.
[9]Karlins and Abelson, op. cit., p. 26.
[10]See *Factors Affecting Credibility in Psychological Warfare Communications* (Washington, D.C.: Human Resources Research Office, George Washington University, 1956).

Again, observation of businessmen's reactions to scores of *Harvard Business Review* articles advocating controversial measures convinces me that the most influential articles have been those that have acknowledged the shortcomings, weaknesses, and limitations of their arguments. When an author wants to sell a new idea to a sophisticated audience, he or she should be candid about the soft spots in his argument.

5. Win Respect by Making Your Opinion or Recommendation Clear.

Although strategy may call for a two-sided argument, this does not mean you should be timid in setting forth your conclusions or proposals at the end. We assume here that you have definite views and seek to persuade your audience to adopt them. The two-sided approach is a *means* to that end; it does not imply compromising or obfuscating your conclusions. An official at Armour & Company once criticized many reports from subordinates to bosses on the ground that, after presenting much data, they concluded, in effect, "Here is what I found out and maybe we should do this or maybe we should do that." The typical response of a boss to such a memorandum, he noted, was to do nothing. Hence, the time taken both in writing and reading was wasted.[11]

6. Put Your Strongest Points Last If the Audience Is Very Interested in the Argument, First If It Is Not So Interested.

This question is referred to by social scientists as the "primacy-recency" issue in persuasion. The argument presented first is said to have primacy; the argument presented last, recency. Although studies of the question have produced inconsistent findings and no firm rules can be drawn, it appears that if your audience is deeply concerned with your subject you can afford to lead it along from the weakest points to the strongest. The audience's great interest will keep it reading, and putting the weaker points at the start tends to create rising reader expectations about what is coming. When you end with your strongest punch, therefore, you do not let readers down.

If your audience is not so concerned with the topic, on the other hand, it may be best to use the opposite approach. Now you cannot risk leading readers along a winding path. They may drop out before you reach the end. So grab their attention right at the beginning with your strongest argument or idea.

In any case, put the recommendation, facts, or arguments you most want the reader to *remember* first or last. Although experiments by social scientists on the primacy-recency issue are inconclusive, there is a firm pattern on the question of recall. The ideas you state first or last have a better chance of being remembered than the ideas stated in the middle of your appeal or case.

---

[11]John Ball and Cecil B. Williams, *Report Writing* (New York: Ronald, 1955).

7. Don't Count on Changing Attitudes by Offering Information Alone.

"People are hostile to big business because they don't know enough facts about it," businesspeople are heard to say. Or, "If customers knew the truth about our costs, they would not object to our prices." Companies have poured large sums into advertising and public relations campaigns on this assumption; civic organizations have often based their hopes on it.

"The trouble with the assumption," states Karlins and Abelson, "is that it is almost never valid. There is a substantial body of research findings indicating that cognition—knowing something new—increasing information—is effective as an attitude change agent only under very specialized conditions."[12]

Social scientists do concede, however, that presentations of facts alone may strengthen the opinions of people who already agree with the persuader. The information reassures them and helps them defend themselves in discussions with others.

8. "Testimonials" Are Most Likely to be Persuasive If Drawn from People with Whom Readers Associate.

It is well known that a person's attitudes and opinions are strongly influenced by the groups to which he or she belongs or wants to belong—work units in a company, labor unions, bowling teams, social clubs, church associations, ethnic associations and so on. To muster third-party support for your proposal or idea, therefore, you would do well to cite the behavior, findings, or beliefs of groups to which your readers belong. In so doing, you allay any feelings of isolation readers might have if tempted to follow your ideas. You suggest that they are not alone with you, that there is group support for the points being made.

As every school child learns, the predominant attitudes of a group toward individuals or regarding standards of behavior, performance, or status influence an individual member's perceptions. For instance, a study of boys at a camp demonstrated that their ratings of various individuals' performances at shooting and canoeing were biased by their knowledge of the status of the rated individuals in the camp society. Thus a boy generally regarded as a leader was seen as performing better with the rifle or canoe than was a boy generally regarded as a follower, even though the first boy's performance was not actually superior.[13]

In addition, it seems fair to say that as modern television, radio, records, and cassettes have brought national celebrities into the home and automobile, these people, too, have been stamped with approval or disapproval by millions of groups across the country.

---

[12]Karlins and Abelson, op. cit., p. 33.
[13]Ibid., p. 50.

Accordingly, if your readers are young, dissident, or "long hairs," refer to a Richard Dreyfuss or a Joan Baez for supporting statements, not to a Gerald Ford or an Arnold Palmer. If your readers are electrical engineers, quote well-regarded scientific sources as your authority, not star salespeople or public relations people. Take into account also that the more deeply attached your readers are to a group, the greater the influence of the group norms on them. For instance, one experiment by social scientists showed that the opinions of Catholic students who took their religion seriously were less influenced by the answers of nonserious Catholics than were the opinions of Catholic students who placed little value on their church membership.[14]

9. Be Wary of Using Extreme or "Sensational" Claims and Facts.

Both research in behavioral science and common sense confirm this rule.[15] Do not be misled by the fact that flashy journalists make successful use of extreme and bizarre cases to dramatize a story. The situation in business and professional writing is different from that in journalism.

When you seek the confidence and cooperation of your readers—and typically you do in the kinds of communications we deal with in this book—it is best to write in terms of the real world as you and they perceive it. Observable, believable, realistic statements carry more weight than any other kind. Although you want reader attention, you do not want to shock your audience with outlandish examples or arguments. These may help you to succeed in making the reader sit up—but they will also provoke distrust and suspicion.

Examples are common in the letters sections of newspapers. A writer who identified himself as a former vice-president of a well-known bank opposed a large power company's plan to build a new plant in a rural area near his town. His letter began as follows: "A great many of us . . . are profoundly disturbed by the proposal now being considered to disrupt and destroy the marvelous little valley southwest of [name of town], in order to build bigger and better power plants. This would be a devastating blow to the last unspoiled bit of country left in Connecticut. . . ."[16]

Like a batter who hits the first two pitches foul and quickly gets two strikes against him, this writer managed to distort the first two sentences he wrote. The proposed plant, though a very large one, would not "disrupt and destroy" the valley — only a small section of the valley area would be affected. Moreover, the valley was not "the last unspoiled bit of country" in the state—it was only a small parcel of the state's beautiful countryside. These exaggerations might have

---

[14]See H. Kelley, "Salience of Membership and Resistance to Change of Group-Anchored Attitudes," *Human Relations,* August 1955, pp.255–289. Cited in Karlins and Abelson, op. cit., p. 58.

[15]See, for example, *Building Opposition to the Excess Profits Tax* (Princeton: Opinion Research Corporation, August 1952), and R. Weiss, "Conscious Technique for the Variation of Source Credibility," *Psychological Reports,* Vol. 20, 1969, p. 1159. Both cited in Karlins and Abelson, op. cit., pp. 36–37.

[16]*Lakeville Journal,* April 2, 1970, p. 11.

drawn cheers from rabid foes of the project, but the writer wasn't interested in appealing to them; he wanted to win uncommitted readers. At the very beginning, however, he antagonized them with hyperbole.

10. Tailor Your Presentation to the Reasons for Readers' Attitudes, If You Know Them.

Your chances of persuading readers are better if you can plan your appeal or argument to meet the main feelings, prejudices, or reasons for their beliefs. For instance, if reader beliefs are the result of their wanting to go along with certain groups they like or associate with, your best bet (as indicated earlier) is to show the acceptability of your point to these groups. If their attitudes reflect personal biases, such as an old grudge against someone in power, it is best to tailor your presentation to that prejudice. And so on.

Summarizing the implications of several behavioral studies, Karlins and Abelson present the example of three people who say they are against private ownership of industry. How should their reasons for this position influence one's choice of strategy or persuasion? The authors explain:

> One of them feels that way because he has only been exposed to one side of the story and has nothing else on which to base his opinions. The way to change this man's opinion may be to expose him to facts, take him to visit some factories, meet some workers and supervisors. A second person is against private ownership because that is the prevailing norm or social climate in the circles in which he finds himself. His attitudes are caused by his being a part of a group and conforming to its standards. You cannot change this fellow just by showing him facts. The facts must be presented in an atmosphere which suggests a social reward for changing his opinion. Some kind of status appeal might be a start in that direction. A third person may have negative attitudes toward private industry because by making business the scapegoat for all his troubles, he can unload his pent-up feelings of bitterness and disappointment at the world for not giving him a better break. . . . Trying to change this third person with facts may actually do more harm than good. The more evidence shows how wrong he is, the more he looks for reasons to support his beliefs. This kind of person can sometimes be influenced by helping him to understand why he has a particular attitude.[17]

11. Never Mention Other People without Considering Their Possible Effect on the Reader.

Other people may, as we saw earlier, be introduced for the sake of "testimonials." More commonly, however, other people's names are mentioned in the course of explaining a situation, narrating an event, or completing the format of a message. This use of names, too, may affect the power of your message.

A reference to the actions of another person—however simple and unobtrusive it may seem to you the writer—may alter your relationship with read-

---

[17]Karlins and Abelson, op. cit., p. 92.

ers. If readers consider that person a friend or enemy, their natural reaction is to begin thinking of the possible bearing of your communication on their friendship or antagonism. This reaction can have significant implications for your approach.

To illustrate, a doctoral student who had failed to meet his school's program requirements tried to muster faculty opinion in support of his petition for re-admission by appearing daily at the entrance to the dining hall and handing out leaflets to faculty members. One such leaflet contained these words: "I am very unhappy about the strain my case has created for Professor [name of the program director]. I am distressed if last Friday's handout . . . created the impression that I was harping on his mistakes. I have told him and I tell you that I could understand his actions and decisions. . . ." The leaflet went on at some length to explain the doctoral student's feelings about the problem.

What this writer did not realize was the impact of the professor's name on his communication strategy. Almost everyone who received the leaflet was a col-league of the professor in question. Therefore the leaflet made it necessary for them to think of their relationship with the professor when they made up their minds about the petition. And their relationship with the professor was more important to them than their relationship with the doctoral student.

If the doctoral student considered it essential to mention the professor, he could have elected to: (1) try to win readers over while convincing them that their relations with the professor would not be affected, or (2) show that the pro-fessor was so far off base that readers were morally bound to risk their relation-ship with him. In the latter case, the leaflet should have contained ready-to-use arguments that readers could draw on in explaining to the professor why they sympathized with the doctoral student. Since the leaflet did neither of these things, it was a failure in persuasion.

Don't overlook the possible effect of distribution. Letters often go to third parties, with "cc" typed at the bottom followed by the names of those people. A memorandum often contains the names of several addressees in the "To" line at the top. Covering letters with reports may indicate several groups of readers. All this may affect your strategy. The background information that you could omit if writing only to Jones may be quite necessary if Brown, too, is an important reader; and the rather offhand treatment you give to a certain test or episode if writing to Jones and Brown might not be fitting at all if Larabee also is an intended reader. Many times the wise manager or professional rewrites part of a letter or memo after deciding to send a copy of it to an additional person who was not considered when the first draft was made.

Many people have strong feelings about "blind copies," that is, copies sent to persons other than those indicated after "cc" at the end of a letter or in the "To" line of a memorandum. Some people feel that blind copies never should be sent. Others feel that since a letter or memo is the property of the writer, he or she can distribute it at will. Although the latter view is legally correct, only an obtuse writer will distribute copies thoughtlessley if the content is in any way confiden-tial, personal, or politically sensitive.

SIZING UP YOUR READERS

We have a tendency to abstract written communications from real life, to act as if the customary ground rules of influence and persuasion don't apply to a message that is in writing. We act with a naiveté almost unheard of in our face-to-face relationships. Not seeing readers, we act as if they weren't real people. "If we write the information clearly, accurately, and correctly," we think wishfully to ourselves, "surely that satisfies the requirements of a piece of paper." But Josh Billings's puckish maxim, "As scarce as truth is, the supply has always been in excess of the demand," applies to truth on paper as well as truth in conversation.

Think of your intended readers as the real people they will be when they take your letter or report out of the "in-box." Only then can you decide intelligently what information and ideas to emphasize and in what order to present them. To help you think of readers as three-dimensional people, ask yourself some questions about their situation and relationships with you. Are they:

- Deeply or only mildly interested in the subject of your communication?
- Familiar or unfamiliar with your views, competence, and feelings about them?
- Knowledgeable or ignorant of your authority in the area discussed, your status, and your associations of possible importance to them?
- Committed or uncommitted to a viewpoint, opinion, or course of action other than the one you favor in your letter, report, or other document?
- Likely or unlikely to find your proposal, idea, finding, or conclusion threatening or requiring considerable change in their thought or behavior?
- Inclined or uninclined to think and feel the way they do about the subject because of identifiable reasons, prejudices, or experiences?
- Associated formally or informally with groups or organizations involved in some way with the idea or proposal you deal with?

With answers to questions like these in mind, you will not see your readers as shadows on the wall. They will sit across from you. You can write as if talking *with* them, not talking to them.

Philip C. Kolin

Proposals

Philip C. Kolin, who has published widely in the fields of literature and communications, is professor of English at the University of Southern Mississippi.

GUIDELINES FOR WRITING A SUCCESSFUL PROPOSAL

Regardless of the type of proposal you are called on to write, the following guidelines will help you to persuade your audience to approve your plan. Refer to these guidelines both before and while you formulate your plan.

1. **Approach writing a proposal as a problem-solving activity.** Your goal is to solve a problem that affects the reader. Do not lose sight of the problem as you plan and write your proposal. Everything in your proposal should relate to the problem, and the organization of your proposal should reflect your ability as a problem-solver. Psychologically, make the reader feel confident that you can solve the problem.
2. **Regard your audience as skeptical readers.** Even though you offer a plan that you think will benefit readers, do not be overconfident that they will automatically accept it as the best and only way to proceed. To determine the feasibility of your plan, readers will question everything you say. They will withhold their approval if your proposal contains errors, omissions, or inconsistencies. Consequently, try to examine your proposal from the readers' point of view. . . .
3. **Research your proposal carefully.** A winning proposal is not based only on a few well-meaning, general suggestions. All your good intentions and enthusiasm will not substitute for the hard facts readers will demand. Concrete examples persuade readers; unsupported generalizations do not. To make your proposal complete and accurate, you will have to do a lot of homework; for example, reading

previous correspondence or research about the problem, doing comparative shopping for the best prices, verifying schedules and timetables, interviewing customers and/or employees, making site visits. . . .

4. **Prove that your proposal is workable.** The bottom-line question from your reader is "Will this plan work?" Your proposal should be well thought out. It should contain no statements that say "Let's see what happens if we do X or Y." By analyzing and, when possible, by testing each part of your proposal in advance, you can eliminate any quirks and revise the proposal appropriately before readers evaluate it. What you propose should be consistent with the organization and capabilities of the company. It would be foolish to recommend, for example, that a small company (fifty employees) triple its workforce to accomplish your plan.

5. **Be sure that your proposal is financially realistic.** This point is closely associated with and follows from guideline 4. "Is it worth the money?" is another bottom-line question you can expect from readers. Do not submit a proposal that would require an unnecessarily large amount of money to implement. For example, it would be unrealistic to recommend that your company spend $20,000 to solve a $2,000 problem that might not ever recur. Study the ecomonic climate, too—are you in an economic slump or in a boom time?

6. **Package your proposal attractively.** Make sure your proposal is letter-perfect, inviting, and easy to read (e.g., use plenty of headings and other visual devices). . . . The appearance as well as the content of your proposal can determine whether it is accepted or rejected. Remember that readers, especially those unfamiliar with your work, will evaluate your proposal as evidence of the type of work you want to do for them. Take advantage of any software programs dealing with desktop publishing . . . that may be available to you. This software will allow you to prepare a proposal that looks as if it were done by a professional printer.

INTERNAL PROPOSALS

The primary purpose of an internal proposal, such as the one included in Figure 1, is to offer a realistic and constructive plan to help your company run its business more efficiently and economically. On your job you may discover a better way of doing something or a more efficient way to correct a problem. In the world of work, typical problems for which proposals are written focus on money, personnel, outdated technology, health concerns, and organizational communications. You believe that your proposed change will save your employer time, money, or further trouble. (Tina Escobar and Oliver Jabur in Figure 1 have identified and researched a more effective and less costly way for Community Federal Bank to do business and to satisfy its customers.) You decide to notify your department head, manager, or supervisor, or your employer may ask you for specific suggestions to solve a problem he or she has already identified. . . .

EQUAL HOUSING
LENDER

COMMUNITY FEDERAL BANK

FDIC/DIFM

Powell
584-5200

Monroe
413-6000

Langston
796-3009

TO: Michael L. Sappington, Executive Vice President
Dorothy Woo, Langston Regional Manager

FROM: Tina Escobar, Oliver Jabur, ATM Services

DATE: June 2, 1994

RE: A Proposal to Install an ATM at the Mayfield Park Branch

<u>PURPOSE</u>

Clearly states why proposal is being sent

We are writing to propose a cost-effective solution to what is a growing problem at the Mayfield Park branch in Langston: inefficient servicing of customer needs and rising personnel costs. We recommend that you approve the purchase and installation, within the next three to four months, of an ATM at Mayfield. Such action is consistent with Community's goals of expanding electronic banking services and promoting our image as a self-serve yet customer-oriented institution.

<u>THE PROBLEM WITH CURRENT SERVICES AT MAYFIELD PARK</u>

Identifies problem by giving reader necessary back-ground information

Currently, we employ four tellers at Mayfield. However, we are spending too much on personnel/salary for routine customer transactions. In fact, as determined by teller activity reports, nearly 25 percent of the 4 tellers' time each week is devoted to routine activities easily accommodated by ATMs. Here is a breakdown of teller activity for the month of May:

Teller #	Total Transactions	Routine Transactions
1	6,205	1,551
2	5,989	1,383
3	6,345	1,522
4	6,072	1,518
	24,611	5,974

FIGURE 1 An Internal Unsolicited Proposal

Divides problem into parts— volume, financial, personnel, customer service

Clearly, we are not fully using our tellers' sales abilities when they are kept busy with routine activities. To compound the problem, we expect business to increase by at least 25 percent at Mayfield in the next few months, as projected by this year's market survey. If we do not install an ATM, we will need to hire a fifth teller, at an annual cost of $20,800 ($15,500 base pay plus approximately 30 percent for fringes), for the additional 6,000 transactions we project.

Verifies that problem is widespread

Most important, customer needs are not being met efficiently at Mayfield. Recent surveys done for Community Federal by Watson-Perry demonstrate that customers are dissatisfied about not having the convenience of an ATM at Mayfield. Seventy-seven percent of the respondents to the Watson-Perry questionnaire pointed to the lack of an ATM as Mayfield's biggest drawback. Customers are unhappy about long waits in line to do simple banking business, such as deposits, withdrawals, and loan payments, and about having to drive to other branches to do their after-hours banking. Conversations we had with manager Rachael Harris-Ignara at Mayfield confirm that customers regularly complain to tellers and loan officers about not having an ATM.

Ultimately, the lack of an ATM at Mayfield Park hurts Community's image. With ATMs available to Mayfield residents at local stores and other banks, our institution risks having customers and potential customers disassociate Community from their banking needs. We not only miss the opportunity of selling them on our other services but also risk losing their business entirely.

A SOLUTION TO THE PROBLEM

Relates solution to individual parts of the problem

The purchase and installation of an ATM at Mayfield Park will initially result in significant savings in personnel costs and time. We will not have to hire a fifth teller at $20,800 for the anticipated increase in transactions. Thanks to an ATM, we will also be able to allocate more effectively the talents of the four existing tellers at Mayfield. These tellers will then be able to assist customers with questions and transactions that cannot be handled through an ATM, such as purchase of savings bonds, traveler's checks, or foreign currency. Mayfield tellers, therefore, will have more opportunities for greater involvement in customer services and can spend more time cross-selling our services. As a result, we will get additional personnel achievements without reducing the level of customer service. In fact, we will be improving that service.

The increase in teller availability will inevitably lead to greater customer satisfaction. An ATM will allow customers the option of meeting their banking needs electronically or through a teller. Customers in a hurry can

FIGURE I   *(Continued)*

easily make a withdrawal with their ATM cards, while those who need more personal attention can take advantage of window service. As customer frustration is eased, so too will be the stress on tellers because of shorter lines and fewer complaints.

Shows problem can be solved and stresses how

It is feasible to install an ATM at Mayfield. This location does not pose the difficulties we faced at some older branches. Mayfield offers ample room to install a drive-up ATM in the stubbed-out fourth drive-up lane. This location is away from the heavily congested area in front of the bank, yet it is easily accessible from the main driveway and the side drive facing Cornith Avenue.

Judging from the ATM vendor's past work, the ATM could be installed and operational within three to four months. That is the amount of time it took to install ATMs at the first two locations in Powell and for Archer Avenue in Langston. Moreover, by authorizing the expenditure at Mayfield within the next month, you will ensure that ATM service is available before the Christmas season.

COSTS

Itemizes costs

The costs of implementing our proposal are as follows:

Diebold Drive-up ATM	$28,000.00
Installation fee	2,000.00
Maintenance (1 year)	1,500.00
	$31,500.00

Interprets costs for reader

This $31,500.00, however, does not truly reflect into our annual costs. We would be able to amortize, for tax purposes, the cost of installation of the ATM over five years. Our annual expenses would, therefore, look like this:

$30,000 (28,000 + 2,000) ÷ 5 years =
$6,000.00 + 1,500 (maintenance) or $7,500 per year.

Compared with the $20,800 a year the bank would have to expend for an additional fifth teller at Mayfield, the annual depreciated cost for the ATM ($7,500) reduces by nearly two-thirds the amount of money the bank will have to spend for much more efficient customer service.

FIGURE 1 *(Continued)*

CONCLUSION

Stresses benefits
for reader and
bank as a whole

Authorizing an ATM for the Mayfield Park branch is both feasible and cost-effective. Adoption of this proposal will save our bank more than $13,000.00 in teller services annually, reduce customer complaints, and increase customer satisfaction and approval. We will be happy to discuss this proposal with you anytime at your convenience.

FIGURE 1   *(Continued)*

Generally speaking, your proposal will be an informal, in-house message, so a brief (usually one- or two-page) memo should be appropriate.

An internal proposal can be written about a variety of topics, such as:

- purchasing new or more advanced equipment: word processors, transducers, automobiles
- hiring new employees or training current ones to learn a new technique or process
- eliminating a dangerous condition or reducing an environmental risk to prevent accidents—for employees, customers, or the community at large
- improving communication within or between departments of a company or agency
- revising a policy to improve customer relations (eliminating an inconvenience, speeding up a delivery) or employees' morale (offering vanpooling; adding more options for a schedule).

As this list shows, internal proposals cover almost every activity or policy that affects the day-to-day operation of a company or agency.

Your Audience and Office Politics

Writing an internal proposal requires you to be aware of and sensitive to office politics. To be successful, your internal proposal should be written with the needs and likes of your audience in mind. Remember that your boss will expect you to be very convincing about both the problem you say exists and the solution you propose to correct it. You cannot assume that your reader will automatically agree with you that there is a problem or that your plan is the only way to solve it.

Your reader in fact may feel threatened by your plan. After all, you are advocating a change. Some managers regard such changes as a challenge to their administration of an office or department. Or your reader may be indifferent, not even wanting to give your work serious consideration. Or your manager-reader may have certain "pet" projects or ways of doing things that you must take into account. To surmount these and other obstacles, show that the change you propose is in everyone's best interest. Do not overlook the possibility that your boss may have to take your proposal up the organizational ladder for commentary and, eventually, approval. . . . Writing a proposal may mean working as a team with your boss—putting his or her name on the document, too, to get notice and credit.

Before you write an internal proposal, consider the implications of your plan for your boss and for other offices or sections in your company. A change you propose for your department or office (transfers; new budgets or schedules) may have sweeping and potentially disruptive implications for another office or division within your company. It is wise to discuss your plan with your boss before you put it in writing. Then you might provide your boss with a draft, asking for his or her revisions or feedback. . . .

Never submit an internal proposal that offers an idea you think will work but relies on someone else to supply the specific details on *how* it will work. For example, do not write an internal proposal that says the payroll, community relations, maintenance, or advertising department can give the reader the details and costs he or she needs about your proposal. That pushes the responsibility onto someone else, and your proposal could be rejected for lack of concrete evidence.

The Organization of an Internal Proposal

A short internal proposal follows a relatively straightforward plan of organization, from identifying the problem to solving it. Internal proposals usually contain four parts, as shown in Figure 1. Refer to this proposal as you read the following discussion.

The Introduction

Begin your proposal with a brief statement of why you are writing to your boss: "I propose that. . . ." State why you think a specific change is necessary now. Then succinctly define the problem and emphasize that your plan, if approved by the reader, will solve that problem. Where necessary, stress the urgency to act—within the next week? month?

Background of the Problem

In this section prove that a problem does exist by documenting its importance for your boss and your company. As a matter of fact, the more you show how the problem affects the boss's work (and area of supervision), the more likely you are to persuade him or her to act. And the more concrete evidence you cite, the easier it will be to convince the reader that the problem is significant and that action needs to be taken now.

Avoid vague (and unsupported) generalizations such as "we're losing money each day with this procedure (or piece of equipment)"; "costs continue to escalate";

"the trouble occurs frequently in a number of places"; "numerous complaints have come in"; "if something isn't done soon, more difficulty will result."

Instead, provide readers with quantifiable details about the number of dollars a company is actually losing per day, week, or month. Emphasize the financial trouble so that you can show how your plan (described in the next section) offers an efficient and workable solution. Indicate how many employees (or work-hours) are involved or how many customers are inconvenienced or endangered by a procedure or condition. Notice how Escobar and Jabur include this information in their proposal in Figure 1. Verify how widespread a problem is or how frequently it occurs by citing specific occasions. Rather than just saying that a new word processor would save the company "a lot of money," document how many work-hours are lost using other equipment for the routine jobs your company now has employees perform.

The Solution or Plan

In this section describe the change you want approved. Tie your solution (the change) directly to the problem you have just documented. Each part of your plan should help eliminate the problem or should help increase the productivity, efficiency, or safety you think is possible.

Your reader will again expect to find factual evidence. Do not give merely the outline of a plan or say that details can be worked out later. Supply details that answer the following questions: (1) Is the plan workable—can it be accomplished here in our office or plant? and (2) Is it cost-effective—will it really save us money in the long run or will it lead to even greater expenses?

To get the boss to say yes to both questions, supply the facts that you have gathered as a result of your research. For example, if you propose that your firm buy a new piece of equipment, do the necessary homework to locate the most efficient and cost-effective model available, as Tina Escobar and Oliver Jabur do in Figure 1. Supply the dealer's name, the costs, major conditions of service and training contracts, and warranties. Describe how your firm could use this equipment to obtain better results in the future. Cite specific tasks the new equipment can perform more efficiently at a lower cost than the equipment now in use.

It is also wise to raise alternative solutions, before the reader does, and to discuss their disadvantages. Notice how Tina Escobar and Oliver Jabur do this in Figure 1, in their discussion of the solution to the bank's customer-service problem. They show why installing an ATM is more necessary than hiring a fifth teller. Their discussion is based on a strong persuasive strategy, demonstrating to their readers that they have examined all alternatives and chosen the best.

If you are proposing that your company hire or reassign employees, indicate where these employees will come from, when they will start, what they will be paid, what skills they must have, where they will work and for how long. If you propose to assign current employees to a job, keep in mind that their salary will still have to be paid by your company. Just because they are coworkers does not mean they will work for nothing. Note again in Figure 1 how Tina Escobar and

Oliver Jabur raise and resolve the problem of new and profitable responsibilities for the tellers at Community Federal Bank.

A proposal to change a procedure must include the following details:

1. How the new (or revised) procedure will work
2. How many employees or customers will be affected by it
3. When it will go into operation
4. How much it will cost the employer to change procedures
5. What delays or losses in business might be expected while the company switches from one procedure to another
6. What employees, equipment, or locations are available to accomplish this change.

As these questions indicate, your boss will be concerned about schedules, working conditions, employees, methods, locations, equipment, and the costs involved in your plan for change. The costs, in fact, will be of utmost importance. Make sure that you supply a careful and accurate budget so that your reader will know what the change is going to cost. Moreover, make those costs attractive by emphasizing how inexpensive they are as compared to the cost of not making the change, as Escobar and Jabur do in the section labeled "Costs." Double-check your math.

The Conclusion

The conclusion to your internal proposal should be short—a paragraph or two at the most. Your intention is to remind the reader that the problem is serious, that the reason for change is justified, and that you think the reader needs to take action. Select the most important benefits and emphasize them again. In Figure 1, Escobar and Jabur again emphasize the savings that the bank will see by following their plan as well as the increase in customer satisfaction. Also indicate that you are willing to discuss your plan with your reader.

SALES PROPOSALS

A sales proposal is the most common type of external proposal. Its purpose is to sell your company's products or services for a set fee. Whether short or long, a sales proposal is a marketing tool that includes a sales pitch as well as a detailed description of the work you propose to do. Figures 2 and 3 contain sample sales proposals.

The Audience and Its Needs

Your audience will usually be one or more business executives who have the power to approve or reject a proposal. Unlike readers of an internal proposal, your audience for a sales proposal may be even more skeptical since they may not know you or your work. Your proposal may also be evaluated by experts in

REYNOLDS INTERIORS
250 Commerce Avenue, S.W.
Portland, OR 97204–2129

503–555–8733   FAX 503–555–1629

January 21, 1994

Mr. Floyd Tompkins, Manager
General Purpose Appliances
Highway 41, South
Portland, OR 97222

Dear Mr. Tompkins:

In response to your Request 7521 for bids for an appropriate floor covering at your new showroom, Reynolds Interiors is pleased to submit the following proposal. After carefully reviewing your specifications for a floor covering and inspecting your new facility, we believe that the Armstrong Classic Corlon 900 is the most suitable choice. I am enclosing a sample of the Corlon 900 so that you can see how it looks.

Corlon's Advantages

Guaranteed against defects for a full three years, Corlon is one of the finest and most durable floor coverings manufactured by Armstrong. It is a heavy-duty commercial floors 0.085-inch thick for protection. Twenty-five percent of each tile consists of interface backing; the other 75 percent is an inlaid wear layer that offers exceptionally high resistance to everyday traffic. Traffic tests conducted by the Independent Floor Covering Institute repeatedly proved the superiority of Corlon's construction and resistance.

Another important feature of Corlon is the size of its rolls. Unlike other leading brands of similar commercial flooring—Remington or Treadmaster—Corlon comes in 12-foot-wide rather than 6-foot-wide rolls. This extra width will significantly reduce the number of seams in your floor, thus increasing its attractiveness and reducing the danger of tile split.

Installation Procedures

The Classic Corlon requires that we use the inlaid seaming process, a technical procedure requiring the services of a trained floor mechanic. Herman Goshen, our floor mechanic, has over fifteen years of experience working with the inlaid seam process. His professional work has been consistently praised by our customers.

Installation Schedule

We can install the Classic Corlon on your showroom floor during the first week of March, which fits the timetable specified in your request. The tile will take three and one-half days to install and will be ready to walk on immediately. We recommend, though, that you not move equipment onto the floor for twenty-four hours after installation.

FIGURE 2   An External Solicited Proposal

Costs

The following costs include the Classic Corlon tile, labor, equipment, and tax:

750 sq. yards of Classic Corlon at $12.50/yard	$ 9,375.00
Labor (28 hr. @ $15.00/hr.)	420.00
Sealing fluid (10 gal. @ 10.00/gal.)	100.00
Total	9,895.00
Tax (5 percent)	494.75
GRAND TOTAL	**$10,389.75**

Our costs are $250.00 under those you specified in your request.

Reynolds' Qualifications

Reynolds Interiors has been in business for more than twenty-eight years. In that time, we have installed many commercial floors in Portland and its suburbs. In the last year, we have served more than sixty customers, including the new multi-purpose Tradex plant in Portland.

Conclusion

Thank you for the opportunity to submit this proposal. We believe you will have a great deal of success with an Armstrong floor. If we can provide you with any further information, please call us.

Sincerely yours,

Sharon Scovill

Sharon Scovill
Sales Manager

Jack Rosen

Jack Rosen
Installation Supervisor

FIGURE 2   *(Continued)*

National Business Equipment

470 Rodgers Rd.   ■   Camden, NJ 08104–0826   ■   (201) 555–1100

September 20, 1994

Ms. Denise Taylor
Business Manager
Madison Tool and Die Company
3400 Veterans Boulevard
Camden, NJ 08104

Dear Ms. Taylor:

While we were servicing your Piko 2500 copier last week, it occurred to me that you might be interested in purchasing a newer model that will give you state-of-the-art features to make your copying work more efficient and reliable. In the past three years since you purchased your 2500, copier technology has advanced tremendously, giving users many benefits at surprisingly low cost.

Based on our assessment of Madison Tool and Die Company's needs for the most reliable office equipment available, we recommend that you purchase the Piko 4000. We believe that this copier will satisfactorily meet all of your copying needs for the present and well into the future.

ADVANTAGES OF THE PIKO 4000

With the Piko 4000, your cost per single copy will be reduced to only 1 1/2 cents, compared with approximately 3 cents per copy with your present machine. The 4000 model is designed to make as many as **75,000 copies per month,** almost double the recommended work load of your current copier. The following description of the main features of the Piko 4000 will show you its other advantages over your current copier.

Printing Technology

The Piko 4000's **printing rate of 80 copies per minute** will save you valuable time. The automatic duplexing capability allows you to turn single-sided documents into double-sided ones, thus reducing paper costs. In addition to the standard paper sizes—8 1/2 × 11. 8 1/2 × 14, and 11 × 17—**the 4000 functions effectively with other paper sizes and shapes,** including heavyweight sheet stock suitable for specifications and drawings. The Piko 4000 has original-size sensing, which can copy varying sizes automatically.

FIGURE 3   An External Uunsolicited Proposal

Copier Control Security System

The Piko 4000's **electronic ID keypad** allows only authorized users, with individual entry codes, to run the copier. This keypad can control access, limit copies, and provide tabulated records of use for your accounting purposes.

Feeding/Sorting Capacity

The 400 includes as standard equipment both a recirculating automatic document feeder and an automatic sorter/stapler. These features **can reduce the time spent on large copying jobs by as much as 50 percent.**

Printing Quality

The new Piko 4000 offers top-quality, high resolution reproduction **in four colors** in addition to black and white. Since seeing is believing, I am enclosing a copy of this proposal made on the 4000 as well as a copy of a blueprint to show you how both copied documents look.

Reduction/Enlargement Function

The Piko 4000's zoom capability allows you to increase or decrease the size of your copy within a range of 50–200 percent. Since these changes are made in increments of 1 percent, your copy is always the exact size you want.

Size

The Piko 4000 measures 4' × 3' × 4', **occupying one-quarter less space** than your present copier. Once the machine is installed, you will find this extra space useful for your storage needs.

INSTALLATION, TRAINING, AND SERVICE

We will deliver and install your new Piko 4000 within one week of receiving your purchase order. The installation requires approximately forty-five minutes.

Our local sales representative, Darlene Simpson, is available to instruct your employees in the operation and routine maintenance of the 4000. A phone call will allow us to set up a mutually convenient date for such instruction. If a problem occurs, **we offer customers the latest in remote diagnostics.** A modem installed in the Piko 4000 allows us to diagnose specific problems (and needed parts) before dispatching a service representative. We save time, so you save time. In addition, our "hot line" is available to you on business days from 8:00 A.M. to 5:00

FIGURE 3 *(Continued)*

P.M. We guarantee that a service representative will arrive at your office **within two business hours** from the time of your call.

COSTS

Below is the price of the Piko 4000 and appropriate initial supplies:

Piko 4000	$18,295.00
Toner (tube)	23.00
Paper (5,000 sheets of medium grade 8 1/2 × 11)	25.00
Service contract—optional	
(per year, parts and labor included)	2,300.00
Total	$20,643.00

We install your new Piko 4000 **free of charge.**

NATIONAL'S REPUTATION

Over the past 22 years, National has supplied copiers to more than 90 companies in the greater Camden area, including Biscayne Industries, Northeast Manufacturing, and Teunissen Accounting, Inc. In addition to providing quality products, we are **dedicated to giving our customers fast and efficient service long after a sale.**

We appreciate your using **National Business Equipment** for your repair needs and hope that you will decide to purchase the new Piko 4000. Please call me if you have any questions about the Piko 4000 or National.

If this proposal is acceptable, please sign and return a copy of this letter.

Sincerely yours,

Marion Copely

Marion Copely

Encls. Copier samples

I accept this proposal made by National Business Equipment.

_____

for Madison Tool and Die Company

FIGURE 3 *(Continued)*

other fields employed by your prospective customer. Readers of a sales proposal will evaluate your work according to (1) how well it meets their needs and (2) how well it compares with the proposals submitted by your competitors. Your proposal must convince readers that you can provide the most appropriate work or service and that your company is more reliable and efficient than any other firm.

The key to success is incorporating the "you attitude" . . . throughout your proposal. Relate your product, service, or personnel to the reader's exact needs as stated in the RFP [request for proposal] for a solicited proposal or through your own investigations for an unsolicited proposal. You cannot submit the same proposal for every job you want to win and expect to be awarded a contract. Different firms have different needs. The most important question the reader will raise, therefore, about your work is "How does this proposal meet our company's special requirements?" Some other fairly common questions readers will have as they evaluate your work include the following:

- Does the writer's firm understand our problem?
- Can the writer's firm deliver what it promises?
- Can the job be completed on time?
- What assurances does the writer offer that the job will be done exactly as proposed?

Answer each of these questions by demonstrating how your product or service is tailored to the customer's needs.

Organizing a Sales Proposal

A sales proposal can have the following parts: introduction, description of the proposed product or service, timetable, costs, qualifications of your company, and conclusion.

Introduction

The introduction to your sales proposal can be a single paragraph in a short sales proposal or several pages in a more complex one. Basically, the introduction should prepare readers for everything that follows in your proposal. The introduction itself may contain the following sections, which sometimes may be combined.

1. **Statement of purpose and subject of proposal.** Tell readers why you are writing and identify the specific subject of your work. If you are responding to an RFP, use specific code numbers or cite application dates, as the proposal in Figure 2 does. If your proposal is unsolicited, indicate how you learned of the problem, as Figure 3 does. Briefly define the solution you propose.
2. **Background of the problem you propose to solve.** Show readers that you are familiar with their problem and that you have a firm grasp of the importance and implications of the problem. In a solicited proposal, such as the one shown in

Figure 2, a section outlining the problem is usually unnecessary, because the potential client has already identified the problem and just wants to know how you would handle it. In that case, just point out how your company would solve the problem, mentioning your superiority over your competitors (see the third paragraph of Figure 2). In an unsolicited proposal, you need to describe the problem in convincing detail, identifying the specific trouble areas. However, if it is an external proposal to a current customer, such as the one in Figure 3, it would be unwise to point out past problems your client may have had with your company's services or products. In Figure 3, note how the problems of the Piko 2500 model are described mainly as a way to sell the advantages of the Piko 4000.

Description of the Proposed Product or Service

This section is the heart of your proposal. Before spending their money, customers will demand hard, factual evidence of what you claim can and should be done. Here are some points that you should cover.

1. **Carefully show your potential customers that your product or service is right for them.** Stress specific benefits of your product or service most relevant to your reader. Blend sales talk with descriptions of hardware.
2. **Describe your work in suitable detail—what it looks like, what it does, and how consistently and well it will perform in the readers' office, plant, hospital, or agency.** You might include a brochure, picture, or, as the writers of the proposals in Figures 2 and 3 do, a sample of your product for customers to study. Convince readers that your product is the most up-to-date and efficient one they could select. Note how the proposals in Figures 2 and 3 emphasize this.
3. **Stress any special features, maintenance advantages, warranties, or service benefits.** Highlight features that show the quality, consistency, or security of your work. For a service, emphasize the procedures you use, the terms of that service, and even the kinds of tools you use, especially any "state-of-the-art" equipment. Be sure to provide a step-by-step outline of what will happen and why each step is beneficial for readers.

Timetable

A carefully planned timetable shows readers that you know your job and that you can accomplish it in the right amount of time. Your dates should match any listed in an RFP. Provide specific dates when the work will begin, how long it will continue, and when you will be finished installing equipment, testing equipment, or training employees to use equipment. For proposals offering a service, specify how many times—by the hour, week, month—customers can expect to receive your help; for example, spraying three times a month for an exterminating service or making deliveries by 10:00 A.M. for a trucking company. Indicate whether follow-up visits or service calls will be provided.

Costs

Make your budget accurate, complete, and convincing. Accepted by both parties, a proposal is a binding legal agreement. Don't underestimate costs in the hope that a low bid will win you the job. You may get the job but lose money doing it, for the customer will rightfully hold you to your unrealistic figures. Neither should you inflate prices; competitors will beat you in the bidding.

Give customers more than the bottom-line cost. Show exactly what readers are getting for their money so they can determine if everything they need is included. Itemize costs for specific services, equipment, labor (by the hour or by the job), transportation, travel, or training you propose to supply. If something is not included or is considered optional, say so—additional hours of training, replacement of parts, and the like. If you anticipate a price increase, let the customer know how long current prices will stay in effect. That information may spur them to act favorably now.

Qualifications of Your Company

Emphasize your company's accomplishments and expertise in using relevant services and equipment. You might list previous work you have done that is identical or similar to the type of work you are proposing to do for the customer. You may even want to mention the names of a few local firms for whom you have worked that would be pleased to recommend you. Never misrepresent your qualifications or those of the individuals who work with or for you. Your prospective client may verify if you have in fact worked on similar jobs for the last five to six years.

Conclusion

This is the "call to action" section of your proposal. As with the conclusion in an internal proposal, encourage your reader to approve your plan. Stress the major benefits your plan has for the customer. Notice how the last paragraphs of Figures 2 and 3 do this effectively. Offer to answer any questions the reader may have. Some proposals end by asking readers to sign and return a copy of the proposal thus indicating their acceptance of it, as the proposal in Figure 3 does.

Part 5

Resumes and Other Written Materials for a Job Search

The written materials that are part of an application for a job face one of the most difficult audiences imaginable: experienced recruiting managers. Because the selections that follow in this section of *Strategies* offer extensive advice on how to prepare and write resumes and cover letters, I will limit my remarks here to some general comments.

First, there is no *one* way to write a resume or a cover letter. Slavishly following some model that an applicant mistakenly thinks represents the ideal can defeat the purpose of writing a resume and a cover letter: to get an interview. The shelves of bookstores and libraries across the country are filled with guides offering advice on how to write resumes and cover letters. These guides offer advice, not commandments. Job applicants should consult as many of these guides as they want, read the selections that follow in *Strategies,* but, in the end, let common sense be their guide when they write what are essentially advertisements for themselves.

The best advice on how to write effective resumes and cover letters quite naturally comes from the recruiting managers who read them. Several years ago, my own university's placement bureau sponsored a panel featuring the recruiting managers from the companies that traditionally have hired the greatest number of our graduates. All the managers agreed that they looked at resumes and cover letters to determine as quickly as possible—sometime merely by scanning candidate's application materials—what preparation and experience candidates had in the following skills and areas:

- written and oral communication skills
- MIS or PC skills
- interpersonal skills, as demonstrated by the ability to work as a member of a team
- self-reliance and initiative, as demonstrated by the ability to work alone
- a sense of what the world of work demands in terms of professionalism and deadlines
- specific skills in at least one business or technical area supplemented by secondary skills in a variety of related areas
- a sense of business and personal ethics
- the ability to manage time, set priorities, and work under stress.

While recruiting managers do not expect job candidates to excel in all these areas or possess all these skills from the start, the list does provide some general guidelines for communicating with recruiting managers by resumes and cover letters—or, for that matter, in person during an interview.

This section of *Strategies* begins with John L. Munschauer's detailed discussion of how to write a resume and cover letter. Based on his many years of experience working at the Cornell University Career Center, Munschauer places the writing of resumes and cover letters within the context of how they are read by busy recruiting managers. Steven Graber follows with a no nonsense discussion of the basics of the cover letter. Richard H. Beatty then discusses the advantages and disadvantages of using a functional resume rather than a more traditional chronological resume in applying for a job, and Burton Jay Nadler concludes this section of *Strategies* with some suggestions about composing and sending electronic resumes.

John L. Munschauer

Writing Resumes and Letters in the Language of Employers

The late John L. Munschauer was Director of the Cornell University Career Center.

"I am sorry, Father O'Mega, I can't let you in."

"But, St. Peter, I did everything I was supposed to. I changed the Mass from Latin to English. I had the communicants hold their hands the new way when they took communion. Everyone, so far as I knew, genuflected properly, and even Mrs. O'Reilley's Protestant husband stood when I read the gospel. I can't think of a thing I did that was wrong."

"The message, Father O'Mega, what about the message?"

"The message? What message?"

"The message of the Lord, Father. Don't you remember? The Ten Commandments? The Beatitudes? The Golden Rule? It's the *message* that gets people in here, not the ritual. The ritual was supposed to help deliver the message, but you made the ritual the message and the meaning got lost."

Meanwhile, back on earth . . . in Tucumcari, New Mexico, and all over the United States, job hunters are at their typewriters and computers pushing words around to make their resumes look like the ones they have seen in books. Whether to have it professionally printed or not, that is the question. Does it have enough action verbs?

In Peoria, Illinois, and elsewhere, employers are scanning resumes. Some are beautiful to look at; a few are even printed commercially on expensive paper. But the resumes are laid aside. Employers are reading them with one thought in mind: What can the candidates do for us? The message they are looking for isn't there.

WHY USE A RESUME?

The purpose of a resume is to convey a message, a purpose easily forgotten in the ritual of preparing it. At every turn, you will get conflicting advice about how to conduct the ritual:

- You must have a resume to get a job.
- The purpose of a resume is to get an interview.
- Every resume must have a job objective.
- A resume should never, ever be longer than one page.
- On a resume, list experience chronologically.

On the other hand, others advise:

- Don't use a resume if you are looking for an executive position or merely seeking information.
- Resumes typecast you and narrow your options.
- Interview for information first, determine what an employer wants, and then decide whether offering a resume will suit your purposes.
- Two or more pages present no problem if your resume follows a logical, easy-to-read outline.
- List your experience by function, not chronologically.

The more opinions you get, the more confused you become, but you finally work up something. You send it out. You get little or no response. You change your resume from one page to two pages—or from two pages to one. The results are no better. Somebody says to try pink paper; *that* will get attention. You decide you should have used blue. You fiddle and fiddle with your resume, trying to find the magic formula that will get you what you want. You are caught up in the ritual, forgetting that the purpose of a resume is to send a message.

It is hard not to be distracted from the message. You concentrate on developing a format, forgetting the message, instead of concentrating on the message and *then* working on a format that will convey it best. But what is your message? To find the answer, use your imagination to step out of yourself and become an employer. Think about what the employer is trying to accomplish and the talent required to get it done. Now, from your imaginary employer's chair, look back at yourself and ponder the answer. If you can't come up with one, you will have difficulty writing a resume that will say to the employer, "I have something to offer you."

Resumes are not tickets to a job. They are just one of several ways to court employers. And, as in any courtship, sometimes it's just as well not to put everything in print so the other party can't draw the wrong conclusion. Take the case of Cheryl Fender, an alto who learned that the Springfield Opera Company was auditioning for an alto to sing *Carmen*. Cheryl was well-qualified, and her voice teacher was well-known as a coach of only the most gifted singers. On the other hand, her resume, incomplete and with gaps in her history, showed extensive experience as a secretary. Wisely, she did not send the resume, which might have established her as a secretary who wanted to sing rather than as a singer who had supported herself by being a secretary. Instead, she asked herself what was important to the director of an opera company. Voice, of course, so she outlined her training in a letter. But dramatic ability also mattered, so she included pictures of herself on stage in roles that beautifully illustrated her dramatic ability. She got an audition. Cheryl followed a good marketing rule: Don't confuse customers by flaunting things that don't speak to their needs.

GIVING YOUR MESSAGE

While the language of employment for you is "I want" and for employers it is "I need," you can create resumes and letters in your language that will be read by employers in theirs. I don't mean statements such as "I have analyzed my qualifications and feel confident that they fit your needs." If you were one of thousands of employers who read this kind of thing, would you ask what qualifications and what needs? If the answer isn't clear—and it rarely is—there won't be a message.

It isn't all that difficult to create a statement that is effective. Start with the written word, with prose. Even if you are going to be approaching employers in person, go through the exercise of writing letters. Otherwise, like most people, you can get caught up in the resume ritual and neglect to develop the words that can be more effective in telling your story. People sweat over resumes, then dash off letters without much thought. Perfect resumes arriving in the mail won't even be read if the cover letters don't impress and engage employers.

I recall being asked to review a proposed application letter for a job on the staff of a yachting magazine. The letter was beautifully written, but it left the impression of being all "I": "I want to write so very much, and I am sure I can learn if you give me a chance. . . . I got straight A's in English. . . . I love boats. . . . I am a sailor. . . . I was captain of the sailing team in college. . . ."

What do you say to a person who has written a letter like that? He had obviously spent hours composing it. In its appearance and use of language, it met the highest standards. I could not squelch the young man's hopes, so I had him read *Book Publishing,* a pamphlet Daniel Melcher wrote when he was president of R. R. Bowker Company. Although the pamphlet is concerned only with book publishing, I thought its message would be useful for someone interested in magazines as well.

Let me paraphrase a portion of Melcher's pamphlet:

I like publishing and you might like it too. It is only fair to warn you, however, that publishing attracts a great many more people than the industry can possibly absorb. Sometimes it seems as though half of the English majors in the country besiege publishing offices for jobs each year. While we hope that you will have something to offer us, you might as well face it. Publishers are experts in the art of the gentle brush-off.

The interviewer hopes that you have what he needs, but it turns out that you have never looked into any of the industry's trade journals nor read any books about the industry. You haven't even acquainted yourself with the work of your university press. You tell the interviewer that you are willing to start anywhere, but it develops that a file clerk's job would not interest you, you do not want to type, you don't think you can sell, and you know nothing about printing.

The fact is, all you have thought about is what you want, but it is his needs that create jobs, and you must address yourself to needs.

Your problem, therefore, is to learn as much as possible about the industry before you go looking for a job. Only in this way will you be able to put yourself in the publisher's place and talk to him about his needs rather than about your wants.

After reading Melcher's advice, the fellow said he got the point, thanked me, and left, returning about three weeks later to show me his revised letter. It began in much the same way as his earlier effort—with "I's"—"I majored in English. . . . I have done considerable sailing. . . ."—but after a few short sentences, he suddenly changed his tack. A new three-sentence paragraph began like this: "With my interests, naturally I want to work for you. But more to the point is not what I want but what you need." He followed this with three words that many women refuse to use and that men almost never think to use: "I can type." Right away, he showed that he knew he must be useful.

So much for the easy part of the letter. Next came the difficult part. Although he had had no experience and had never submitted an article to a magazine or even written for a student newspaper, he still had to come up with something that would interest the editor. He found the solution. The letter continued in this manner:

. . . In looking into the field of journalism, I visited the editor of our alumni magazine, and I talked to magazine space salesmen and to executives responsible for placing ads in magazines. I also visited a printer who has contracts with magazines. In addition, I have been reading trade journals and several books on the industry. As I looked into publishing, it occurred to me that, of all the things I have done, the one I could most closely relate to the field was, strangely enough, an experience I had as a babysitter.

Immediately, he had the editor's full attention. How could babysitting fit in with publishing? During the summer of his junior year in college, he had taken a job as a sailing instructor, tutor, and companion to the children of a wealthy family that summered on the coast of Maine. The parents often went away for a week or more at a time, leaving a governess in charge, with a cook, chauffeur, maid, and gardener to do the chores and the student to keep the children busy. While the parents were on a cruise, the governess suffered a stroke, sending the cook into a tizzy, the maid into tears, and the chauffeur and gardener to the local bar. Only the student could cope, and he took charge and managed the estate for the rest of the summer.

In his letter of application, he described the crisis and its subsequent problems and told how he had met them. Then he related those experiences to the problems that he had learned editors, advertisers, printers, and others encounter in the publishing industry. Reading his letter, you could picture the student working for a publisher. There would be no slipups with the printers. Advertisers and authors would be handled with tact, yet he would get them to turn in their copy on time. He came through as someone with ingenuity, energy, and reliability; and the letter itself testified to his writing ability.

Did he get the job? Yes and no. He got an offer, but because of the publisher's urgent need, the job had to be filled immediately, and he could not accept it. He was teaching school at the time and felt that in fairness to his pupils he should finish the school year. However, a while later, the letter surfaced again when a group of editors at a meeting chatted during lunch. The subject of good editorial help came up. It followed the usual theme: "They don't make 'em like they used to." The editor of another sailing magazine complained that he had been looking for an assistant, but despite hundreds of applicants he had found no one suitable. At this point, the editor of the yachting magazine described the young man's letter and agreed to share it. The upshot was, again, a job offer. This time the timing was right, and the young man took the job. There is nothing quite like the staying power of a well-written letter. It is remembered.

Later, I complimented the young man, telling him I had never read a better letter. "It was easy," he said. "The first letter was the tough one. I didn't have anything to say other than 'I have a good record. Please give me a chance,' but I knew everyone else was saying the same thing, so I would have to say it better. I struggled with every word, trying to make an ordinary message extraordinary, but even elegant words can't make something out of fluff. I didn't know anything about publishing, so I didn't have anything to say to publishers; nor could I be really convincing without knowing enough about the work to decide whether I wanted that kind of job or not. But publishing sounded exciting, so I thought I would give it a fling."

The Importance of Knowing What the Job Is All About

"When I looked into the field in depth," the young man continued, "I became confident that I had something to offer. Thanks to a few good high school teachers and college professors, I knew where to place a comma and a colon, so I had a technical skill to offer publishers. And I knew sailing. But the big need I saw was one I discovered when I was teaching school and again when I was an assistant manager at a McDonald's—a need for people who can get things done. Such a quality is hard to describe without an analogy. I could have alluded to any number of jobs I had held, but I chose the baby-sitting job because I thought it would be different and would introduce an element of surprise. Apparently, the analogy worked. I got the job. More important, I wanted it. If I had received a job offer after sending the first letter, before I had really investigated publishing, I would have taken the job with an attitude of 'teach me.' That's a passive role, an observer's role, and observers tend to be critical. The chances are fifty-fifty that I would not have known how or where to contribute and would have quit after a while. Instead of offering my employer solutions to his problems, I would have become one of his problems."

I asked the young man if he had used any other supporting documents, such as a resume, to help him get the job. "I had a resume," he replied, "but I held it back, because my task was to transfer the qualities I had demonstrated as a baby-sitter to the needs of a publisher, and I couldn't seem to do that in a resume. A resume is a good way to outline facts, but I had to use prose to develop the analogy.

"I did think about including a writing sample, but when I studied the magazine I realized that most of the articles were written by contributors rather than staff. Their job was to select and edit articles. If I presented myself as a writer, I wouldn't have received an offer; the magazine didn't hire authors. When a friend of mine saw an advertisement for an editing job, he made a list of the required qualifications, then presented his case point by point, right on target. Then the dope attached a resume that said loud and clear, 'I want to be a writer.' He either should have skipped the resume or prepared a new one."

Are these examples intended to be good arguments for not using resumes? No. They simply emphasize that it is important to determine the best way to get a message across. There are times when there is no substitute for a resume. When employers advertise and list the qualifications they seek, there is no better way to respond than to send a resume outlining qualifications.

LETTERS OF APPLICATION

Sometimes, however, it is hard to figure out how to make the letter of application you send with your resume more effective than those of the hundreds of other people who are responding to the ad. Imagine being on the receiving end of applications at General Motors or Exxon.

To find out about the effectiveness of letters and resumes, I visited corporations and asked employment managers for their comments. "Here," said one employment manager, as he picked up an 18-inch stack of letters and handed it to me. "This is my morning's mail. Read these letters and you'll have your answer."

"I can't," I protested. "I have only 2 hours, and there's a day's reading here."

"Yes, you can," he replied. "Unfortunately, you'll get through the pack in half an hour, because a glance will tell you that most are not worth reading."

It was hard to believe that the letters could be that bad, but he was right. The typical letter was an insult. Among the letters that I did not finish reading was one that was obviously a copy—the machine must have run out of toner thirty copies back. It began:

Dear Sir:

I am writing to the top companies in each industry and yours is certainly that. I want to turn my outstanding qualities of leadership and my can-do abilities to. . . .

Enough of that. Also, the applicant hadn't even bothered to type in the employment manager's name, which he could easily have found in reference books such as the CPC Annual, published by the College Placement Council, and Peterson's Job Opportunities, published by Peterson's Guides.

The next letter was written in pencil on notepaper. There may have been an Einstein behind that one, but I can't imagine anyone taking the time to find it out. Many other letters were smudged and messy. Some applicants tried to attract attention with stunts, such as putting cute cartoons on their letters. One piece of mail contained a walnut and a note that read, "Every business has a tough nut to crack. If you have a tough nut to crack and need someone to do it, crack this nut." Inside the nut, all wadded up, was a resume. Cute tricks and cleverness don't work at the General Mammoth Corporation.

At the same time, the good letters stood out like gold. Five letters, only five letters in that pile of hundreds, were worth reading. They had this in common:

- They looked like business letters. Their paragraphing, their neatness, and their crisp white 8 1/2" X 11" stationery attracted attention like good-looking clothing and good grooming.
- They were succinct.
- There were no misspellings or grammatical errors.

As I read them, I heard a voice—the voice of a fusty old high school English teacher—commanding out of the past:

- If you can't spell a word, look it up in a dictionary.
- Use a typewriter. Pen and ink are for love letters.
- Clean your typewriter. Avoid fuzzy type. You wouldn't interview in a dirty shirt, so don't send a dirty resume or letter.
- For format, use a secretarial manual. If you don't have one, borrow one from the library. What do you think libraries are for?

How that English teacher would have loved the following advice given by the late Malcolm Forbes when he was editor-in-chief of *Forbes* magazine:

Edit ruthlessly. Somebody ~~has~~ said that words are ~~a lot~~ like inflated money—the more ~~of them that~~ you use, the less each one ~~of them~~ is worth. ~~Right on~~. Go through your entire letter ~~just~~ as many times as it takes. ~~Search out and~~ **A**nnihilate all unnecessary words ~~and~~ sentences—even ~~entire~~ *paragraphs*.

The following letter is typical of those I saw that day. Give it the Forbes treatment, and see what you can do with it. You may need a scissors as well as a pencil.

Dear Mr. Employer:

I am writing to you because I am going to be looking for employment after I graduate which will be from Michigan where I studied Chemical Engineering and I will be getting a Bachelor of Science degree in June. The field in which I am interested and hope to pursue is process design and that is why I am writing your company to see if you have openings like that. I think I have excellent qualifications and you will find them described in the resume which I have attached to this letter.

The five letters that stood out favorably were characterized by their simplicity. Here is a letter, fictitious of course, but enough like the letter I remember to give you an idea of the ones that created a favorable impression:

February 1, 1991

Mr. Paul Boynton
Manager of Employment
The United States Oil Company
1 Chicago Plaza
Chicago, Illinois 60607

Dear Mr. Boynton:

This June I will receive a Bachelor of Science in Chemical Engineering from the University of Michigan, and I hope to work in process design or instrumentation. I saw your description in *Peterson's Job Opportunities for Engineering, Science, and Computer Graduates* soliciting applicants with my interests. Enclosed is a resume to help you evaluate my qualifications.

While I find all aspects of refining interesting, my special interest in process design and instrumentation developed while working as a laboratory assistant to Professor Juliard Smith, who teaches process design. I wrote my senior thesis on the subject of instrumentation under him, and part of what I wrote will be used in a textbook he is writing and editing.

Would it be possible to have an interview with you in Chicago during the week of March 1? To be even more specific, could it be arranged for 10 A.M. on Tuesday, March 3? I am going to be in Chicago that week, and this time and date would be best for me, but of course I would work out another time more convenient for you. In any event, I will call your office the week before to determine whether an interview is possible.

Sincerely yours,

Charles C. Thompson

Thompson's effective letter, and three of the four others, followed a similar pattern:

1. The first paragraph stated who the writer was and what he wanted.
2. The second paragraph, sometimes the third, and in one case a fourth paragraph, indicated why the writer wrote to the employer and mentioned areas of mutual interest, special talents that might be of interest to employers, or other factors relating to qualifications that could be better described in a letter than in a resume.
3. A final paragraph suggested a course of action.

The fifth letter covered the same points in a different order. I remember it because it complemented the good but not outstanding resume shown in Figure 1. In that resume, a perceptive employer could see a person he might like, someone who was energetic and personable. Yet it didn't quite hang together, because the work history and activities didn't seem to support what the writer wanted to do. But look at the letter that "made" the resume:

LANCE ZAROTE

<u>Campus Address</u>
101 Morril Hall
University of Puget Sound
Tacoma, Washington 95840
(206) 555-6206

<u>Permanent Address</u>
25 The Byway
Provincetown, Massachusetts 05840
(617) 555-6026

GOAL: A SALES CAREER

EDUCATION: UNIVERSITY OF PUGET SOUND
                   Bachelor of Arts 1991
                   Philosophy Major

                   <u>Business-Related Courses</u>
                   Statistics   Introduction to Computers
                   Economics   Calculus
                   Accounting   English

EMPLOYMENT: UNIVERSITY OF PUGET SOUND (Work-Study Program)
                   Dining hall supervisor   1990–91
                   Kitchen helper   1987–89
                   OTHER WORK while in college
                   Ma and Pa Motel, Tacoma—night clerk   1990
                   Joe's Bar and Grill—weekend waiter
                       & bartender   1989–1991
                   Baby-sitting, gardening, house cleaning   1985–1991
                   WATCHEE OUTEY SUMMER CAMP, Nome, Alaska
                   Sailing coach and waterfront director   1990
                   Counselor   1986–1989

ACTIVITIES: CAMPUS
                   Chair, Campus Chest Drive
                   Intramural hockey, tennis, and volleyball
                   LIVING UNIT
                   House Manager, $25,000 budget
                   Secretary
                   Membership Committee
                   CIVIC
                   Reader-companion in nursing home
                   Big Brother-Sister Program, Southside Youth Center
                   CHURCH
                   Choir
                   Youth leader
                   Sunday school teacher

INTERESTS: Skiing   Music
                   Chess   Dancing

FIGURE I   An ordinary resume can become effective when attached to a powerful letter.

Mr. Paul Boynton, Manager of Employment
The United States Oil Company
1 Chicago Plaza
Chicago, Illinois 60607

March 10, 1991

Dear Mr. Boynton:

This June, following my graduation from the University of Puget Sound, I want to pursue a career in sales. Between April 10 and 23, I plan to call on leading companies whose products I would like to sell. The purpose of this letter is to determine whether you would like to have me include you in my itinerary.

Let me tell you why I believe I can sell. It seems to me that I am always selling. As a camp counselor, I persuaded the director to buy a fleet of small sailboats so I could start a sailing program. When our college housing co-op needed painting, I persuaded the members to give up a vacation to do the job. In thinking about how I enjoyed selling these and other projects, I decided to look into a sales career. I persuaded several sales representatives to let me spend a day or more traveling with them to see what it was like.

While with them I realized something more about myself that further convinced me I belong in sales. The best sales representatives were well organized, had high energy levels, and used their time efficiently, qualities I feel I have. As evidence, I have enclosed a resume that outlines my accomplishments in college and during vacations.

I hope to hear from you. United States Oil is in the top group of employers on my list.

Sincerely,

Lance Zarote

Encl: Resume

Hard Work and Attention to Detail Make for a Good Letter

While only the five letters I have mentioned were effective, the rest of the correspondents could have done as well. The point is they didn't. Most people won't. Therein lies your opportunity, because, like Thompson and Zarote, you can write letters that set you apart. You don't have to create a literary masterpiece; just don't knock off a letter hastily with thoughts that wander all over the page. Write it and rewrite it, following Forbes's advice. Unless you are an exceptional typist, you are not good enough to type it yourself. Hire a professional or use a word processor, but be sure the print is letter quality. Also, get an English teacher or someone in the word business to check your spelling, punctuation, and grammar.

Don't Delegate the Job of Letter Writing

More important than style, however, is the thought process used in preparing letters and resumes. Don't shortchange yourself by delegating your thinking to someone else. Be sure it is *your* letter. Somehow, a ghost-written letter always has a phony ring to it. When you write to employers, think about their needs; then think about yourself and what you offer, and relate this to what you would like to do. Putting your thoughts on paper—thoughtfully—will make you sort out your ideas and interrelate them. When you see them on paper they will talk back to you, at times to suggest better ideas, at other times to tell you that you are off the mark. To organize your ideas, create an outline. In other words, prepare a resume even if you decide not to use it. *The value of a resume is frequently more in its preparation than in its use.*

RESUME PREPARATION

When you do give an employer your resume, make it a testimony to your ability to organize your thoughts. Remember, too, it must look sufficiently attractive to get an employer to read it. Unfortunately, most of the resumes I have seen on employers' desks were just as unattractive as the letters; they had sloppy, crowded margins, were poorly organized, and were badly reproduced. At least 30 percent of the resumes had been put aside with hardly a glance because their physical appearance was so awful. The rest got a 20-second scan to see if they were worth studying.

Following are two resumes that pass the appearance test with flying colors. Let's see how they fare during a 20-second scan and beyond.

Nancy Jones—A Good Resume Made Better

Nancy Jones's resume has arrived at the desk of a laboratory director who needs an assistant to help run a quality-control laboratory in a pharmaceutical company (Figure 2). With candidates far outnumbering openings in biology, the advertisement for the job has brought in hundreds of applications, and the director is wearily scanning them one by one to find the few that will be of interest to him. Conditions are not favorable for Nancy. She has to catch his eye with impressive qualifications, or she is not going to get anywhere.

The director picks up Nancy's resume. Immediately, he is impressed, because it looks attractive. He thinks the resume reflects an orderly mind. Most resumes he has looked at just do not put it all together.

He begins to read. The job objective annoys him; it strikes him as being long-winded. Why couldn't she simply say she is interested in applied biology? What is this business about working with people? Is that there because she has doubts about biology?

If resumes are supposed to say only what needs to be said, what about this line?

Born January 6, 1969   5'7"   135 lbs.   Single   Excellent health

Does it say anything about her ability to do the job? The biologist doesn't think so.

His eyes move down the page:

NANCY O. JONES

Present Address
105 Belleville Place
Ames, Iowa 50011
Phone: 515-555-6674

After June 1, 1991
1212 Centerline Road
Old Westbury, New York 11568
Phone: 516-555-7664

Born January 6, 1969   5'7"   135 lbs.   Single   Excellent health

Career Objective	Research and development in most areas of applied biology, with an opportunity to work with people as well
Education	Iowa State University, Ames, Iowa Bachelor of Science, June 1991 Major: Biology Concentration: Physiology GPA: 3.3 on a 4.0 scale

Major Subjects
Mammalian Physiology
Vertebrate Anatomy
Histology
Genetics

Minor Subjects
Qualitative Analysis
Quantitative Analysis
Organic Chemistry
Biochemistry

Scholarships and Honors	University Scholarship: $2850/year Iowa State Science and Research Award Dean's List two semesters
Activities	Volunteer Probation Officer, 1988–89 Probation Department, Ames, Iowa Tutor, Chemistry and Math, 1988–91 Central High School, Ames, Iowa Member, Kappa Zeta social sorority Women's Intercollegiate Hockey Team
Special Skills	Familiarity with Spanish; PL/C and FORTRAN computer languages; typing
Work Experience	Teaching Assistant and Laboratory Instructor Freshman Biology School year, 1990–91 Waitress, Four Seasons Restaurant Catalina Island Summers, 1989–90

FIGURE 2 The resume of Nancy O. Jones does not communicate her career-related experience.

<u>Education</u>   Iowa State University, Ames, Iowa
                    Bachelor of Science, June 1991
                    Major: Biology   Concentration: Physiology

                    GPA: 3.3 on a 4.0 scale

                    <u>Major Subjects</u>   <u>Minor Subjects</u>
                    Mammalian Physiology   Qualitative Analysis
                    Vertebrate Anatomy   Quantitative Analysis
                    Histology   Organic Chemistry
                    Genetics   Biochemistry

She has used the outline form well, so he is able to take in a great deal of information in one look. Her education impresses him.

Double spacing above and below her grade point average makes it stand out. However, if her average had not been quite as good, and if she had not wanted to feature it, she could have used single spacing to make it less conspicuous, like this:

<u>Education</u>   Iowa State University, Ames, Iowa
                    Bachelor of Science, June 1991
                    Major: Biology   Concentration: Physiology
                    GPA: 2.7 on a 4.0 scale

Now that she has told him about her grades and her courses she wants to drive home the point that she was no ordinary student. She flags him down with the headline "Scholarships and Honors," and he sees her financial aid and science awards, which make a favorable impression. Two times on the dean's list may not be important enough to set apart by double spacing, but it makes a modest impression.

Up until now, Nancy has made a favorable impression overall. The director is ready to take in the next batch of information:

<u>Activities</u>   Volunteer Probation Officer, 1988–89
                    Probation Department, Ames, Iowa

                    Tutor, Chemistry and Math, 1988–91
                    Central High School, Ames, Iowa

                    Member, Kappa Zeta social sorority

                    Women's Intercollegiate Hockey Team

She almost loses him by featuring her work as a probation officer. That is not of primary interest to him, but through good spacing and placement, she draws his eye to the next item, which states that she has tutored chemistry and math. Unfortunately, Nancy now loses him permanently by ranking tutoring along with the sorority and the hockey team. He guesses she has made her major statement about biology, and so he turns to the next resume.

Good as it is, Nancy Jones's resume could be improved by reorganization. Her tutoring chemistry and her two years as a teaching assistant and laboratory instructor in biology reveal a more-than-academic interest in biology. They should be featured. Everything related to biology and any other information of possible use to the employer should be put under a new marginal headline and statement, as follows:

Career-Related Experience	Biology Lab Instructor and Teaching Assistant Freshman Biology (1990–91)
	Chemistry and Mathematics Tutor Central High School, Ames, Iowa (1988–91)

Skills and Interests	Microscopy	Computer Languages:
	Electron Microscopy	FORTRAN, PL/1,
	Histology	COBOL
	Spectrum Analysis	Statistics
	Small-Animal Surgery	

Now, the director is able to see the things she can do. "Good," he says to himself, "she can use an electron microscope. We need someone with that skill."

By no means should Nancy eliminate mention of the hockey team, but since she is applying for a professional job, the first bait to throw out is credentials; they testify to her ability to do the work. After that, what may sink the hook is how the employer sees her as a person. He may have picked out bits here and there that testify to her diligence, but the picking out depends on chance reading; it would be best not to leave anything to chance. With a slightly different presentation, she might ensure that he receives an impression of her diligence.

There are other areas she could strengthen, things she has underplayed or totally neglected to mention. Sticking the waitress job in down at the bottom of the page is almost an apology for it. She may also have had other jobs, such as baby-sitting, household work, or door-to-door sales, that she belittles in her mind and has not even mentioned. Mention of such things might make a good impression on an employer looking for somebody who is not afraid to work and who is mature for her years.

If we quizzed Nancy, we might find that she could put something like this on her resume:

Scholarships and Financial Support	90% self-supporting through college as follows:
	University Scholarship: $2850/year Iowa State Science and Research Award
	Waitress, Four Seasons Restaurant Catalina Island (Summers, 1989–90)
	Teaching and instructing, baby-sitting, home mainte- nance, selling

And this, because her activities tell something about her as a person:

Activities	Volunteer Probation Officer (1988–89) Kappa Zeta social sorority Women's Intercollegiate Hockey Team Skiing, sailing, singing, tennis

Double spacing has been cut down so as not to overemphasize the less important items, yet a string of other things not terribly important in themselves has been inserted to support the impression of an active, interesting person. Sometimes you want to leave an impression, at other times you want to emphasize a qualification. For example, Nancy wanted to feature her studies, and the way she brought them out by listing them in a column was good. If she had listed them like this, they would not have stood out:

> Major Subjects: Mammalian Physiology, Vertebrate Anatomy, Histology, Genetics
> Minor Subjects: Quantitative Analysis, Qualitative Analysis, Organic Chemistry, Biochemistry

In Figure 3, you will see the Nancy Jones resume as she might have revised it to emphasize the strong points that would have been of interest to that director who was looking for a resourceful assistant. The added emphasis might have made the difference that would have landed the job for her.

Nancy's revised resume is pure outline, devoid of prose. It works well for her. When she states that she has studied quantitative and qualitative analysis and knows PL/1 and FORTRAN, a scientist reading her resume knows what this means.

Janet Smith—The Proper Use of Headlines

Janet Smith, whose resume is shown in Figure 4, has a different problem in presenting her qualifications. She needs to *describe* what she did in order to tell an employer about their qualifications, so her resume calls for a mixture of key words and prose to get her message across. Yet prose can destroy the effect of an outline. The solution lies in imitating newspaper editors, who use headlines and subheadlines to attract readers. Like a newspaper, a resume should lend itself to skimming so the reader can quickly pick up a good overview of what is important. Then the reader can select specific things of interest and read further. There is an art to using headlines, but Janet Smith hasn't mastered it, at least not in the resume she used to apply for a job with Hermann Langfelder, a hardbitten old hand with thirty years in personnel and labor relations in the machinery business. He picked up her resume and read:

| CAREER OBJECTIVE | A challenging position in personnel administration requiring organizational ability and an understanding of how people function in business and industry. |

That was pure baloney, and he choked on it. He thought of the hours he had spent in meetings, listening to a lot of hot air. "Challenging, my foot!" he muttered. Then he read the bit about her organizational ability and her understanding of how people function in business and industry. "I've been at this business for thirty years," he groused to himself, "and I still can't figure out how people function in industry. But she knows all about it."

Beware of Misleading Headlines

"Well, let's see what she's done," he said to himself, and then his eyes fell on "UNIVERSAL METHODIST CHURCH." That did it! He didn't want to bring any do-gooder into his factory to preach. He rejected her. And the fault was hers, in using the headline.

NANCY O. JONES

Present Address
105 Belleville Place
Ames, Iowa 50011
Phone: 515-555-6674

After June 1, 1991
1212 Centerline Road
Old Westbury, New York 11568
Phone: 516-555-7664

Career Objective   Research and development

Education   Iowa State University, Ames, Iowa
Bachelor of Science, June 1991
Major: Biology   Concentration: Physiology

GPA: 3.3 on a 4.0 scale

Major Subjects   Minor subjects
Mammalian Physiology   Qualitative Analysis
Vertebrate Anatomy   Quantitative Analysis
Histology   Organic Chemistry
Genetics   Biochemistry

Career-Related   Biology Lab Instructor and Teaching Assistant
Experience   Freshman Biology (1990–91)

Chemistry and Mathematics Tutor
Central High School, Ames, Iowa (1988–91)

Skills and Interests   Microscopy   Computer Languages:
Electron Microscopy   FORTRAN, PL/1,
Histology   COBOL
Spectrum Analysis   Statistics
Small-Animal Surgery

Scholarships   90% self-supporting through college as follows:
and Financial
Support   University Scholarship: $2850/year
Iowa State Science and Research Award

Waitress, Four Seasons Restaurant
Catalina Island (Summers, 1989–90)

Teaching and instructing, baby-sitting, home
maintenance, selling

Activities   Volunteer Probation Officer (1988–89)
Kappa Zeta social sorority
Women's Intercollegiate Hockey Team
Skiing, sailing, singing, tennis

FIGURE 3   The revised resume of Nancy O. Jones effectively features her career-related experience in a separate section.

JANET V. SMITH
111 Main Street
North Hero, Vermont 05073
802-555-1234

CAREER
OBJECTIVE

A challenging position in personnel administration requiring organizational ability and an understanding of how people function in business and industry.

EDUCATION

PURDUE UNIVERSITY
Hammond, Indiana
Master of Industrial Relations   June 1991

SMITH COLLEGE
Northampton, Massachusetts
Bachelor of Arts, magna cum laude   June 1986

WORK
EXPERIENCE

UNIVERSAL METHODIST CHURCH
1 Central Square
New York, New York 10027
Assistant Personnel Officer. 1988–90
Responsible for interviewing applicants for clerical positions within the organization and for placing those who demonstrated appropriate skills, for accepting and dealing with employees' grievances, and for developing programs on career advancement.

CORTEN STEEL COMPANY
10 Lake Street
Akron, Ohio 44309
Assistant, Personnel Office. 1986–88
Responsible for all correspondence of Personnel Director and for interviewing some custodial applicants and referring them to appropriate supervisors for further interviewing.

BORG-WARNER, INC.
Ithaca, New York 14850
Assembly-line worker. Summer 1985
Assembled parts of specialized drive chains in company with thirty other men and women.

COMMUNITY
SERVICE

PLANNED PARENTHOOD
North Hero, Vermont 05073
Counselor. Summers, 1983–84
Explained various aspects of family planning and provided birth control information to clients of Planned Parenthood. Made referrals to other counselors and physicians where appropriate.

AUXILIARY
SKILLS

French: Fluent. Knowledge of office procedures.
Experience with mainframe computes and PCs.

FIGURE 4   The resume for Janet V. Smith misleads the reader with irrelevant headings.

Janet's job at the church was administrative, not ministerial, and the church didn't care whether she was Jewish, Catholic, or agnostic. Only in its ministerial work does the Methodist Church need Methodists. But people—Langfelder and the rest of us—respond to symbols and make snap judgments on the basis of symbols. Janet had put the symbol of the church—its name—in a heading, when she could have done something better.

When you lay out your resume, think of symbols. Imagine you are a newspaper editor who wants to put a story across. As an editor planning headlines, you must imagine yourself in the position of the reader. Ask yourself, "What words will catch the reader's eye? What word will put the reader off?"

Use words that fit the job in question, and play down those that can lead an employer to think of you in terms that don't relate to the job. Ask yourself, "Does this say something to the employer?" Janet Smith missed Langfelder, and some of her duties would have yielded key words to send appropriate messages. The following arrangement would have been more effective for her:

WORK
EXPERIENCE

ASSISTANT PERSONNEL OFFICER, 1988–90
Universal Methodist Church
1 Central Square
New York, New York 10027
Interviewing, placement, grievances, and training of applicants and employees in the clerical and support services of the church organization. Developed programs for the career advancement of employees.

ASSISTANT, PERSONNEL OFFICE, 1986–88
Corten Steel Company
10 Lake Street
Akron, Ohio 44309
Interviewing and referring as an assistant to Personnel Director. Responsible for all correspondence and for interviewing some custodial applicants and referring them to supervisors for further interviews.

Let's suppose that Janet has a chance to send Langfelder the revised resume shown in Figure 5. This time, she uses headlines that pinpoint the ideas she most wants to get across, so that he makes it past the Universal Methodist Church and gets down to the assembly line. "Hey, now, look at that!" he thinks. "She's worked out there on the floor. That means she's heard all the language and knows the gripes and the tedium. We have a lot of women in this factory, and it might be a good thing to have a down-to-earth, smart woman on my staff." (His sexism may have been showing, but this kind of employer is alive and kicking somewhere out there, and you may have to deal with him.)

What if Janet Smith wanted a job in the computer industry and had had ten years' experience with IBM? Employers are impressed by "graduates" of companies

JANET V. SMITH
111 Main Street
North Hero, Vermont 05073
802-555-1234

CAREER INTERESTS	Personnel Administration and Labor Relations
EDUCATION	PURDUE UNIVERSITY June 1991 Hammond, Indiana Master of Industrial Relations SMITH COLLEGE June 1986 Northampton, Massachusetts Bachelor of Arts, magna cum laude
WORK EXPERIENCE	ASSISTANT PERSONNEL OFFICER. 1988–90 Universal Methodist Church 1 Central Square New York, New York 10027 Interviewing, placement, grievances, and training of applicants and employees in the clerical and support services of the church organization. Developed programs for the career advancement of employees. ASSISTANT, PERSONNEL OFFICE. 1986–88 Corten Steel Company 10 Lake Street Akron, Ohio 44309 Interviewing and referring as an assistant to Personnel Director. Responsible for all correspondence and for interviewing some custodial applicants and referring them to supervisors for further interviews. ASSEMBLY-LINE WORKER. Summer 1985 Borg-Warner, Inc. Ithaca, New York 14850 Factory work experience on an assembly line. Worked with a team of thirty other men and women.
COMMUNITY SERVICE	COUNSELOR. Summers, 1983–84 Planned Parenthood North Hero, Vermont 05073 Explained various aspects of family planning and provided birth control information to clients of Planned Parenthood. Made referrals to other counselors and physicians where appropriate.
AUXILIARY SKILLS	French: Fluent. Knowledge of office procedures. Experience with mainframe computers and PCs.

FIGURE 5   The revised resume for Janet V. Smith stresses what she did rather than the less important point of where she did it.

like IBM that are known as leaders in their fields. Ten years with them is significant. It might make sense to present her experience this way:

EXPERIENCE 1981–91
IBM Corporation
Binghamton, New York
Assistant Personnel Officer

Were she an engineer after a technical job that could use her IBM experience, highlighting the name of the company would have been a good idea. Imagine employers giving the resume a 20-second scan. What words should you use and how should they be placed to catch the eye and make employers want to read further?

Mark Meyers—The Functional Resume

Janet Smith and Nancy Jones were lucky. Their training and experience translated into satisfactory headlines to highlight their experience, but that doesn't always work. Mark Meyers, whose resume appears in Figure 6, adopted a different technique to help him get a job in community recreation. He got his message across by creating a resume based on functions.

When he began to write his resume, he tried time and time again to get his message across in a conventional form in which he first listed his education, then his experiences in chronological order, and finally his activities, hobbies, and interests. But writing it conventionally raised all sorts of problems. He wanted to highlight his public relations and promotion experience, some of which he had been paid for and some not. Some of it had also been secondary to a primary assignment. Dividing up this experience and placing bits and pieces of it in various parts of the resume to make it conform to a conventional style diluted its impact. Also, his athletic ability and experience would mean a great deal to an employer in his field, but how could he show it effectively? Some of it had been gained as a participant, some through training, and some as a coach. Could he expect an employer to sift through the various sections of the resume to find out all he had done in athletics? (Remember, you can only count on an employer giving a resume a quick scan before deciding whether or not to study it more fully.)

His solution, as seen in Figure 6, was to feature the functions of the job he wanted and then describe things he had done that pertained to each area. Thus, under each function he developed the equivalent of a mini-resume.

Preparing a Resume for a Specific Job

Mark stated his case well, but you can't get blood from a stone. He found he had to look outside his field, because jobs in it were virtually nonexistent due to cutbacks in government funding. In his search, he ran across the following job listing from the publisher of a magazine for parents:

EDITORIAL SECRETARY

BA. in Liberal Arts
Interested in childhood training. Well organized, outstanding language skills. Typing and
clerical skills, potential to use electronic text-
editing equipment. Reporting to Coordinating
Editor, Happy Days magazine. Assist in all editorial functions. Evidence of creativity essential. Entry-level position with career potential.

Can you put yourself in his shoes, analyze the job, and devise a resume that
speaks to the stated needs of this employer? The clue to doing it is to go through
the job description and step by step take your cue from the employer. Right off,
you will hit a bit of a snag because the employer has specified a B.A., while Mark
has a B.S. His drama minor might give him appropriate credentials, however. If
he shows his education as in the following example it might reflect the liberal
background the employer apparently prefers:

EDUCATION   HOBOKEN UNIVERSITY, B.S., 1987
                                   Major: Recreation   Minor: Drama

                                   Humanities courses:
                                   Introduction to Dramatic Literature
                                   British Drama to 1700
                                   History of Theater
                                   Playwriting
                                   Introduction to Poetry
                                   Shakespeare
                                   English History

Next, the job calls for an interest in childhood training, then language skills,
and so on. Each specification suggests a headline for a resume. Mark faces
another stumbling block when it comes to demonstrating an interest in childhood training, since his experience has been with older youths only. Since a
resume makes points by stating facts, he cannot demonstrate an interest in childhood training in his resume because he lacks the appropriate experience. However, he *can* describe his interest in the letter that usually goes hand in glove with
a resume. He has a good basis for doing so, because the field of recreation certainly has much to do with the entire range of human development from childhood on. He should be able to point out correlatives in his education and
experience with the work being done in childhood training, and he could check a
library for information about childhood training to help develop the correlatives.
Above all, he should read the magazine and try to tie in as many of his experiences as possible with the purpose of the magazine.

With his interest in childhood training brought out in a letter to complement
his resume, he might then proceed to develop his outline as follows:

EXPERIENCE

HUMAN DEVELOPMENT

Dover Youth Bureau. Planned and implemented programs in drama, photography, athletics, health. Summer 1986

Research Project. Studied recreational needs of a Hoboken neighborhood. Interviewed residents, developed questionnaire. Project provided an insight into the family life and problems of parents in a neighborhood setting. Fall semester 1986

Outdoor Leadership Training. Practical experience in human development through a knowledge of nature and survival skills. Summer 1983.

LANGUAGE SKILLS AND CREATIVITY

Writing. Wrote releases, developed advertising, and prepared radio announcements for Drama Club, men's sports programs, and other events. Wrote report on Hoboken research project. 1984–87

Lecturing. Gave talks and led tours at Atlantic County (Delaware) Park nature trail and visitors' center. Summer 1985

Audiovisual. Prepared slides for audiovisual presentation for visitors' center. 1985.

Promotional. Created innovative techniques—such as costume parade of cast—to arouse interest in avant-garde staging of "Alice in Wonderland," which gained campuswide attention. 1984–85.

Design. Designed posters, fliers, and other graphics for sporting events, plays, and other campus events. 1984–87.

Constructed nature exhibits at nature center. 1985.

CLERICAL AND ADMINISTRATIVE

Clerical. Typing and use of office machinery. McGary's Department Store. Summer 1982.

Administration. Responsible for equipment, scheduling of programs, and coordinating of competitions. Hoboken University, 1984–86.

The functional resume allows you to develop a different message for each job or type of job you wish to apply for. Different functions can be highlighted, depending on what the job requires, and your specific experiences rearranged under different headings. It gives you the flexibility you need if your experience has been diverse.

Almost every resume ought to have something of the functional resume about it. With computers making it so easy to change a text, there is no reason why each resume can't be slanted to appeal to the particular employer, even if it's the resume of a generalist like Bruce Gregory Robinson, who hopes to get into a training program.

MARK MEYERS

1414 South Harp Road
Dover, Delaware 19901
302-555-4444

CAREER INTEREST
Community recreation.

RECREATION PROGRAMMING EXPERIENCE
Planned and implemented programs in stagecraft and drama; assisted with pro-
gramming in ceramics, photography, and physical fitness for Dover Youth
Bureau summer program. Summer 1986.
Lectured and led tours at Atlantic County (Delaware) Park nature trail and visi-
tors' center. Prepared slides to illustrate lecture; helped in construction of
nature exhibits. Summer 1985.
Coordinated men's intramural sports competitions for Hoboken University.
Had responsibility for equipment and scheduling. 1984–86.

PUBLIC RELATIONS AND PROMOTION EXPERIENCE
Directed publicity efforts of University Drama Club for several productions.
Developed innovative techniques—such as a costumed cast parade to arouse
interest in an avant-garde staging of "Alice in Wonderland", which gained
campuswide attention. 1984–85.
Advertised in various media and became familiar with advertising methods,
including writing news releases, taping radio announcements, designing
graphics for posters and fliers. Drama Club, 1985–87; men's sports,
1984–87.

LEADERSHIP AND ATHLETIC ABILITIES
Trained in outdoor leadership and survival skills by Outward Bound.
Summer 1983.
Coached hockey and basketball. Dover Youth Bureau, 1986.
Participate in hockey, swimming, backpacking.

RESEARCH AND EVALUATION ABILITIES
Prepared college research project on the recreational needs of residents of a
Hoboken neighborhood. Designed questionnaire to solicit residents' own
perceptions of their needs; interviewed residents and local officials. Fall
semester 1986.
Reported on effectiveness of Dover Youth Bureau programming. Summer 1986.

OFFICE SKILLS
Dealt with unhappy and irate customers at McGary's Department Store, Dover.
Worked at service desk, tracking down problems and rectifying errors.
Summer 1982.
Typing and use of office machinery.

EDUCATION
Bachelor of Science, June 1987   Major: Recreation
Hoboken University, Hoboken, New Jersey   Minor: Drama

FIGURE 6   The resume of Mark Meyers illustrates the functional style.

Bruce Gregory Robertson—A Resume Reflecting an Active Mind and Body

Put yourself in the chair of a merit employer, someone who is interested in candidates not so much for what they know as for what they can learn. Imagine yourself as a banker, a merchant, or a manufacturer. The position you have to fill can be learned easily on the job. What interests you are candidates' traits—their energy, intelligence, leadership qualities—the things that tell you the candidate can grow in your employ and eventually become an executive in the company. You are looking for a resume that reflects an active person with an active mind. Bruce Robertson is such a person, and his resume shown in Figure 7 has been designed accordingly.

Now make yourself a textbook publisher. Again, you are a merit employer and promote from within. Your company hires people with a good liberal education and little or no postgraduate experience. The men and women you hire will travel from college to college either soliciting manuscripts from professors or showing them texts that might fit their courses. Impressive candidates should have a high energy level and be able to show that they have used time efficiently and worked independently. How does Bruce look to you?

Now change the scenario slightly. This time you are a publisher of texts for primary and secondary school. To get your texts adopted, your representatives will have to make presentations to state regents, school superintendents, and other educational groups. Remember that experience and college major don't matter, but do you think it might help Bruce's cause if he did a bit more to highlight his interest in public speaking and his honors as a debater? He has a computer. It would be a cinch for him to change his resume slightly. Can you think how he might do it?

Next, put yourself in the chair of an employer on Main Street. You need someone in customer relations, a person to follow up on complaints and misunderstandings about credit arrangements or bills. You want to hire someone who has had experience of this kind. Still, the work easily could be learned on the job . . . if the right person came along. In response to your classified advertisement, Bruce's resume arrives accompanied by a routine cover letter. He recounts people problems he has resolved and describes his administrative experience in a way that indicates he is a team worker. What's your decision? What if his letter is more inspired?

Michelle Trio—The Curriculum Vitae

A curriculum vitae (literally, "course of life" in Latin), sometimes called a C.V. or vita, is a resume for academic positions and as such does not need a statement of goals or interest. While there is merit in keeping nonacademic resumes brief by focusing on employers' needs, a faculty tends to select colleagues not just to teach but for the prestige they will bring to the department, especially in the long run. An eminent faculty attracts eminent associates. Publications, research, memberships, and honors all contribute to telling what a candidate is like; hence

BRUCE GREGORY ROBERTSON

Home Address
105 Comstock Drive
Pierre, South Dakota 57501
605-555-5445

College Address
5 Erasmus Drive
St. Paul, Minnesota 55101
612-555-4554

CAREER INTERESTS	— The marketing of products or services in industries such as banking, publishing, and retailing.
EDUCATIONAL BACKGROUND	— Macalester College, St. Paul, Minnesota A.B., May 1991
	Major: English Minor: Economics
	Honors: Dean's List, 1990–91
EMPLOYMENT	— Entrepreneurial—Organized a company to paint road markings in parking lots; rented necessary equipment; employed two additional workers; had personal contact with mall managers, store owners, etc., to promote the company's services: summers, 1989–90.
	— Administrative and Clerical—Temporary Help, Inc. Typing and other business machine operations, bookkeeping, clerking, complaint adjustments; short-term assignments with auto dealers, banks, real estate operations, schools, and similar employers: summer 1988; part-time 1989.
	— Miscellaneous—Camp counselor, stock clerk in grocery store, baby-sitter, newspaper boy. From high school on have earned money for clothes, travel, and purchase and maintenance of an automobile.
ACTIVITIES	— Vice President, Seven-Come-Eleven investment club
	— Varsity basketball
	— Coach and tutor, St. Paul Concordia Boys Club
	— Debate Club
	— Book and Bottle literary club
SKILLS	— Typing
	— PL/C and FORTRAN computer languages
	— German: fluent
HONORS	— Dakota Interstate Scholarship
	— Hubert H. Humphrey First Prize, Minnesota Intercollegiate Debate
INTERESTS	— Public speaking and debate
	— Investments
	— Writing
	— Parachute jumping

FIGURE 7 A resume designed more to reflect an active, energetic personality than specific experience.

long vitae that reflect many achievements are traditional. The same candidate applying for an industrial position wouldn't list a raft of publications, for example, if those publications didn't relate to the job in question.

The C.V. that is shown in Figure 8 lays out in a logical way all the essentials of a good vita. Note the correct way to list publications. In academia, those who list them incorrectly are jeopardizing their chances of being hired.

The Job Objective

"Do I have to have a job objective?" According to my calculations, as of this writing I have been a career counselor for 2,288 weeks, and I have been asked this question at least six times a week, except for the 132 weeks when I was on vacation. When I answer, "No, I don't like the heading JOB OBJECTIVE," the sigh of relief is audible. The job seekers think I have let them off the hook for one of the most important parts of a resume. I haven't. With rare exceptions, a resume *should* open with an objective—it's the way it is stated that can be changed.

I show them the preceding Robertson, Meyers, and Smith resumes and tell them that I prefer the headline CAREER INTEREST, because it leads to a simple and direct way of stating the purpose of the resume. For example, I suggest they try rewriting the Robertson and revised Smith statements of interest as job objectives to see if they don't find it awkward. After a struggle they come up with the same kind of baloney found in the unrevised Jones and Smith resumes.

They listen politely. I may have helped them with a minor problem of phrasing, but I know that I really haven't dealt with their question. Then they tell me what I already know: They don't want to state a goal because they don't know what they want to do. I ask them to imagine themselves as the employers reading their resumes. If theirs is like the revised Nancy Jones resume, which is so obviously slanted toward biology, then employers can figure out what the resume is for without a stated objective. However, I wouldn't fool with a resume that didn't tell me at the outset what it was all about. When I have to start studying a resume to guess what the writer wants, I throw it in the wastebasket.

Next I point to the Zarote and Robertson resumes, which wouldn't make any sense without a statement of purpose. And I can't get up much enthusiasm for a letter as a substitute for an objective. A letter stating a purpose accompanied by a resume without purpose is a wasted letter.

I remember a young woman who, on short notice, got a chance to be interviewed by a recruiter from a large department store. Knowing very little about merchandising, she headed for the nearest department store, where several managers were kind enough to give her information and the loan of their trade journals. In three hours of investigation, her eyes were opened to an industry in which people pursued careers in training, employee relations, promotion, credit, public relations, merchandising, and other occupations. Several of these looked interesting, so thanks to word processing, she easily changed a few things on her resume, then listed her objective like this:

CAREER INTERESTS: Training, promotion, and public relations in a retail setting.

MICHELLE TRIO

Department of English
Athens College
Athens, NY 14850
607-555-4567

406 East Bates Street
Athens, NY 14850
607-555-7654

EDUCATION

Ph.D., 1984, Cornwall University.

Dissertation: "The Anatomy of Sin: Violations of *Kynde* and *Trawþe* in
                  Cleanness," directed by L. E. Cooper (DAI 40/09, p. 5046-A).
Major Subject: Old and Middle English language and literature.
Minor Subject: Medieval philology.
Courses: Old English; *Beowulf;* seminars on the Junius
                  Manuscript, the Exeter Book, and hagiography;
                  Middle English literature; Chaucer; *Piers Plowman;*
                  Medieval Latin; paleography; Old French; Middle High
                  German; Old Icelandic; Dante.

A.M., 1979, with High Honors, Boston University.

B.A., 1977, *magna cum laude,* Honors in English, State University of New York at Stony
Brook.

PROFESSIONAL EXPERIENCE

1985–present: Assistant Professor, Medieval and Renaissance Literature, Department of
English, Athens College. Position includes: Engl 325 Chaucer, Engl 323 Triumph of
English (a course that I instituted on the history of English), Engl 232 Medieval
Literature, Engl 420 Shakespeare Seminar, Engl 231 Ancient Literature, Engl 107
Introduction to Literature: Myth, Legend, and Folktale.

1985: Assistant Professor, Department of English, Cornwall University Summer Program.
One section of Practical Prose and Composition and training of a graduate teaching
assistant.

1984–85: Lecturer, Department of English, Cornwall University. Two courses in
Freshman Seminar Program: Shakespeare and Politics, Practical Prose Composition. I
was Co-director of the latter, with responsibility for planning the syllabus, training and
evaluating graduate teaching assistants, and leading staff meetings on problems and
goals of teaching composition.

PUBLICATIONS

"Heroic Kingship and Just War in the Alliterative *Morte Arthure*," to be published in
Acta, 11.

Articles on the Beatitudes, the Handwriting on the Wall, the Parable of the Marriage
Feast, Sarah, and Sodom and Gomorrah in *Dictionary of Biblical Tradition in English
Literature,* ed. David L. Jeffrey, to be published by Oxford University Press and W. B.
Eerdmans.

"On Reading *Bede's Death Song:* Translation, Typology, and Penance in Symeon of
Durham's Text of the *Epistola Cuthbert: de Obitu Bedae.*" *Neuphilologische Mitteilungen,*
84 (1983), 171–81.

FIGURE 8   A Curriculum Vitae

PROFESSIONAL SERVICE AND ACTIVITIES

1. At Athens College
   1989–90: English Department Library Representative.
   1989–90: Member, Athens College Faculty Enrichment Committee.
   1988–90: English Department Personnel Committee.
   1987–89: English Department Committee on the London Center.

2. Elsewhere
   1985: Organizer and Chair, Latin Section, Northeast Modern Language Association. Topic: "Eschatology and Apocalypticism."
   1980–82: Organizer and Chair, *Quodlibet: The Cornwall Medieval Forum.*
   Memberships in MLA, Medieval Academy of America, International Arthurian Society.

HONORS

Charles A. Dana Fellowship for Excellence in Teaching, Athens College.

Goethe Prize in German Literature, Cornwall University.

George Lincoln Fellowship in Medieval Studies, Cornwall University.

Teaching Assistantship, Medieval Studies program, Cornwall University ("Medieval Literature in Translation").

Teaching Fellowship, English Department, Boston University ("Freshman Rhetoric and Composition").

LANGUAGES

Reading knowledge of Latin, Old French, Spanish, Italian, Middle High German, and Old Icelandic, in addition to the usual French, German, and Old and Middle English.

CREDENTIALS

Dossier may be obtained from the Educational Placement Bureau, Barnes Hall, Cornwall University, Athens, NY 14853.

FIGURE 8 *(Continued)*

Like most of us, her range of aptitudes was wide, so, like a chameleon, she showed the recruiter only those that seemed to match retailing.

One Page or Two?

If a resume can be kept to one page, so much the better. The length depends on the message. In reading resumes in which everything is jammed on one page with none of the white space or headings that can make them attractive and readable, I have wearied of trying to find information and have given up. On the other hand, it has never been the least bit tiring to lift a piece of paper and turn to a second page when scanning an interesting resume. A resume is an outline. It needs a white space. It needs headings that stand out. Don't sacrifice them for some arbitrary notion about a one-page maximum.

Additional Advice About Resumes

No matter how you develop your message, test it before you send it to employers. Get friends to give you a critique of your resume or vita, especially if they are in an occupation in which you hope to find a job. A word of caution, however. Unless you guide them, their critique may relate more to the ritual than to the message. One job hunter had modeled a resume after Janet Smith's. The critic took a red pencil and put all the dates of employment in the left-hand margin. That would have been a good idea if it had indicated long years of experience with a company such as 3M, signifying considerable experience with one of the best-managed companies in the country. But the dates in question referred to short-term summer jobs that were of no consequence to the message and cluttered the margin with information that distracted from the headlines.

One way to get a resume criticized is to hold it up a few feet from the reader and ask for comments on its appearance. Does it look neat? Is the layout pleasing? Does it look easy to read? Is the print good-looking?

Next, give the critics the resume to read. Let them make all the comments they want. You may pick up valuable ideas for improving its style and layout, but be careful you don't get caught up in inconsequentials. What you really want is to have your critics look at the resume as if they didn't know you. You might even show them a resume with an alias, then ask:

- What qualifications does this person have?
- What do you see this person doing with these qualifications?
- What kind of an employer would want to hire this person?
- Does the resume project an image of a certain kind of person? What kind? Aggressive? Thoughtful? Energetic? What?

In other words, ask your critics the most important question about your resume: "What message do you get about me?"

Steven Graber

The Basics of a Cover Letter

Steven Graber is the former managing editor of the JobBank *series and the* Adams Almanac *series.*

Your cover letter, like your resume, is a marketing tool. Too many cover letters are merely an additional piece of paper accompanying a resume, saying "Enclosed please find my resume." Like effective advertisements, effective cover letters attract an employer's attention by highlighting the most attractive features of the product. Begin by learning how to create an effective sales pitch. As with resumes, both the format and the content of your cover letter are important.

FORMAT

Before reading a word of your cover letter, a potential employer has already made an assessment of your organizational skills and attention to detail simply by observing its appearance. How your correspondence looks to a reader can mean the difference between serious consideration and dismissal. You can't afford to settle for a less than perfect presentation of your credentials. This chapter outlines the basic format you should follow when writing a cover letter and shows you how to put the finishing touches on a top-notch product.

The Parts of a Letter

Your cover letter may be printed on the highest-quality paper and typed on a state-of-the-art computer, but if it isn't arranged according to the proper format, you won't come across as a credible candidate. Certain guidelines apply when composing any letter.

Either of two styles may be used for cover letters: business style (sometimes called block style) or personal style. The only difference between them is that in business style, all the elements of the letter—the return address, salutation, body, and complimentary close—begin at the left margin. In personal style, the return address and complimentary close begin at the centerline of the page, and paragraphs are indented.

Return Address

Your return address should appear at the top margin, without your name, either flush left or beginning at the centerline, depending on whether you're using business style or personal style. As a rule, avoid abbreviations in the addresses of your cover letter, although abbreviating the state is acceptable. Include your phone number if you're not using letterhead that contains it or it doesn't appear in the last paragraph of the letter. The idea is to make sure contact information is on both the letter and the resume, in case they get separated in the hiring manager's office (this happens more often than you would expect!).

Date

The date appears two lines below your return address, either flush left or centered, depending on which style you're using. Write out the date; don't abbreviate. *Example:* October 12, 2000.

Inside Address

Four lines beneath the date, give the addressee's full name. On subsequent lines, give the person's title, the company's name, and the company's address. Occasionally, the person's full title or the company's name and address will be very long and can appear awkward on the usual number of lines. In this case, you can use an extra line.

The text of the letter below the date should be centered approximately vertically on the page, so if your letter is short, you can begin the inside address six or even eight lines down. If the letter is long, two lines is acceptable.

Salutation

The salutation should be typed two lines beneath the company's address. It should begin "Dear Mr." or "Dear Ms." followed by the individual's last name and a colon. Even if you've previously spoken with an addressee who has asked to be called by his or her first name, never use a first name in the salutation. In some cases, as when responding to "blind" advertisements, a general salutation may be necessary. In such circumstances, "Dear Sir or Madam" is appropriate, followed by a colon.

Length

Three or four short paragraphs on one page is ideal. A longer letter may not be read.

Enclosure

An enclosure line is used primarily in formal or official correspondence. It's not wrong to include it in a cover letter, but it's unnecessary.

Paper Size

As with your resume, use standard 8½- by 11-inch paper. A smaller size will appear more personal than professional and is easily lost in an employer's files; a larger size will look awkward and may be discarded for not fitting with other documents.

Paper Color and Quality

Use quality paper that is standard 8½ by11 inches and has weight and texture, in a conservative color like white or ivory. Good resume paper is easy to find at stores that sell stationery or office products and is even available at some drugstores. Use *matching* paper and envelopes for both your resume and cover letter. One hiring manager at a major magazine throws out all resumes that arrive on paper that differs in color from the envelope!

Do not buy paper with images of clouds and rainbows in the background or anything that looks like casual stationery you would send your favorite aunt. Do not spray perfume or cologne on your cover letter. Also, never use the stationery of your current employer.

Typing and Printing

Your best bet is to use a word processing program on a computer with a letter-quality printer. Handwritten letters are not acceptable. You will generally want to use the same typeface and size that you used on your resume. Remember that serif typefaces are generally easier to read.

Don't try the cheap and easy ways, like photocopying the body of your letter and typing in the inside address and salutation. Such letters will not be taken seriously.

Envelope

Mail your cover letter and resume in a standard, business-sized envelope that matches your stationery. Unless your handwriting is *extremely* neat and easy to read, type your envelopes. Address your envelope, by full name and title, specifically to the contact person you identified in your cover letter.

CONTENT

Personalize Each Letter

If you are *not* responding to a job posting that specifies a contact name, try to determine the appropriate person to whom you should address your cover letter. (In general, the more influential the person, the better.) Try to contact the head of the department in which you're interested. This will be easiest in mid-sized and small companies, where the head of the department is likely to have an active role in the initial screening. If you're applying to a larger corporation, your application will probably be screened by the human resources department. If you're instructed to direct your inquiry to this division, try to find out the name of the senior human resources manager. This may cut down on the number of hands through which your resume passes on its way to the final decision-maker. At any rate, be sure to include your contact's name and title on both your letter and the envelope. This way, even if a new person occupies the position, your letter should get through.

Mapping It Out

A cover letter need not be longer than three or four paragraphs. Two of them, the first and last, can be as short as one sentence. The idea of the cover letter is not to repeat what's in the resume. The idea is to give an overview of your capabilities and show why you're a good candidate for the job. The best way to distinguish yourself is to highlight one or two of your accomplishments or abilities. Stressing only one or two increases your chances of being remembered.

Be sure it's clear from your letter why you have an interest in the company—*so many candidates apply for jobs with no apparent knowledge of what the company does!* This conveys the message that they just want any job. Indicating an interest doesn't mean you should tell every employer you have a burning desire to work at that company, because these statements are easy to make and invariably sound insincere. Indicating how your qualifications or experience meet their requirements may be sufficient to show why you're applying.

First paragraph. State the position for which you're applying. If you're responding to an ad or listing, mention the source. *Example:* "I would like to apply for the position of research assistant advertised in the *Sunday Planet*" (or "listed on the Internet").

Second paragraph. Indicate what you could contribute to this company and show how your qualifications will benefit them. If you're responding to an ad or listing, discuss how your skills relate to the job's requirements. Don't talk about what you can't do. Remember, keep it brief! *Example:* "In addition to my strong background in mathematics, I offer significant business experience, having worked in a data processing firm, a bookstore, and a restaurant. I am sure that my courses in statistics and computer programming would prove particularly useful in the position of trainee."

Third paragraph. If possible, show how you not only meet but exceed their requirements—why you're not just an average candidate but a superior one. Mention any noteworthy accomplishments, high-profile projects, instances where you went above and beyond the call of duty, or awards you've received for your work. If you have testimonials, commendations or evaluations that are particularly complimentary, you may want to quote a sentence from one or two of them. *Example:* "In a letter to me, Dewayne Berry, president of NICAP Inc., said, 'Your ideas were instrumental to our success with this project.'"

Fourth paragraph. Close by saying you look forward to hearing from them. If you wish, you can also thank them for their consideration. Don't ask for an interview. If they're interested, they'll call. If not, asking won't help. Don't tell them you'll call them—many ads say "No phone calls." If you haven't heard anything in one or two weeks, a call is acceptable.

Complimentary close. The complimentary close should be two lines beneath the body of the letter, aligned with your return address and the date; Keep it simple—"Sincerely" followed by a comma, suffices. Three lines under this, type your full name as it appears on your resume. Sign above your typed name in black ink.

Don't forget to sign the letter! As silly as it sounds, people often forget this seemingly obvious detail. An oversight like this suggests you don't take care with your work. To avoid this implication if you're faxing the letter and resume directly from your computer, you can type your name directly below the complimentary close, without any intervening space. Then follow up with a hard copy of the resume and the signed letter, with your name typed in the traditional place under the signature.

TIPS FOR SUCCESSFUL COVER LETTERS

What Writing Style Is Appropriate?

Adopt a polite, formal style that balances your confidence in yourself with respect for the employer. Keep the style clear, objective, and persuasive rather than narrative. Don't waste space boasting instead of presenting relevant qualifications.

Example: "In addition to a Bachelor of Arts degree in Business Administration, I recently received a Master's, *cum laude,* in International Marketing from Brown University. This educational experience is supported by two years' part-time experience with J&D Products, where my marketing efforts resulted in increased annual product sales of 25 percent."

Tone: Reserved Confidence Is Always in Style

Think of how you'd sell your qualifications in a job interview. You'd probably think harder about what to say and how to say it than in an informal conversation. Above all, you'd want to sound polite, confident, and professional. Adopt a similar tone in your cover letter. It should immediately communicate confidence in your abilities. The trick is to sound enthusiastic without becoming melodramatic. Take, for example, the candidate who expressed his desire to enter the advertising field as "the single most important thing I have ever wanted in my entire twenty-three years of existence." The candidate who was actually offered the position began her letter as follows: "My extensive research into the industry, coupled with my internship and education, have confirmed my interest in pursuing an entry-level position in advertising."

Emphasize Concrete Examples

Your resume details the duties you've performed in your jobs. In contrast, your cover letter should highlight your most significant accomplishments. Instead of stating something like "My career is highlighted by several major achievements," use concrete examples:

"While Sales Manager at Shayko Chicken, I supervised a team that increased revenues by 35 percent in 18 months."

"I published four articles in *The Magical Bullet Newsletter*."

"At MUFON Corporation, I advanced from telephone fundraiser to field manager to canvassing director within two years."

List tangible, relevant skills rather than personal attributes. A sentence like "I am fluent in C++, Pascal, and COBOL" is a good substitute for a vague statement like "I am a goal-oriented, highly skilled computer programmer." Avoid using "etc."—don't expect a potential employer to imagine what else you mean. Either describe it or leave it out.

Use Powerful Language

Your language should be hard-hitting and easy to understand. Your message should be expressed using the fewest words possible. As with your resume, make your letters interesting by using action verbs like "designed," "implemented," and "increased," rather than passive verbs like "was" and "did." Use simple, common language and avoid abbreviations and slang. Change "Responsible for directing" to "Directed" if appropriate. Also steer clear of language that's too technical or jargon-heavy. The first person who reads your cover letter may not possess the same breadth of knowledge as your future boss.

Avoid Catchphrases

In the course of a job search, it's tempting to use catch phrases you've picked up from advertisements or reference materials, phrases that sound as though they *should* go in a resume or cover letter. Many people are tempted to reach for expressions like "self-starter," "excellent interpersonal skills," and "work well independently or as part of a team."

Improve on these descriptions by listing actual projects and goals. For example, rephrase "Determined achiever with proven leadership skills" as follows: "Supervised staff of fifteen and increased the number of projects completed before deadline by 10 percent." Once you begin working, employers will discover your personal attributes for themselves. While you're under consideration, concrete experiences are more valuable than vague phrases or obscure promises.

Mention Personal Preferences?

Candidates often worry if, and how, they should include salary requirements and availability to travel or relocate. Refrain from offering salary information unless the advertisement you are responding to requires it. If you must include salary requirements, give a salary range rather than a number. Another option is to simply indicate that salary concerns are negotiable.

If you're applying to an out-of-state firm, indicate a willingness to relocate; otherwise, a hiring manager may question your purpose in writing and may not take the initiative to inquire.

Proof with Care

Mistakes on resumes and cover letters are not only embarrassing, they will often remove you from consideration (particularly if something obvious, like your name, is misspelled). No matter how much you paid someone else to type, write, or typeset your resume or cover letter, *you* lose if there is a mistake. So proofread it as carefully as possible. Get a friend to help you. Read your draft aloud as your friend checks the proof copy. Then have your friend read aloud while you check. Next, read it letter by letter to check spelling and punctuation.

If you're having it typed or typeset by a resume service or a printer and you don't have time to proof it, pay for it and take it home. Proof it there and bring it back later to get it corrected and printed.

If you wrote your cover letter with a word processing program, use the built-in spell checker to double-check for spelling errors. Keep in mind that a spell checker will not find errors like "to" for "two" or "wok" for "work." Many spell-check programs don't recognize missing or misused punctuation, nor are they set to check the spelling of capitalized words. It's important to still proofread your cover letter for grammatical mistakes and other problems, even after it's been spell-checked.

If you find mistakes, do not fix them with pen, pencil, or white-out! Make the changes on the computer and print out the letter again.

COVER LETTER BLUNDERS TO AVOID

The following discussion focuses on examples that have been adapted from real-life cover letters. Although some of these blunders may seem obvious, they occur far more often than one might think. Needless to say, none of the inquiries that included these mistakes met with positive results.

Unrelated Career Goals

Tailor your cover letter to the position you're applying for. A hiring manager is only interested in what you can do for the company, not what you hope to accomplish for yourself. Convey a genuine interest in the position and a long-term pledge to fulfilling its duties.

Example A (wrong way): "While my true goal is to become a professional dancer, I am exploring the option of taking on proofreading work while continuing to train for the Boston Ballet's next audition."

Example B (right way): "I am very interested in this proofreading position, and I am confident of my ability to make a long-term contribution to your capable staff."

Comparisons and Clichés

Avoid clichés and obvious comparisons. These expressions detract from your letter's purpose: to highlight your most impressive skills and accomplishments.

Examples of what not to do:

"My word processor runs like the wind."

"I am a people person."

"Teamwork is my middle name."

"Your company is known as the crème de la crème of accounting firms."

"I am as smart as a whip."

"Among the responses you receive for this position, I hope my qualifications make me leader of the pack."

Wasted Space

Since cover letters are generally four paragraphs long, every word of every sentence should be directly related to your purpose for writing. In other words, if you are applying for a position as a chemist, include only those skills and experiences most applicable to that field. Any other information weakens your application.

Examples of what not to do:

"As my enclosed resume reveals, I possess the technical experience and educational background to succeed as your newest civil engineer. In addition, I am a certified gymnastics instructor who has won several local competitions."

"I am writing in response to your advertisement for an accounting clerk. Currently, I am finishing an associate degree at Peacock Junior College. My courses have included medieval architecture, film theory, basic home surgery, and nutrition."

Form Letters

Mass mailings, in which you send a form letter to a large number of employers, are not recommended. This approach doesn't allow you to personalize each application. Every cover letter you write should be tailored to the position you're seeking and should demonstrate your commitment to a specific industry and familiarity with each employer. Mass mailings may indicate to a hiring manager that you're not truly interested in joining that organization.

Inappropriate Stationery

White and ivory are the only acceptable paper colors for a cover letter. Also, don't rely on graphics to "improve" your cover letter; let your qualifications speak for themselves. If you're a cat enthusiast, don't use stationery with images of favorite felines. If you're a musician, don't send a letter decorated with a border of musical notes and instruments.

"Amusing" Anecdotes

Imagine yourself in an interview setting. Since you don't know your interviewer, you wouldn't joke with him or her until you determined what demeanor was appropriate. Similarly, when writing, remain polite and professional.

Erroneous Company Information

If you were the employer, would you want to hire a candidate who confuses your company's products and services or misquotes recent activities? To avoid such errors, verify the accuracy of any company information you mention in your cover letter. On the other hand, if you haven't researched the company, don't bluff. Statements like "I know something about your company" or "I am familiar with your products" signal to an employer that you haven't done your homework.

Desperation

In your cover letter, sound determined, not desperate. While an employer appreciates enthusiasm, he or she may be turned off by a desperate plea for employment. However, a fine line often separates the two.
   Examples of what not to do:
   "I am desperately eager to start, as I have been out of work for six months."
   "Please call today! I'll be waiting by the phone."
   "I really, really need this job to pay off medical bills."
   "I AM VERY BADLY IN NEED OF MONEY!"

Personal Photos

Unless you're seeking employment in modeling, acting, or other performing arts, it's inappropriate to send a photograph.

Confessed Shortcomings

Some job seekers mistakenly call attention to their weaknesses in their cover letters, hoping to ward off an employer's objections. This is a mistake, because the letter emphasizes your flaws rather than your strengths.
   Examples of what not to do:

"Although I have no related experience, I remain very interested in the management consultant position."

"I may not be well qualified for this position, but it has always been my dream to work in the publishing field."

Misrepresentation

In any stage of the job-search process, never, *ever,* misrepresent yourself. In many companies, erroneous information contained in a cover letter or resume will be grounds for dismissal if the inaccuracy is discovered. Protect yourself by sticking to the facts. You're selling your skills and accomplishments in your cover letter. If you achieve something, say so, and put it in the best possible light. Don't hold back or be modest—no one else will. At the same time, don't exaggerate to the point of misrepresentation.

Examples of what not to do:

"In June, I graduated with honors from American University. In the course of my studies, I played two varsity sports while concurrently holding five jobs."

"Since beginning my career four years ago, I have won hundreds of competitions and awards and am considered by many to be the best hairstylist on the east coast."

Demanding Statements

Your cover letter should demonstrate what you can do for an employer, not what he or she can do for you. For example, instead of stating "I am looking for a unique opportunity in which I will be adequately challenged and compensated," say "I am confident I could make a significant contribution to your organization, specifically by expanding your customer base in the northwest and instituting a discount offer for new accounts." Also, since you're requesting an employer's consideration, your letter shouldn't include personal preferences or demands. Statements like "It would be an overwhelmingly smart idea for you to hire me" or "Let's meet next Wednesday at 4:00 P.M., when I will be available to discuss my candidacy further" come across as presumptuous. Job candidates' demands are rarely met with an enthusiastic response.

Missing Resume

Have you ever forgotten to enclose all the materials you refer to in your cover letter? This is a fatal oversight. No employer is going to take the time to remind you of your mistake; he or she has already moved on to the next application.

Personal Information

Do not include your age, health, physical characteristics, marital status, race, religion, political/moral beliefs, or any other personal information. List your personal interests and hobbies only if they're directly relevant to the type of job you're seeking. If you're applying to a company that greatly values teamwork, for instance, citing that you organized a community fundraiser or played on a basketball team may be advantageous. When in doubt, however, leave it out.

Choice of Pronouns

Your cover letter necessarily requires a thorough discussion of your qualifications. Although some applicants might choose the third person ("he or she") as a creative approach to presenting their qualifications, potential employers sometimes find this disconcerting. In general, using the first person ("I") is preferable.

Example A (wrong way): "Bambi Berenbeam is a highly qualified public relations executive with over seven years of relevant experience in the field. She possesses strong verbal and written communication skills, and has an extensive client base."

Example B (right way): "I am a highly qualified public relations executive with over seven years of relevant experience in the field. I possess strong verbal and written communication skills and have an extensive client base."

Tone Trouble

Tone problems are subtle and may be hard to detect. When reading your cover letter, patrol for tone problems by asking yourself, after each sentence, "Does this statement enhance my candidacy? Could a hiring manager interpret it in an unfavorable way?" Have a second reader review your letter. If the letter's wording is questionable, rewrite it. A cover letter should steer a middle course between extremely formal, which can come across as pretentious, and extremely informal, which can come across as presumptuous. Try to sound genuine, not stilted. When in doubt, err on the side of formality.

Gimmicks

Gimmicks like sending a home video or a singing telegram to replace the conventional cover letter may seem attractive. No matter how creative these ideas may sound, the majority of employers will be more impressed with a simple, well-crafted letter. In the worst-case scenario, gimmicks can even work against you, eliminating you from consideration. Examples include sending a poster-sized cover letter by courier service or a baseball hat with a note attached: "I'm throwing my hat into the ring!" Avoid such big risks; most hiring decisions are based on qualifications, not gimmicks.

Typographical Errors

It's easy to make mistakes in your letters, particularly when you're writing many in succession. But it's also easy for a hiring manager to reject any cover letter that contains errors, even those that seem minor. Don't make the mistake that one job-hunting editor made, citing his attention to detail while misspelling his own name! Here are a few common technical mistakes to watch out for when proof-reading your letter:

Misspelling the hiring contact's name or title in the address or salutation or on the envelope.

Forgetting to change the name of the organization you're applying to each time it appears in your application, especially in the body of the letter. For example, if you're applying to Boots and Bags, don't express enthusiasm for a position at Shoe City.

Indicating application for one position and mentioning a different position in the body of the letter. For instance, one candidate applying for a telemarketing position included the following statement: "I possess fifteen years experience related to the marketing analyst opening." Another mistake here is that the applicant didn't use "years" as a possessive: ". . . fifteen years' experience. . . ."

Messy Corrections

Your cover letter should contain *all* pertinent information. If, for any reason, you forget to communicate something to your addressee, retype the letter. Including a supplementary note, either typed or handwritten, will be viewed as unprofessional or, worse, lazy. For example, one candidate attached a "post-it" note to his cover letter, stating his willingness to travel and/or relocate. This and all other information must be included in your final draft. Also, avoid using correction fluid or penning in any corrections.

Omitted Signature

However obvious this may sound, don't forget to sign your name neatly in blue or black ink. Far too many letters have a typed name but no signature. Also, don't use a script font or a draw program on your word processor.

COVER LETTERS FOR SPECIAL SITUATIONS

Writing a cover letter can seem like an even more formidable task when you find yourself in what we call "special situations." Perhaps you lack paid job experience, have been out of the workplace to raise children, are concerned about possible discrimination due to age or disability, or are trying to enter a field in which you have no practical experience. The key to improving your cover letter in these special situations is to emphasize your strengths. Focus on your marketable skills (whether they were acquired in the workplace or elsewhere), and highlight impressive achievements, relevant education and training, and/or related interests. And, of course, you should take care to downplay or eliminate any information that may be construed as a weakness.

For example, if you're a "displaced homemaker" (a homemaker entering the job market for the first time), you can structure your cover letter to highlight the special skills you've acquired over the years while downplaying your lack of paid experience. If you're an older job candidate, use your age as a selling point. Emphasize the depth of your experience, your maturity, your sense of responsibility, and your positive outlook. Changing careers? Instead of focusing on your job history, emphasize the marketable skills you've acquired that are considered valuable in the position you're seeking. For example, let's say your career has been real estate and, in your spare time, you like to run marathons. Recently, you heard about an opening in the sales and marketing department at an athletic shoe manufacturer. What you need to do is emphasize the skills you have that the employer is looking for. Not only do you have strong sales experience, you're familiar with the needs of the company's market, and that's a powerful combination!

RESPONSE TO A "BLIND" ADVERTISEMENT

A form of classified advertisements, "Blind" advertisements do not list employer information and generally direct inquiries to a post office box rather than a company's address. Since you're not provided with a company name in a blind ad, your cover letter should sharply define your knowledge of the industry, position (if mentioned) and how your qualifications specifically match up to the stated requirements. In other words, tailor your letter to any information given. For example, consider a blind ad that reads:

> *Large-size law firm in need of paralegal with experience in legal research, writing briefs, and office administration.*

You need to target everything in your response: what you know of the operations of large-size firms; why you want to be and remain a paralegal; how much experience you have in legal research and writing; and exactly what office skills you have. Avoid longwinded passages that don't follow these guidelines. Without knowing your readers, you've caught their attention. They're more likely to invite you for an interview, and suddenly you're one step closer to getting the job.

COLD LETTERS

With a "cold" cover letter, you can directly contact potential employers without a referral or previous correspondence. Job-seekers most commonly use this type of letter to advertise their availability to hiring managers or personnel departments. Presumably, after researching your field, you will have devised a list of the top employers you would like to work for and gathered basic company information for each.

BROADCAST LETTERS

With a broadcast letter, well-qualified candidates can advertise their availibility to top-level professionals in a particular field. The candidate attempts to entice the potential employer to consider his or her impressive qualifications for available positions. Although the broadcast letter discusses a candidate's background in detail, a resume is usually included. Since this type of letter is used primarily by seasoned executives, its tone should reflect the candidate's experience, knowledge, and confidence in his or her capabilities.

A candidate using the broadcast letter format might begin, "Are you in need of a management accountant who, in her most recent association, contributed to productivity improvements resulting in an annual savings of $20 million?" This attention-grabbing opening is effective only if the reader understands the significance of such an accomplishment. For this reason, broadcast letters are not recommended for those candidates conducting widespread job searches, where cover letters may end up in the human resources department rather than in the hands of a fellow industry executive.

LETTER TO AN EMPLOYMENT AGENCY

When searching for a job, many candidates rely on the help of employment agencies. These agencies offer services to a wide range of job-seekers, primarily for clerical or support staff positions. Letters addressed to employment agencies should focus on who you are, what type of position you are looking for and in what specific industry, and some of your strongest skills related to that field. For the agency to place you in an appropriate position, mention personal preferences, including geographic and salary requirements.

LETTER TO AN EXECUTIVE SEARCH FIRM

Although executive search firms actively recruit candidates for client companies, don't let this discourage you from writing. A well-crafted cover letter can alert an otherwise unknowing recruiter to your availability. Highlight your most impressive accomplishments and attributes and briefly summarize all relevant experience. If you have certain preferences, like geographical location, travel, and salary, mention them in your cover letter.

NETWORKING LETTERS

For the most part, networking letters refer to a third-party industry contact to garner the reader's attention and induce him or her to assist you in your job search. It is essential to achieve the right tone in your networking letters. Unless you are familiar with a contact, word your correspondence in a businesslike manner. In other words, do not use your addressee's first name or rely on an overly casual writing style. Likewise, if you have been in contact with this person recently, it could be useful to remind him or her, "It was great seeing you at the Chicago Writers' Convention last month" or "It's been several months since we bumped into each other on that flight to London. How are you?"

Many networking letters are written to an addressee whom the candidate has not met but has been referred to by a mutual acquaintance. In this case, immediately state the name of the person who referred you, such as "Jean Rawlins suggested I contact you." It is generally more effective to ask a contact with whom you are unfamiliar for assistance and names of people to contact than it is to ask for a job. Chances are, if your letter is politely persuasive, people will be interested in talking with you.

THANK YOU LETTERS

Your correspondence doesn't end with cover letters. Other types of letters, such as thank you letters, are often appropriate, even obligatory. It's acceptable to handwrite your thank you letter on a generic blank note card (but *never* a postcard). Make sure handwritten notes are neat and legible. If you're in doubt, typing your letter is always a safe bet. If you met with several people, it's fine to send each an individual thank you letter. Call the company if you need to check on the correct spelling of their names. Remember to keep the letters short, proofread them carefully, and send them *promptly*.

Richard H. Beatty

The Functional Resume

Richard H. Beatty, a nationally recognized authority on job searches, is currently president of the Brandywine Consulting Group, a corporate outplacement consulting firm.

Although, as a seasoned employment professional, I must admit to not being a strong advocate . . . [the functional] resume, I will quickly add, however, that it does occasionally have its place in the employment process. As an indication of this fact, I would estimate that approximately 5 to 10 percent of all resumes received by corporations make use of the functional format. This makes the functional format the second most popular resume style (second only to the chronological format which is estimated to account for 90 to 95 percent of all resumes). . . .

ADVANTAGES OF THE FUNCTIONAL RESUME

Advocates of the functional resume are quick to point out that it makes use of an excellent marketing strategy. This resume format is designed to capture the interest of the resume reader right from the beginning. This is accomplished by positioning the candidates's most salable strengths at the beginning of the resume in the form of some brief statements that capsulize the candidate's major accomplishments and most salable professional experience.

The marketing premise upon which this resume design is based is that most employment professionals are not thorough in their reading of employment resumes. Further, advocates of this style of resume would tell you that many employment professionals in fact read only the first paragraph or two of a resume. If these initial paragraphs do not capture their interest, the reader simply discards the resume and moves on to the next.

Although there may be some employment professionals who are not particularly thorough and conscientious in their resume reading, my experience suggests that this is certainly not the case with most. In my opinion, the majority of employment professionals discover early in their professional career that thorough resume reading is a must. In having read thousands of resumes as an employment manager, I have found that the specific experience and/or skills sought by my employer were sometimes not described until the second or third page of the applicant's resume. This experience has conditioned me to read resumes with a fair degree of thoroughness. Most professional employment managers are likewise sufficiently concerned with finding well-qualified employees for their organizations that they are equally thorough in their resume reading approach.

This is not to say that there are not those who are less diligent in their reading. To the contrary, I am sure there are some who are much less thorough. Such individuals may be superficially impressed with these opening paragraphs and may elect to invite the candidate in for an employment interview with a less than a complete screening of the candidate's qualifications. In the better companies, however, this superficial approach to resume screening will quickly attract attention, and the employment professional will soon be exposed as a nonprofessional. Poor resume screening will assuredly result in unnecessary interviews, wasting the valuable time of line managers and costing the company unnecessary money in the form of candidate travel reimbursement. Such inefficiencies will not long be tolerated, and careless employment managers may well find themselves in the unemployment line.

Of course there are well-qualified candidates who without professional guidance sometimes elect to use the functional employment resume. It is up to the employment manager, therefore, to thoroughly read the *entire* resume to determine whether or not the candidate possesses the prerequisite skills and qualifications sought by his or her company.

SOME CAUTIONS

Perhaps the biggest drawback of the functional resume is that it seems to be the format most frequently chosen by individuals who wish to disguise some flaw in their credentials. In fact, as an employment professional who has read several thousand resumes, I would estimate that in seven out of every ten cases functional resumes have intentionally been chosen by the employment candidate to camouflage such problems. Awareness of this history of deception causes most employment professionals to become suspicious when this particular format is used. The tendency is thus to read these resumes as if one were on a witch hunt. Energy and focus are directed toward finding out what is wrong with the candidate instead of being directed toward whether the candidate possesses the desired qualifications.

I have discussed my feelings concerning the functional resume with several other personnel executives and have frequently found that their feelings are generally similar to my own. In general, the consensus appears to be that if an employment candidate uses the functional resume format, he or she is probably attempting to hide one or more of the following facts:

1. *Job Hopper.* The applicant has worked for an abnormally high number of employers in a relatively short period of time and would therefore be a high employment risk.

2. *Older Worker.* The applicant is older and is attempting to hide this fact.
3. *Employment Gap.* There is an undesirable or unexplained gap or break in the candidate's employment history. Since this gap is frequently unexplained, it leads the employment executive to conjure up all kinds of undesirable explanations (for example: major illness, marital problems, alcoholism). Most of these spell high risk in the minds of the employment professional.
4. *Educational Deficit.* The applicant lacks the requisite educational credentials normally required for the position for which he or she is applying.
5. *Minimal Experience.* The candidate has little if any meaningful experience related to his or her job objective.

CHOOSING WHICH RESUME FORMAT TO USE

The rule of thumb for choosing between the chronological (both classical and linear) and the functional resume is fairly simple. If you have good credentials and a solid work history, I strongly recommend that you use a form of the chronological resume. Why cast unnecessary doubt on otherwise excellent credentials?

The reciprocal of the above statement is also true. If you have some good experience and major credentials, but also have one or more of the aforementioned problems, you may well want to seriously consider use of the functional resume format.

Resume Format Test

If you are still in doubt, perhaps the following practical test may be of assistance to you in deciding whether to use the functional resume format.

Answer each of the following questions with a yes or no.

1. Have you worked for four or more employers in the last 10 years? Yes _____ No _____
2. If employed for more than 10 years, have you averaged less than three years of service per employer? Yes _____ No _____
3. Have you been employed with more than seven companies during your professional career? Yes _____ No _____
4. Are you age 50 or older? Yes _____ No _____
5. Have you been unemployed (or substantially underemployed) for a period of more than one year in the last 10 years? Yes _____ No _____
6. Have you been unemployed (three months or longer) more than once in the last five years? Yes _____ No _____
7. Do you have the necessary educational qualifications normally required by most employers for your occupation? Yes _____ No _____
8. Do you lack the experience normally required by most employers for your profession? Yes _____ No _____

If you answered yes to one or more of the first three questions, by most standards you would be considered a job hopper. Use of the chronological resume would highlight this problem early in the resume and, in many cases, would result in your being screened out. Use of the functional resume under these circumstances is to your distinct advantage.

Although illegal, the screening out of candidates because of age continues. Some inroads have been made in this area, and it appears that most employers are now paying less attention to age than they are to the qualifications of the individual. Unfortunately, however, age 50 appears to be a bench mark of sorts for many employers. Rightly or wrongly, many employers assume that a candidate who is age 50 or older lacks the necessary energy and vitality to be a productive worker. This fact suggests that candidates who are in this age group should attempt, where possible, to conceal this fact. The functional resume format can be particularly effective in accomplishing this objective.

Questions 5 and 6 have to do with gaps in the employment history. Should you have such a gap or should you have been substantially underemployed for a period of time, you may wish to employ the functional resume to focus on your strengths and accomplishments and to draw attention away from this employment gap.

If you answered no to question 7, chances are that use of the chronological resume may cause you to be screened out on the basis of educational credentials. The functional resume, on the other hand, gives you the opportunity to highlight your strengths and accomplishments early in the resume and may generate interest sufficient to cause the prospective employer to disregard your educational deficiency.

If question 8 was answered yes, you will probably do better in adopting the functional resume. Here again the reader's attention will be focused on what you consider your strengths and accomplishments rather than on your lack of specific job experience. . . .

FUNCTIONAL RESUME COMPONENTS

Although there are several different styles or versions of the functional resume, the most frequently used format is comprised of the following components:

1. Heading
2. Qualification summary
3. Major accomplishments
4. Work history
5. Education

The order in which these components are presented on the resume can vary, but the sequence represented above is the most commonly used, accepted, and recommended format. The most frequent deviation from this recommended sequence, however, is the positioning of the education and work experience components. Note that the sample resume shown in Figure 1 has been positioned such that education follows work experience.

BRUCE B. CHAMBERS
813 Locust Lane
Lake City, Texas 09413
Phone: (217) 855-3939

SUMMARY

Marketing executive with over 20 years experience in sales and marketing management. Excellent reputation as a creative, innovative manager capable of successfully revitalizing old product lines and introducing new. Full range of marketing and sales experience to include: market research, market planning & analysis, advertising & promotion, sales, and sales management.

MAJOR ACCOMPLISHMENTS

Market Research

- Investigated and analyzed European market for U.S. lumber export with resultant successful market entry.
- Worked closely with R&D in development of consumer mini-pocket calculator. Careful design of test market and resultant market feedback assured successful product development and subsequent market entry ($5 million sales in two years).
- Developed market research computer model to forecast ten-year market projection for microwave ovens.

Advertising & Promotion

- Coordinated company efforts with major New York City consumer advertising agency to develop effective campaign to revitalize failing product line (fishing reels). Campaign expenditure ($2 million) yielded annual increase in sales of $4.1 million in one year.
- Developed creative special value coupon and supportive advertising campaign that increased annual sales volume for photographic film product line by 35% over two-year period.

Marketing & Sales Management

- Managed national sales organization of 95 employees (ten regional managers and 75 salespersons) in the sale of consumer photographic film to wholesale and retail trade.
- Directed Corporate Marketing Staff (25 employees) in the development of all marketing plans and strategies for manufacturer of consumer hardware (annual sales volume $500 million).
- Successfully organized, trained and motivated new national sales organization of 45 employees for manufacture of consumer calculators. Sales reached $20 million in four years.

FIGURE I   Sample Resume

WORK HISTORY

U.S. PAPER & WOOD PRODUCTS   1997 to Present
Manager of Corporate Marketing

PHOTO FILMS INTERNATIONAL   1995 to 1997
National Sales Manager

SELF EMPLOYED   1993 to 1995
Marketing Consultant

MICRO OVENS, INC.   1992 to 1993
Manager of Market Research

SELF EMPLOYED   1991 to 1992
Marketing Consultant

NATIONAL CALCULATORS, INC.   1977 to 1991
Director of Marketing   (1989 – 1991)
National Sales Manager   (1985 – 1989)
Regional Sales Manager   (1982 – 1985)
Salesperson   (1977 – 1982)

EDUCATION

B.A., Utah State University, 1977
Major: Business Administration

FIGURE I   *(Continued)*

The general rule for deciding whether to position education before or after work experience on the functional resume is this: Always position education after work history unless it will clearly serve to enhance your marketability to do otherwise.

In cases where the candidate has exceptionally strong educational credentials and this would be clearly recognized as the case by prospective employers, it would be in the candidate's best interest to position education prior to work history on the resume. Conversely, a candidate whose work history is extensive but whose educational credentials would be considered somewhat light by most employers' standards should definitely position education following the presentation of work history.

Additionally, it is recommended that younger candidates with good educational credentials and little work experience list education first. By contrast, older workers with considerable experience and persons who do not have strong educational credentials are generally best served by placing education after work experience.

For the older worker, placing education first in the resume can often draw attention to his or her age—something that should be avoided. This makes it easy for those employers who are in defiance of federal law, still practicing age discrimination, to weed you out. If you are successful in getting such employers to review your work experience and major accomplishments first, and thereby convince them that you have something of value to contribute to their organization, perhaps less attention will be paid to your age.

You are now ready to be in the step-by-step process of preparing the functional resume. The approach that we follow examines each resume component in some detail and provides some practical exercises that will assist you in developing each of these components. . . . [If] you follow the step-by-step directions provided, you will conclude this . . . [selection] having prepared a logical, professional resume that will effectively portray and market your qualifications to prospective employers.

Heading

The resume heading consists of three parts: your full name, complete address, and home telephone number. Some sample resume headings follow:

DAVID. C. JOHNSON
18 Smith Road
Richmond, Ohio 19870
Phone: (815) 852–4310

CAROLYN A. CRISWELL
401 East 7th Street
Knoxville, Tennessee 19760
Phone: Home (315) 769–6588
Office (315) 766–4233

Your name should be typed in capital letters and in bold type so that it stands out from the rest of the heading. Your address and telephone number, on the other hand, are typed in lower case. It is advisable to exclude your office telephone number from your resume unless you are in a position to accept employment-related telephone calls at your office. Should you be on a confidential voice mail system where you are the only one who can access your messages, this is likely not a problem. Listing your office telephone number on your resume, however, sometimes raises suspicion on the part of a prospective employer that you are "on the skids" with your current employer and that your current employer is cooperating with you in your job search. It is therefore advisable to avoid listing your office telephone number on your employment resume unless by excluding it you would be making it difficult for prospective employers to reach you.

Most employment managers do not mind placing telephone calls in the evening to prospective candidates. These managers can be encouraged to do so by inclusion of a simple request to call you at home in the evenings. Such a request can easily be inserted into your cover letter. For example, you might include a statement similar to the following in the cover letter that accompanies your resume:

> I can usually be reached at my home telephone number during the evening hours after 8:00 P.M. In the event that you find this inconvenient and wish to contact me during business hours, my husband, George, will be pleased to relay your message. George can be reached at (315) 972–8051.

Using the preceding instructions, try your hand at developing your own heading.

Summary

The purpose of the Summary section of the functional resume is twofold. First, it is intended to convey the breadth and scope of your experience as a professional. Second, it is intended to provide you with the opportunity to sell your key strengths. The overall intent is to hook the resume reader so that he or she will be encouraged to read further.

Let's now examine a few summary statements.

1. *Summary.* Marketing executive with over 20 years of experience in sales and marketing management. Excellent reputation as a creative, innovative manager capable of revitalizing old product lines and introducing new ones. Full range of marketing and sales experience to include: market research, market planning and analysis, advertising and promotion, sales, and sales management.
2. *Summary.* Accomplished human resources professional with over 25 years experience in all phases of human resources management. Excellent reputation as an individual who truly understands the relationship between human resource utilization and profitability. Full range of human resources experience incudes: human resource planning, internal and external staffing, organization design and development, compensation and benefits, employee and labor relations, and public affairs.
3. *Summary.* Creative design engineer with 15 years experience in innovative design of high speed electromechanical devices. Well known for innovative, practical, and cost-effective designs that work. A real contributor to bottom-line results.

In reviewing these summary statements, you will note certain similarities. First, the initial sentence of each summary statement gives the writer's career/professional area and indicates the number of years of experience The second sentence is used to market particular strengths possessed by the writer. The third sentence, as in the case of the marketing executive and human resources professional, can be used to further communicate the breadth of the writer's experience. As an option to this, as shown in the designer's summary statement, the resume writer can use this third sentence to further market some unique skill or fact that would likely prove valuable to the prospective employer.

Another similarity between these summary statements is that they are fairly concise. In preparing your summary statement, you should avoid lengthy, rambling paragraphs that will only serve to dilute your message to the prospective employer. The key to this section is to be concise while stating sufficient positive information to entice the reader to continue. Additionally, your summary statement should convince the prospective employer that you have something of value to contribute to their organization.

Note that these sample summary statements are not written in complete sentences. These statements are simply descriptive phrases that are intended to convey strong meaning in as few words as possible. They are short "bullets" much the same as you would find in magazine or newspaper advertising. As with such advertising copy, each is intended to convey value and to compel the reader to respond favorably.

Review of the sample summary statements will also reveal the lack of the articles "a" and "the." Each sentence begins with an adjective followed by a noun. Following this type of format will force you to write concise, brisk statements that have considerably more marketing impact than the normal sentence.

Now try your hand at writing your own sample summary statements. Try writing this statement three times with the idea of bringing about quantum improvement with each successive writing.

Major Accomplishments

The Major Accomplishments section is intended to point out specific accomplishments that you have realized during your professional career and to highlight these accomplishments in discrete, functional areas. The functional areas under which these accomplishments are highlighted are normally major subfunctions or subactivities of your profession. Some examples of these functional areas would be as follows:

Marketing Executive

1. Market research
2. Advertising and promotion
3. Marketing and sales management.

Human Resources Executive

1. Human resource management
2. Labor negotiations
3. Employment
4. Compensation and benefits.

Research Engineer

1. Conceptual design
2. Prototype development
3. Prototype testing
4. Redesign
5. Reduction to practice.

In preparation for this section of your resume, make a detailed listing of all of the subfunctions relating to your career specialty. Think in terms of the major components of your present and past positions—the key functional areas for which you were accountable in each of your past positions. Having done this, you are now ready to develop a listing of your major accomplishments for each of the corresponding functional areas that you have listed.

A good exercise for stimulating your memory is to list each of your past employers in reverse chronological order, including each of the respective positions you have held. Starting with the most recent employer, make a list of the most significant results that you have realized—the most significant contributions you have made to each employer. As you list each of these major accomplishments, be sure to include, wherever possible, some kind of quantitative measurement that will serve to convey to the reader the significance of this result to your past employer. Think in terms of the savings in time and money that your accomplishments provided. Additionally, think in terms of other ways you have positively affected the profitability of your employer. In this regard, actions that you have taken to increase employee productivity, sales volume, and so on should certainly be included in the category of major accomplishments.

Although the procedure described may seem laborious, it will prove to be extremely worthwhile to you as you begin the job-hunting process. In addition to providing you with an excellent basis for the construction of your functional resume, this personal analysis of major accomplishments will provide you with the information and confidence necessary to perform well during the interview phase of your employment campaign. Take time therefore to carefully develop this inventory of major accomplishments.

After you have listed your major accomplishments by functional area, you are ready to begin to rework these for inclusion in the final draft of your functional resume. Before doing this, however, review the major accomplishments section of the sample resume shown in Figure 1. . . . You will note that this area of the resume is comprised of only three functional areas: market research, advertising and promotion, and marketing and sales management. It is important that your resume contain no more than three or four functional areas which you plan to highlight. Listing more than four areas will serve to dilute the impact that is intended when using the functional resume. This intent is to highlight and market your major areas of strength. Listing more than four functional areas with accompanying major accomplishments may serve to give the resume reader the impression that you are a jack of all trades but master of none. The reader will begin to question the credibility of your accomplishments rather than being impressed with a carefully chosen few.

A careful review of your list of past employers and positions held will highlight your major accomplishments and contributions. Scanning these accomplishments will show that several are similar in nature and can fit nicely under the key functions that you have chosen to feature on the resume. Picking what appear to be the most prominent three or four subfunctions, put each of these subfunctions as a heading on a separate piece of paper. Then systematically go back over your employment history, extract each major accomplishment and add it to the page containing the functional heading under which it fits. In following this procedure you will soon see the major accomplishments section of your functional resume taking shape.

Your next step is to decide in what order these key functions should be listed in the major accomplishments section of your resume. There are two schools of thought relative to the order in which these functions should be listed. The first school of thought focuses on the old adage "lead from strength." This adage suggests that you should list that function in which you have made the greatest contributions and have realized the most significant accomplishments. This approach reasons that your most marketable or salient qualifications should be listed first since they are your areas of greatest strength and therefore are those areas that will be of greatest value to any prospective employer. The second school would argue that you should position that function first that most closely parallels your job-hunting objective. For example, if you are a human resources professional and are seeking a position as director of labor relations, you would obviously wish to cite your accomplishments in the area of labor relations first. Prospective employers looking for a director of labor relations are probably less concerned about your expertise in the area of compensation and benefits.

Both of the above schools of thought have their point. This may leave you a bit confused as to which function to list first. As a general rule of thumb, I would suggest listing first that function that most closely relates to your current career objective. However, should your accomplishments in this area be less than stellar, I would suggest listing your area of strongest accomplishment first without respect to job-hunting objective. The remaining functional areas should be listed in a priority sequence based upon the significance of the accomplishments in each of these respective functions.

In this section of your resume, writing style can be extremely important if you wish to have maximum impact on the reader. I have extracted two functions from the major accomplishments section of the resume shown in Figure 1 for your review. The first is market research:

- Investigated and analyzed European market for U.S. lumber export with resultant successful market entry.
- Worked closely with R&D in development of consumer mini pocket calculator. Careful design of test market and resultant market feedback assured successful product development and subsequent market entry ($5 million sales in two years).
- Developed market research computer model to forecast ten-year market projection for microwave ovens.

The second function is advertising and promotion:

- Coordinated company efforts with major New York City consumer advertising agency to develop effective campaign to revitalize failing product line (fishing reels). Campaign expenditure ($2 million) yielded annual increase in sales of $4.1 million in one year.

- Developed creative special value coupon and supporting advertising campaign that increased annual sales volume for photographic film product line by 35% over two-year period.

In reviewing these accomplishments you will note that each statement begins with a verb or action word. By starting most of your statements with a verb or action word, you will be forced to be brief, concise, and to the point in your statements. Additionally, you will be forced to state specific results and accomplishments that you contributed. . . .

Another effective technique to employ in developing the Major Accomplishments section of your resume is to use quantitative descriptions, where these serve to highlight your specific accomplishments and contributions. For example, consider the following:

Wrong	Increased sales significantly in first year.
Right	Increased sales by 50% from $100 to $150 million in first year alone.
Wrong	Developed new equipment design that saved the company a lot of money.
Right	Developed new plastic extrusion equipment design resulting in $1 million company savings in 1995 alone.
Wrong	Developed and installed computerized brand costing system resulting in substantial payroll savings.
Right	Developed and installed computerized brand costing system resulting in elimination of 12 positions and annual payroll savings of $385,000.

You are now ready to develop the Major Accomplishments section of your resume. Use the format illustrated here to carefully develop each of the functions you wish to highlight with specific accomplishments. Remember to begin each of your statements with an action word and to use quantitative descriptions wherever possible. Although you will wish to highlight only three or four functional areas, rewrite each functional area at least twice. Experiment by attempting to upgrade your second writing of each function by being more concise and choosing words that have greater impact.

Functional area: _____

Accomplishments:_____

1. _____

_____

2. _____

_____

3. _____

_____

4. _____

_____

Functional area: _____

Accomplishments: _____

1. _____

_____

2. _____

_____

3. _____

_____

4. _____

_____

Work History

The Work History section of the resume should be organized in reverse chronological order. Shown first should be your most recent employer, second, your second most recent employer, and so forth back to your original employer. Shown on the left should be dates of your employment, when you began and when you left. Adjacent to these dates should be the name of the employer followed by the position or positions held. The following work history section has been extracted from the sample functional resume shown in Figure 1.

Work History

1997 to Present	**U.S. Paper & Wood Products** Manager of Corporate Marketing
1995 to 1997	**Photo Films International** National Sales Manager

1993 to 1995	**Self-Employed** Marketing Consultant
1992 to 1993	**Microwave Ovens, Inc.** Manager of Market Research
1991 to 1992	**Self-Employed** Marketing Consultant
1977 to 1991	**National Calculators, Inc.** Director of Marketing (1989–1991) National Sales Manager (1985–1989) Regional Sales Manager (1982–1985) Salesperson (1977–1982)

There is nothing sacred about the order of position held and employer in this section of the resume. When the candidate has been employed by major corporations and highlighting this fact could prove to the candidate's advantage from a marketing standpoint, Company Name should be listed first with job title to follow on the next line. The answer to sequence is really marketability. List first either job title or employer, whichever you feel consistently gives you the greatest advantage from a marketability standpoint.

Listing position first followed by employer will work well as long as you have held only one position with each employer. In those cases where you have held multiple positions with a single employer, however, you will need to list employer first, followed by the positions held. (Note the National Calculators, Inc. section of the sample Work History from Figure 1.) You will note that National Calculators, Inc. is listed first, with the positions held listed following the company name. You will also note that the dates of the positions held at National Calculator are shown in brackets following each of the job titles of the positions held.

The work history section of the resume is intended to be simple. No additional information other than company name and job title should be included in this section. The following space is provided for you to develop the Work History section of your own functional resume.

From: _____ To: _____   Employer: _____

_____

Job Title: _____

_____

From: _____ To: _____   Employer: _____

_____

Job Title: _____

_____

From: _____ To: _____   Employer: _____

_____

Job Title: _____

_____

From: _____ To: _____   Employer: _____

_____

Job Title: _____

_____

From: _____ To: _____   Employer: _____

_____

Job Title: _____

_____

From: _____ To: _____   Employer: _____

_____

Job Title: _____

_____

Education

The Education section of a functional resume should include the following items:

1. Degree awarded
2. School attended
3. Year of graduation
4. Major field of concentration
5. Grade point average (G.P.A.) (List only if good and education is recent.)
6. Honoraries.

Some sample education statements have been provided for your reference.

1. Education:

Ph.D., Massachusetts Institute of Technology, 1997
Major: Chemical Engineering

M.S., Rochester Institute of Technology, 1995
Major Chemical Engineering
G.P.A. 3.85/4.0

B.S., Rochester Institute of Technology, 1993
Major: Chemical Engineering
G.P.A. 3.75/4.0

2. Education:

M.S., University of Pennsylvania, 1996
Major: Statistics
G.P.A. 3.5/4.0
Dixon Mathematics Scholarship

B.S., Drexel University, 1994
Major: Mathematics
G.P.A. 3.6/4.0
Kettering Award (Mathematics Honorary Award)

3. Education:

M.S., University of Michigan, 1996
Major: Mechanical Engineering
G.P.A. 3.4/4.0
Sloan Engineering Scholarships

B.S., University of Maryland, 1995
Major: Mechanical Engineering
G.P.A. 3.4/4.0
Tau Beta Pi (Engineering Honorary)

4. Education:

B.A., North Carolina State University, 1996
Major: Business Administration

Generally, unless you are a recent college graduate, you should leave little space for the Education section of your employment resume. The general rule to be followed in developing this section of the resume is: "The further along you are in your professional career, the less important are your educational credentials and the more important are your work experience and specific accomplishments." Additionally, since you have chosen the functional resume for the purpose of highlighting your experience and accomplishments, it stands to reason that the Education section of the resume will be de-emphasized.

You now can develop the Education portion of your resume using the instructions provided.

Miscellaneous

In almost all cases, extracurricular activities and hobbies should be completely excluded from the resume unless they are directly related to your qualifications for the position you seek. On occasion, however, there may be some advantage to listing a specific extracurricular activity or hobby, that is, if it further enhances your marketability to a specific employer. For example, if you are applying for a position with a sporting goods manufacturer, your athletic interests may be to your advantage. Likewise, if you are applying to a company that places a lot of stress on providing leadership in your local community, you may want to include such information as (1) President, Parent Teachers Organization, (2) Chairperson, March of Dimes Campaign, (3) President, School Board. The preceding are carefully chosen exceptions to the rule. In general, however, hobbies and extracurricular activities are specifically excluded from the modern resume.

In the case of technical professionals, such things as publications, papers, and patents can be important. This is particularly true for items that are directly related to the professional's employment qualifications. These may help to increase the candidate's marketability to prospective employers. List only those items that are truly important and meaningful, however, and do not list minor or insignificant publications.

Burton Jay Nadler

Your E-Resume

Burton Jay Nadler is the Assistant Dean of the College and Director of the Career Center at the University of Rochester. Previously, he was Manager of College Recruiting and College Relations for Merrill Lynch Consumer Markets.

RESUMES FOR THE E-SUPERHIGHWAY

Almost all resumes today are electronic to some degree. (For instance, we use personal computers and software.) Beyond the means we use to create them, it is important that resumes can be transmitted electronically and that the reviewer can open them easily. After that, our concern as usual is with quality. The aim is to present yourself efficiently, so that no matter who (or what, in the case of a database search engine) opens your resume, it is clear that you are goal-oriented and qualified for the job.

UPLOADING YOUR E-RESUME

Many potential employers use the Internet to list their job postings. Some use their own corporate Web sites. (The *Everything* ® book publisher, Adams Media at *www.adamsmedia.com,* has an Employment link on its home page.) Many also use headhunter sites, where the Web site organizes and displays an employer's advertisement and also makes sure qualified applicants come to the employer's attention.

The process of uploading a resume from your personal computer is simple. Web sites walk job seekers through the process step by step. The process may vary a little, but in all cases you are using your Internet connection to send the electronic file containing your resume to the online job bank.

File Format for the E-Resume

Uploading is easy, and it is a very effective way of making your resume easily available to many potential employers. However, the electronic file that contains your resume must be in the correct format. Your uploaded resume is useless if the database can't store it.

All online job banks specify the file format you should use. The following are some common formats.

- **Microsoft Word.** Almost everyone in almost all professional fields uses Microsoft Word. If you create your resume using another application, you can probably save your file as a Word document.
- **PDF.** This format is also very common on the Web. PDF files retain the formatting of the original file and they tend not to be corrupted in their travels across the Internet. You can create a PDF from almost any file format. PDFs cannot be edited or altered, making them a good way of keeping your content exactly as you created it.
- **Web page input.** Some job banks ask you to create a resume by filling in the fields of a resume form. In this case, you are not really uploading a file at all. The information you enter into the various fields gets saved directly into the job banks database. Instead of storing resume files, this database contains the information directly.

No matter what format you use, you should do a quality check before uploading your file:

1. Make sure your resume file can be opened, and that it is formatted correctly. Word files can be corrupted, and PDFs can substitute fonts and change things like font size or margins. Do not assume that just because you saved the file or created the PDF that it will meet your requirements.
2. Name your file correctly. Check the requirements of the job bank: They might have a system for naming files. In general, your filename should be short and descriptive.
3. If you are filling in a Web form, take your time entering your information in the fields. Keywords here are essential. You might need to edit your resume, aiming to keep things short while using as many keywords as possible.

Keywords Are Essential

Many job seekers upload their resumes to Web job banks. These are true e-resumes: the potential employer only sees this electronic version. Often, before any human being lays eyes on these e-resumes, they are subject to review by a search engine. This tool is loaded with certain keywords that the employer has defined as critical to any applicant's qualifications. Regardless of your abilities, if your qualification summaries do not use these keywords, it is likely that you'll never make even the first cut.

Here is where your research pays off. In defining your goals and drafting your qualification summaries, you learned the terms and phrases used in your chosen field. When possible, in your statement of objectives, you should name a particular title that you know is used in your field. The verbs you use to describe your qualifications and experience should mirror those terms as closely as possible.

Here are two statements of objectives that could be used for the same resume. See which one would yield the more fruitful keyword search:

- **Objective:** Position designing logos and graphics for state-of-the-art Web sites.
- **Objective:** Position as Web Designer for software design firm, using skills in XML, DTML, streaming video, Flash animation, Oracle database design, and Web site maintenance.

Chances are that you are uploading your resume into a job bank in answer to a specific job posting. The smartest thing to do, therefore, is to compare the job description to your resume. Does your resume, particularly your qualification summary section, contain the key verbs, terms, and phrases used in the job description? If this job description is short or vague, search the Web for other descriptions of similar jobs. You will notice that they use a common vocabulary to define job functions and desired qualifications.

Keep an Inventory of Keywords

As you continue to search out new job postings, you will learn new industry-specific ways of phrasing your qualifications and summaries. Rather than rewriting your resume every time you find a new, more appropriate term, keep a list. Update your resume when you have a collection of these terms. In this way, your update will not only be more efficient in terms of time, you may come up with better ways of phrasing whole sections, instead of changing a word at a time.

Check the rules of your online job banks. Most allow you to regularly update your resumes. When you do an update, take the time to upload it so all potential employers always see the best and most accurate depiction of your job skills.

E-MAILING YOUR E-RESUME

Some job postings ask you to e-mail your resume directly to a contact person. The process here is very simple: Your e-mail message is your cover letter, and your resume gets included as a file attachment. An e-mail cover letter should be short and direct. The first line should state your purpose ("I am writing in response to your posting for an experienced Web Designer"). Any subsequent text should state your qualifications bluntly ("I have five years of freelance experience designing Web sites for independent movie studios, sometimes using GoLive and Adobe Design software but mostly writing my own code to incorporate movie clips and other complex elements").

How much info should I include in my e-mail?

It's a good idea to say what made you respond to this posting ("I use links to your site to give visitors access to real-time industry news, and I would love the opportunity to work for you"). E-mails are quick and informal by nature. You know from experience that an overwritten e-mail is hard to read. Be polite, but don't be stuffy.

E-Mail Etiquette

In this case, the question of file formats is particularly important. Pay attention to the stated format requirements, and do not bother submitting a resume that doesn't meet them. You may think PDF is much better than Word, but if Word is what the contact person requests, he or she will not appreciate your superior taste.

The common question with an e-mailed file of any kind is whether the recipient got it in a legible form. If your resume file is huge, with graphics and other special features, it will take a long time to transmit and to open. Nobody appreciates being made to wait, so for e-mailed resumes, it's a good idea to keep things very simple.

Regardless of your care, file attachments do sometimes still get mangled. That gives you an automatic excuse for a follow-up. Ask the recipient to let you know if your file did not come through. Take the opportunity to slip in another quick selling point. There's no harm, as long as you have the contact's address, in sending periodical follow-ups, as long as they are very brief and not demanding in any way.

Addressing the E-Mail

Job postings often ask you to reference a job number in your subject line. Be sure you include this information, as it is unlikely your e-mail (or resume) will reach its destination without it.

Most job postings include a link to the contact person's e-mail. All you have to do to address your e-mail properly is to click on the link. You must also use a personal address, however, in your cover e-mail. Online job postings are notorious for giving little to no information about the contact person. Sometimes all you have is an e-mail address and no name at all. Use whatever information you have. If you don't have a contact name, start the letter immediately after the address, using no salutation at all.

WEB-BASED RESEARCH AND JOB SEARCH

The Web is full of all the information you could ever need to know about your chosen career. Chances are that your dream job is out there somewhere, too. If only you could find it.

In doing research on the Web, the best place to start is with what you know. It doesn't matter whether you start off in the exact right place. The Web is a job seeker's paradise because it is so easy to follow any trail in whatever direction you choose. Here's an example. We'll let our Web Designer friend Chris Smith take a quick look around the Internet to see what's happening in the working world.

1. Chris is just beginning a job search, so she starts at her favorite all-purpose search engine (we'll say that's *www.google.com*).
2. Chris types in her chosen job title, "Web Designer," in the search field and narrows the search just a little by adding "employment."
3. The Google search engine returns more than a quarter of a million Web sites that are related somehow to the phrase "Web Designer employment." Some of these sites belong to other Web Designers looking for employment. Chris ignores these for now, although later they might be a good way of networking with others in her field. She wants actual job postings. She doesn't have to look very hard to find them. The second listing on the page lists a job posting for a Web Designer.
4. In the next five minutes, Chris finds the following information, just by clicking links that look promising:

- Dozens of job postings for Web Designers, containing plenty of keywords.
- Several big job banks, or headhunter sites. These contain not only job postings for Web Designers, they give Chris an idea of what companies are hiring, what industries in her part of the country use Web Designers, and what qualifications employers expect.
- Salary information. Some job postings include a salary range. But these are not often reliable, so Chris clicks around a little and finds salary statistics on an employment-related government Web site.
- Sites of individual companies with pages devoted to current openings, all with contact people listed.
- An online forum of people just like herself, experienced Web Designers of all kinds. Their discussions cover topics like "Freelance Survival" to "Making It Big in the Corporate World."

These are the results of a real Web search, using these keywords, conducted during a quick coffee break. As you can imagine, the real key to getting useful information from the Web is knowing when to stop. Five minutes can easily yield more information than anyone could use in a week. The Web is a huge, chaotic haystack of information; it's true. But as long as you don't worry yourself about hunting for one in particular, you'll find more needles there than you can possibly use.

Vistas for the E-Job Search

Here are a few avenues for beginning your online job search:

- Check with your local librarian to see whether your state sponsors a job bank or whether individual listings can be found online.

- Check the Web sites of companies in your chosen field.
- Use a Web search engine to find the big headhunter sites (like *www.hotjobs.com, www.monster.com,* or *www.headhunter.com*).

Your local newspaper may also list its help wanted ads online. Online listings are usually much easier to search than the printed version. Also, ads often include a link to the company's Web site, where other suitable postings may be found.

REACTIVE AND PROACTIVE E-STRATEGIES

It is easy to think of the Web as a giant fishing pond. You as the job seeker have a certain kind of bait on your pole, and you dangle your line in hope that the right fish is out there, waiting to bite.

This is a reactive job-search strategy, where you the job seeker react to postings and try to model yourself as the best candidate for the job. Sometimes that's all it takes, but a truly effective job-search campaign incorporates proactive techniques as well.

You have already begun using the Web in a proactive manner. When you research your chosen field, for instance, to find the Web sites of the top companies, you are engaging in proactive research. Maybe the company you would most like to work for has several postings that are close to what you're looking for, but nothing that you feel qualified to apply for. You plan to keep checking back, but in the meantime why not write a quick e-mail to the contact person listed for that job. Tell him or her of your interest. Explain briefly how you believe you could be a real asset to the company. Close by saying you will keep in touch, and do it.

Another proactive technique on the employment e-frontier is a twist on our old favorite, networking. Online forums, chat groups, and mail groups cover almost any topic imaginable. They are easy enough to search out (try Yahoo!'s e-group listing, at *www.groups.yahoo.com*) and to join. You probably won't get any job offers, but you can learn a lot about your chosen field and the function you hope to perform in it.

A recent poll of college career services and corporate recruiters came up with fourteen favorite job-search sites. They include America's Job Bank (www.ajb.org), Careerbuilder.com, CareerXRoads.com, Dice.com, Headhunter.net, Hotjobs.com, JobHuntersBible.com, Jobdirect.com, and Job-Hunt.org.

ONLINE APPLICATIONS

Postings and job bank sites and the actual Web sites of employers increasingly grant you the opportunity to apply for particular positions online. Original copy and paste e-forms are being replaced by PDF systems that allow for uploading of resumes formatted in graphically creative ways. You no longer have to remove bolding or italics, nor worry about tabs, justification, or where dates are presented. But, even with these newer systems you should follow some essential advice.

- Always reflect knowledge of job description within qualification summaries.
- Don't hesitate to use actual phrases appearing in these announcements within summary sections that appear very early (after the objective) in your e-resume.
- Use cover letters created in the same document as the e-resume.
- Update and enhance keyword content of e-resumes.
- Follow up with phone calls and with additional e-mails.

Estimates regarding how many resumes are stored on or transmitted via Web-based resources vary. Some believe that the numbers of resumes stored in resume banks and connected to postings via mega job-search sites doubled each year over the past five years. While actual totals are difficult to confirm, almost all agree that millions of resumes are stored annually and tens-of-thousands are transmitted by e-mail daily.

Part 6

And Now a Word (or Two or Three) About Ethics

Turn on the television or radio, read a newspaper, get updates on current events on your PC, and the news is discouragingly the same: corruption, ethical violations, malfeasance in office, major companies ruined by corporate greed, high flying executives, and bankrupt pension funds and corporations. Ethical behavior seems more and more to be a concept foreign to public and business life.

But business and technical writers have moral and ethical obligations, and the essays in this final section of *Strategies* discuss some of the ethical issues we all face in writing for the world of work. The more general topic of ethics in business and industry is beyond the scope of this section or of this entire collection. What follows is meant to lay the groundwork for a continuing discussion in the classroom or the corporate setting.

Dorothy A. Winsor begins by examining the public documents available on the Challenger explosion to show how a "history of miscommunication" contributed to the accident. Her aim is to show how the dynamics at work in the Challenger case can help business and technical writers and managers reduce intentional or unintentional miscommunication in their own companies.

In a now classic piece, Darrell Huff suggests ways writers can misuse graphics and statistics to manipulate the truth. Dan Jones follows with a broader discussion of the relationship between ethics and writing style, and Carolyn Rude concludes this new edition of *Strategies* with an overview of the ethical issues as they impact business and technical writing and editing.

Dorothy A. Winsor

Communication Failures Contributing to the Challenger Accident: An Example for Technical Communicators

Dorothy A. Winsor is on the faculty at Iowa State University and conducts research on the writing of engineers. Her work has won the NCTE (National Council of Teachers of English) Outstanding Research Award on three separate occasions.

A technological failure such as the explosion of the space shuttle Challenger can be puzzling in retrospect. Investigation often reveals that various people in the organization involved knew that the failure was likely and knew how to prevent it, and yet that knowledge was not shared within the organization as a whole. How does it happen that such important knowledge is not communicated? In the case of the Challenger, why did those who knew of the problem with the shuttle's solid rocket boosters not convince those in power to stop the launch?

The answer to this question lies in a complex set of factors, the most important of which seem to be (1) managers and engineers viewing the same facts from different perspectives, and (2) the general difficulty of either sending or receiving bad news, particularly when it must be passed to superiors or outsiders. An analysis of the communication failures that contributed to the Challenger accident is potentially of great interest to engineers and their managers because a large part of an engineer's job is to communicate both good and bad news upward to management for decision making. The Challenger explosion was a horrifying public event, but it resulted from factors that are probably at work more quietly in many other organizations.

The first of these factors—managers and engineers viewing the same facts from different perspectives—suggests that knowledge is not simply seeing facts but rather interpreting them, and that interpretation varies depending upon one's vantage point. Communication, then, is not just shared information; it is shared interpretation. Achieving shared interpretation within an organization is relatively easy if the sender and receiver of communication share the same corporate role and hence the same concerns and values. If sender and receiver are from different corporate subcultures, however (as they often are in technical communication), then achieving shared interpretation is more difficult [1, 2]. In an appendix to the *Report of the Presidential Commission on the Space Shuttle Accident,* for instance, Commission Member R. P. Feynman notes the difference between probability estimates of flight failure and loss of life by managers—1 in 100,000—and engineers—1 in 100 [3, V. 1, p. F-1]. This difference occurred despite the fact that the managers were almost all engineers by training. Presumably, managers and working engineers had much the same background and many of the same facts at their disposal, but they interpreted the facts quite differently because they approached them from different points of view.

By the same token, communication about the solid rocket booster joint that failed was made more difficult because it was bad news. Research has repeatedly shown that bad news is often not passed upward in organizations [4, 5]. Moreover, even when bad news is sent, people are less likely to believe it than good news [6, 7]. In the shuttle disaster, bad news moved up only slowly from engineers to management within NASA; Marshall Space Center, where the shuttle program was headquartered; and Morton Thiokol International (MTI), the contractor responsible for the solid rocket boosters. It also moved slowly among the organizations because they were in a hierarchical relationship, with MTI dependent on Marshall for the contract and Marshall dependent on NASA for funds and career opportunities.

Additionally, the three organizations seemed to view one another as outsiders despite the fact that they were working jointly on the same project and, in the case of NASA and Marshall, were nominally part of the same agency. So the taboo against airing organizational dirty linen in public was added to the general difficulties of bad news transmission [8]. Communication about O-ring problems, then, had to overcome the barriers to moving bad news between engineering and management subcultures, up through organizational hierarchies, and out to other organizations. Under these circumstances, it is hardly surprising that the communication failed.

The following paragraphs explain, first, what failed physically on the shuttle system, and then what failed in the organizations' attempts to understand those physical problems and communicate about them in the two-year period before the January 28, 1986 launch. I hope that this presentation will give engineers and engineering managers some insight into how to minimize the occurrence of events like the Challenger accident.

PHYSICAL CAUSE OF THE ACCIDENT

The physical cause of the Challenger explosion was the failure of a rubber seal in the solid rocket booster. The shuttle system consisted of three parts: the orbiter, which contained crew and experimental equipment; a large tank of liquid fuel, which was used by the orbiter's engines during liftoff; and two solid rocket boosters, which assisted at liftoff and were jettisoned to be recovered and reused on later flights. These solid rocket boosters (SRBs) were made in segments, which were stacked together at the launch site. The joints between the segments were sealed with two O-rings, which were protected from the heat of combustion by putty. The joint was pressure sealed, meaning that during rocket firing, expanding gases from burning fuel pushed the putty into the air space in the joint; this compressed air, in turn, pushed the O-ring into place and held it there. The second O-ring in each joint, in theory, provided redundancy or backup for the primary ring. During the Challenger launch, the O-rings in one of the SRB joints failed to seal, allowing hot gases to escape from the side of the SRB and burn a hole into the nearby liquid fuel tank, which exploded approximately 73 seconds into the flight.

In hindsight, the failure of the O-rings should not have been unexpected. From early 1984 on, postflight evidence increasingly showed that the joint seals were failing to meet design expectations. After every shuttle flight, the SRBs were recovered, disassembled for inspection, and readied for reuse in future flights. The inspections looked for any anomaly indicating the O-rings had not functioned as they should, and specifically for anything such as charred or eroded surfaces, which would indicate that the rings had failed to seal or had come into close contact with the heat of combustion. Before February 1984, only one O-ring anomaly had been found on the first nine flights. Beginning with the tenth shuttle flight, however, launched two years before Challenger on February 3, 1984, anomalies occurred on more than 50 percent of the flights [3, V. 1, p. 155].

We now know that the increased number of O-ring anomalies after early 1984 was probably caused by the use of increased pressure in the leak check done on the SRB joints after assembly at the launch site. The leak check involved blowing air into the joint through a hole located between the two O-rings and testing to see if the joint pressurized or sealed. Flights one through seven were tested at a pressure of 50 psi, flights eight and nine at 100 psi. From flight ten on, tests were done at 200 psi. In May 1985, MTI experiments showed that the increased pressure was likely to have blown holes through the putty that shielded the primary O-ring from the hot blast of ignition [3, V. 1, p. 156]. The holes were particularly damaging, because they not only allowed the hot gases to penetrate but actually focused them, so that they would cause maximum O-ring erosion, which is the eating away of the edge of the O-ring by the hot gases rushing past before the ring seals. Ironically, the pressure at which the leak check was conducted was increased to 200 psi because of concern that the joints were not sealing. Some officials believed that the blow holes were necessary—unless they existed, the putty rather than the O-ring could seal the joints during the test, and a defective O-ring could escape detection [3, V. 1, p. 134].

EARLY RESPONSES TO BAD NEWS: DISBELIEF AND FAILURE TO SEND UPWARD

When O-ring anomalies first began appearing in early 1984, neither engineers nor management at MTI treated them as serious problems in their communications to Marshall. They did not send a grave interpretation of the data upward and, judging by internal documents, did not believe one themselves. Marshall's reactions are more ambiguous, for they treated the O-ring situation as serious when they communicated downward to MTI but as relatively minor when they communicated up to NASA headquarters.

After MTI engineers saw erosion on the February 3, 1984 flight, they filed a problem report, and O-rings were entered into formal problem tracking systems at both Marshall and MTI. In an action that illustrates the difficulty of accepting bad news, MTI claimed (in a subsequent briefing to Marshall) that the problem was not serious because even if the primary O-ring were damaged, the second ring would provide redundancy and seal [3, V. 1, p. 128]. Tests done more than a year earlier had shown the secondary O-rings to be unreliable because of a phenomenon called joint rotation. Joint rotation means that under the pressure of launch, the two sides of the O-ring joint bent apart, widening the gap the O-ring had to seal. Joint rotation was apparently especially hard on the secondary ring, making it likely to pull completely out of its groove and never seal at all.

As a result of these tests, Marshall had changed the joint's classification from criticality 1R (a critical system with backup) to criticality 1 (a critical system without such backup) in December 1982. However, those involved in shuttle design apparently found this change hard to accept. Marshall personnel, for instance, apparently believed that they had redundancy in all but exceptional cases [3, V. 1, p. 128]. At MTI, the difficulty of belief was even more marked. Some MTI engineers and officials told the Presidential Commission that they were not notified of the criticality rating change, although their names appear on distribution lists on MTI documents [3, V. 1, p. 128]. Did these people lie about being notified? It seems more likely that they literally could not keep the bad news in mind.

Marshall's response to MTI's briefing illustrates what would be a pattern in their reaction to the O-ring difficulties. Some people at Marshall were willing to say that there was a serious problem—as long as any failure was perceived as MTI's. On February 28, for instance, John Miller, chief of the solid rocket motor branch at Marshall, wrote to his superior, George Hardy, through project engineer Keith Coates, urging that tests be done to see if the leak checks were causing problems. In an unusual recognition of the seriousness of the matter, Miller said O-ring failure could be "catastrophic" [3, V. 1, p. 245]. The next day, Coates also wrote to Hardy saying that MTI's briefing had minimized the extent of joint rotation possible and thus was too optimistic [3, V. 1, p. 128].

When the O-ring problem had to be claimed as Marshall's, however, and NASA had to be informed, Marshall, too, became optimistic. On March 8, 1984, a Flight Readiness Review for the eleventh shuttle flight was held at Marshall. These reviews were held at four levels, with each level resolving what problems it could and passing unresolved issues on to the next higher one. The March 8 meeting was Level III, meaning that it was a meeting between Marshall and its contractors and that the highest officials present were the Marshall project managers. At the meeting, MTI reported that maximum erosion on the O-rings would be 0.09 inch and that tests had shown that the rings would function with

0.095 inch of erosion. The 0.005-inch difference appears to be an extremely small safety margin. Rather than report a serious problem to a Level II meeting at Johnson Space Center, however, Marshall apparently accepted the margin, because this same information was entered in the Marshall problem assessment report with a note that future flights need therefore not be delayed. The Marshall problem tracking record reads: "Remedial action—none required" [3, V. 1, p. 128].

The 0.005-inch safety margin was also used as a rationale to justify no flight interruptions at the Level I briefing of top NASA personnel on March 27. NASA accepted Marshall's recommendation but wrote to Lawrence Mulloy, SRB project manager at Marshall, asking for further study of the O-rings. Mulloy had Marshall engineer Lawrence Wear ask MTI to identify the cause of erosion, determine its seriousness, and define any necessary changes.

Internal MTI documents show that the contractor was examining the problem but with little sense of urgency, again evidencing the tendency to see the problem in the best light possible. MTI analyzed the erosion history and test data and, on May 4, presented Marshall with a plan for studying the O-rings to produce the information NASA had asked for. The information was not actually produced, however, until a briefing on August 19, 1985—16 months later.

Despite its optimism to NASA, Marshall was apparently uncomfortable with this pace and pressed for prompter action than MTI was giving. On July 2, L. H. Sayers, MTI's Director of Engineering Design, suggested by phone to Marshall's Ben Powers that tests done to date by MTI were sufficient [3, V. 1, p. 134]. This again implies that MTI did not perceive the matter to be crucial or dangerous.

Early signs of serious O-ring problems, then, were generally not believed at MTI, were accepted at Marshall only when it was possible to see the problem as MTI's, and were not sent upward to NASA headquarters.

CONTINUED BAD NEWS REJECTION DESPITE CONTRADICTORY EVIDENCE

The optimistic view of the O-rings persisted at both MTI and Marshall over the 1984–85 period despite mounting evidence that the rings were not functioning well. This evidence had to do with the effect of cold on the rings and the amount of erosion that could occur in an O-ring.

On January 24, 1985, the fifteenth shuttle flight was launched at a temperature of 53° F, the lowest up to that time. It showed much greater O-ring erosion than any previous flight. Tests have since shown that cold reduces O-ring resiliency and increases the time to seal, thereby exacerbating the problems the joint was already experiencing. The damage on the fifteenth flight was severe enough to bring concern about the rings to the fore again. On January 31, Mulloy wrote to Wear asking him to get MTI to prepare information on O-ring erosion for a Level III Flight Readiness Review scheduled for February 8. At that review, MTI personnel mentioned the cold as a factor in the damage but labeled the risk "acceptable," mostly because they assumed the secondary ring would seal if the first one failed.

An optimistic interpretation of the data on cold was held by both managers and engineers at MTI, and one of the primary advocates of continued launch was MTI engineer Roger Boisjoly, who would later be one of the primary opponents to the launch of Challenger. The split between managerial and engineering interpretation of the data did not develop for four or five more months. MTI's claim that the secondary O-rings would seal is a further example of retaining previous theories in the face of contradictory evidence because, as noted above, the redundancy of the secondary ring had been in question since late 1982.

As they had done earlier, Marshall management accepted MTI's rationale, at least in what it told NASA. A Level I Review was held on February 21. At this meeting, the influence of temperature was not mentioned, and only a single reference was made to O-ring erosion, saying redundancy made the risk acceptable [3, V. 1, p. 136]. Although each of the next four flights experienced joint seal problems, neither MTI nor Marshall seemed unduly concerned. Perhaps the very frequency of the problem added to its acceptability because the damage kept occurring with no serious consequences.

On June 25, however, one of the joints from a flight that had been launched April 29 was examined and found to have severe erosion of not only the primary ring, but even the secondary ring, calling its redundancy into question once again. As mentioned above, MTI had predicted maximum erosion of 0.09 inch for the primary seal. The primary O-ring on this joint was eroded 0.171 inch, almost double the predicted maximum and far beyond the 0.095 inch MTI had claimed to know was safe. This was bad news indeed. It could not be ignored, since an engineer from NASA headquarters was present when the damage was discovered. The engineer wrote to Michael Weeks at NASA on June 28, reporting the damage. The joint affected was a nozzle joint—that is, the joint linking the SRB to the flared section at its base. The NASA engineer blamed the damage on the fact that, in contrast to the rest of the SRB joints, which were now pressure-tested at 200 psi, nozzle joints were still being tested at only 100 psi, which might have permitted a defective ring to escape notice [3, V. 1, pp. 137–38].

In July 1985, Mulloy placed a launch constraint on the nozzle joints. This, in theory, meant that no other flights would take place until O-ring erosion at the nozzle joint had been fixed or shown not to be a problem. By including only the nozzle joints in the constraint, Mulloy was taking the most optimistic view possible of the problem. He reasoned that the nozzle joint had failed, not because of defective design, but because of a defective ring that had escaped notice in the nozzle joint's less rigorous leak test. Thus he believed that the leak test, and not the joint, was problematic. As a consequence, although he had assigned the launch constraint, Mulloy waived it for every subsequent flight, including Challenger, believing he was justified because Marshall increased to 200 psi the leak check pressure on the nozzle joint.

The launch constraint was treated as bad news by both MTI and Marshall. MTI officials testifying before the Commission all said they did not know about it, although subsequent MTI documents refer by document number to the report imposing the constraint [3, V. 1, p. 137]. The officials apparently did not take the news in. NASA officials, on the other hand, seem genuinely not to have been informed of the constraint, although regulations required that Level II be told. Marshall seems to have kept the news to itself rather than pass it out and up to NASA [3, V. 1, p. 138].

INTERNAL VERSUS EXTERNAL COMMUNICATION OF CONCERN FROM MTI ENGINEERS

Despite MTI's ignorance of the launch constraint, the damage discovered in late June seems to have galvanized its engineers into action. Among them, at least, there seems to have been increased recognition of the problem's existence and its seriousness. MTI engineer Roger Boisjoly, for instance, became increasingly insistent about the potential danger from the O-rings. On July 22, his activity report predicted loss of the contract or flight failure if no solution was found. On July 31, he sent the following memo to R. K. Lund, MTI's Vice President of Engineering [3, V. 1, pp. 249–50]:

SUBJECT: *SRM O-ring Erosion/Potential Failure Criticality.*

This letter is written to insure that management is fully aware of the seriousness of the current O-ring erosion problem in the SRM joints from an engineering standpoint. The mistakenly accepted position on the joint problem was to fly without fear of failure and to run a series of design evaluations which would ultimately lead to a solution or at least a significant reduction of the erosion problem. This position is now drastically changed as a result of the SRM 16A nozzle joint erosion which eroded a secondary O-ring with the primary O-ring never sealing.

If the same scenario should occur in a field joint (and it could), then it is a jump ball as to the success or failure of the joint because the secondary O-ring cannot respond to the clevis opening rate and may not be capable of pressurization. The result would be a catastrophe of the highest order—loss of human life.

An unofficial team [a memo defining the team and its purpose was never published] with leader was formed on 19 July 1985 and was tasked with solving the problem for both the short and long term. This unofficial team is essentially nonexistent at this time. In my opinion, the team must be officially given the responsibility and the authority to execute the work that needs to be done on a non-interference basis (full time assignment until completed).

It is my honest and very real fear that if we do not take immediate action to dedicate a team to solve the problem with the field joint having the number one priority, then we stand in jeopardy of losing a flight along with all the launch pad facilities.

It is evident from this memo that Boisjoly's interpretation of the data had changed and that he was trying to communicate his new interpretation to his management. This memo does not give much space to new or old factual information, but rather concentrates on what the facts mean. Boisjoly's concern is evidenced both by the way he faults his own company (particularly his own management) and by his use of emotional language unusual in engineering documents. The memo implies, for instance, that MTI management may have an inaccurate understanding of the situation. The company's previous position was "mistakenly accepted." The company had planned to solve the problem but had not actually gotten to work to do so. The situation had changed "drastically," and a "catastrophe" could result. Boisjoly's memo and subsequent similar ones evidently had some effect on their receiver, because on the night before Challenger flew, Lund did at least begin by recommending against launch. Boisjoly's concern, however, was kept within MTI. He marked his memo *COMPANY PRIVATE* at both top and bottom. Although he was sufficiently alarmed to try to reach his superiors, he still attempted to keep bad news from the prying eyes of outsiders.

Bad news went to Marshall only in response to specific questions from them. On August 9, MTI engineer Brian Russell wrote to Marshall's Jim Thomas about results of tests investigating the effect of cold on the O-rings. The tests had been initiated after the January 24 flight showed such severe damage. The tone of Russell's letter makes an instructive contrast to Boisjoly's [3, V. 5, pp. 1568–69]:

SUBJECT: *Actions Pertaining to SRM Field Joint Secondary Seal.*

Per your request, this letter contains the answers to the two questions you asked at the July Problem Review Board telecon.

1. *Question:* If the field joint secondary seal lifts off the metal mating surfaces during motor pressurization, how soon will it return to a position where contact is re-established?

    Answer: Bench test data indicate that the o-ring resiliency (its capability to follow the metal) is a function of temperature and the rate of case expansion. MTI measured the force of the o-ring against Instron plattens, which simulated the nominal squeeze on the o-ring and approximated the case expansion distance and rate.

    At 100° F the o-ring maintained contact. At 75° F the o-ring lost contact for 2.4 seconds. At 50° F the o-ring did not re-establish contact in ten minutes at which time the test was terminated.

    The conclusion is that secondary sealing capability in the SRM field joint cannot be guaranteed.

2. *Question:* If the primary o-ring does not seal, will the secondary seal seat in sufficient time to prevent joint leakage?

    Answer: MTI has no reason to suspect that the primary seal would ever fail after pressure equilibrium is reached, i.e., after the ignition transient. If the primary o-ring were to fail from 0 to 170 milliseconds, there is a very high probability that the secondary o-ring would hold pressure since the case has not expanded appreciably at this point. If the primary seal were to fail from 170 to 330 milliseconds, the probability of the secondary seal holding is reduced. From 330 to 660 milliseconds the chance of the secondary seal holding is small. This is a direct result of the o-ring's slow response compared to the metal case segments as the joint rotates.

Please call me or Mr. Roger Boisjoly if you have additional questions concerning this issue.

Russell and Boisjoly were actually of much the same opinion on the dangers of the joint, and Boisjoly helped Russell write the memo shown here, but Russell is speaking to outsiders and this affects the memo. The memo is not in any way untruthful, but it is not very communicative. In a sense, it is the opposite of Boisjoly's memo. It gives just the facts, providing little interpretation. Its tone is adamantly objective, in contrast to Boisjoly's more emotional one. In retrospect, some of the facts Russell gives should have been frightening. Note, for instance, that if there was joint rotation, the secondary O-ring never sealed when it was tested at 50° F. The conclusion reached from this information was "that secondary sealing . . . cannot be guaranteed." This is a negative wording of the finding, which was that at 50° F or below, MTI could pretty well guarantee no secondary seal.

That this memo did not communicate its intent is shown by the fact that the people who read it were uncertain about what it meant. Thomas copied the memo to be sent to NASA headquarters, but when the memo went through Mulloy's office for his signature, Mulloy returned it to Thomas saying it sounded like old news. The NASA official to whom Thomas was sending it has since said that even had he received the memo, he might not have understood it. "I don't know if anybody at that time understood the joint well enough to realize that the data was crucial," he said. When Mulloy was asked why he had not treated the temperature data as more important, he said he had not realized its significance, adding, "There were a whole lot of people who weren't smart enough to look behind the veil and say, 'Gee, I wonder what this means'" [9]. As can be seen, the urgency in Boisjoly's memo had not been conveyed to Marshall. Marshall officials had been given the facts about the effects of cold on the O-rings. They did not, however, interpret those facts in the same alarmed manner Boisjoly and Russell did, and Russell's memo did not attempt to communicate the more pessimistic interpretation.

Thus, MTI engineers concluded that the O-ring problems were serious before their management did. However, in their written communication, they varied the extent to which they voiced that seriousness, depending on whether their audience was internal or external.

THE SPLIT BETWEEN MANAGERS AND ENGINEERS

As is suggested by Boisjoly's memo above, MTI managers and engineers were beginning to disagree over the seriousness of the O-ring problem, and engineers had a difficult time communicating their view upward. Support from the mini-culture of the task force probably made it easier than it had been previously for engineers to recognize the problem and speak up about it. On October 1, MTI engineers Roger Ebeling and S. R. Stein complained in separate internal MTI memos to management that the O-ring task force was being slowed by administrative delays and lack of cooperation [3, V. 1, pp. 252–53]. Both men complained that, although the task force members regarded their work as urgent, administrators required that all testing and design be done according to routines established for more leisurely long-term development.

On October 3, the team met with Joe Kilminster, MTI's Vice President of Space Programs, to discuss these administrative difficulties, but apparently the members were not successful in convincing him of the gravity of the situation. On October 4, Roger Boisjoly's activity report complained bitterly that "upper management apparently feels" MTI has the SRB contract "for sure and the customer be damned" [3, V. 1, p. 255]. In December, Ebeling actually told fellow task force members that MTI should not ship any more SRBs to Marshall until the problem was solved. However, consistent with patterns of bad news transmittal, he did not tell this to any of his superiors [3, V. 1, p. 142].

In January 1986, final preparation for Challenger's flight began. A launch scheduled for January 27 was cancelled and rescheduled for the next day. The temperature at launch time was 36° F, 17° colder than it had been for any previous launch. When MTI engineers, including Ebeling, Russell, and Boisjoly, heard of the predicted low temperatures, they became alarmed enough to convince Lund, their Vice President of Engineering, to recommend that the launch be delayed until the temperature of the joints reached 53° F, the previous lowest launch temperature. In their argument, they cited the information from Russell's memo and the severe erosion from the 53° F launch the previous January.

In a teleconference involving numerous MTI and Marshall managers and engineers, Lund, MTI's Vice President of Engineering, did recommend delaying launch. Marshall was apparently surprised by MTI's action and, refusing to accept the bad news, they resisted the recommendation. George Hardy, who headed engineering at Marshall, said he was "appalled" [3, V. 1, p. 94]. He later testified to the Presidential Commission that he meant he was appalled at MTI's data, but MTI personnel believed he said he was appalled at their recommendation. Mulloy apparently asked if they expected him to wait until April to launch.

In general, Marshall challenged, not MTI's facts, but the conclusions drawn from them. Mulloy and Hardy conceded that the primary ring might be slower to seal in the cold and that the increased time to seal would allow more erosion to take place. But they argued that the ring would seal eventually and that the joint could sustain three times the worst erosion they had seen to date and still have a good seal. In the time before the primary ring sealed, joint rotation would not yet have taken place; therefore, the secondary ring would still be good. Moreover, they believed that temperature could not be the deciding factor because, although they had had severe erosion at 53° F, they had also had it at 75° F. (It is true that erosion occurred at various temperatures. However, the Commission later pointed out that 15 percent of the flights launched above 65° F had O-ring anomalies, while 100 percent of those launched below that temperature had them.) In the face of Marshall's opposition, Joe Kilminster, MTI's Vice-President of Space Programs, asked that MTI have a private caucus off the phone line.

During the caucus, it became obvious that MTI was split along role lines. The engineers continued to argue against launch. Boisjoly says that, in the caucus, "there was never one comment in favor . . . of launching by any engineer or other nonmanagement person in the room" [3, V. 1, p. 93]. And in words that describe his listeners shutting out bad news, he says he himself argued until it became clear "that no one wanted to hear what I had to say" [3, V. 4, p. 697].

At this point, Jerald Mason, MTI's Senior Vice President, said it was obvious that all present would not reach agreement and that a management decision would have to be made. He polled the other three vice presidents in the room, first asking Lund, who had presented the recommendation not to launch, to take off his "engineering hat" and put on his "management hat" [3, V. 1, p. 94]. When Lund changed his role, he changed his position, and the four managers voted unanimously to launch. Engineer Brian Russell describes an atmosphere in which it was difficult to maintain opposition to the launch, wondering "whether I would have the courage, if asked, . . . to stand up and say no" [3, V. 4, p. 822]. As it happened, no engineer was asked to vote, and MTI went back on the conference line and reversed its earlier warning.

During the time MTI was caucusing, Marshall engineer Ben Powers also told his immediate supervisors that he agreed with MTI's recommendation not to launch, but his supervisors did not pass his view upward to Hardy and Mulloy. Similarly, Hardy and Mulloy did not pass on to NASA officials the fact that MTI engineers were opposed to the launch. At 11:38 the next morning, Challenger took off.

CONCLUSION

Looking at prelaunch miscommunications, then, several factors are apparent. First, no one at MTI or Marshall wanted to believe the growing evidence of O-ring problems. Second, even when MTI engineers came to believe that a problem existed, they had a difficult time convincing their management, with its different perspective on operations, to interpret the facts in the same light. In turn, on the night before launch, MTI personnel were unable to convince Marshall of the situation's gravity, even though they looked at the same facts, because Marshall, too, saw things differently. Finally, both engineers and managers at MTI were especially reluctant to communicate bad news to those outside the company.

All of this suggests a number of precautions engineers and their managers might take in the face of the same kind of pressure-induced miscommunication. From a manager's point of view, one of the most important precautions is to establish an atmosphere in which engineers feel free to communicate bad news as well as good. A number of writers have offered advice on how this can be done [8, 10, 11, 12, 13]. Establishing an open atmosphere takes time and a concerted effort from the whole organization. It cannot be done on short notice when emergencies arise.

In addition, pressures for holding back bad news should be anticipated to be especially strong when contractors are involved. Encouraging bad news transmittal is difficult when the bearer of bad tidings is afraid of losing a contract. Contracts can, perhaps, be designed to lessen this fear, but those issuing the contracts should be alert for any sign of problems, since full disclosure of bad news is unlikely in this situation.

Lastly, managers and engineers alike should anticipate that they are probably erring on the side of optimism in interpreting data bearing on already established designs and programs. Failure to believe bad news is probably caused by a number of factors, including reluctance to admit that one was wrong, fear of practical consequences such as expensive redesign, and a kind of intellectual inertia that makes it easier to persist in an already established belief than to change it. Such optimism can, however, have disastrous consequences, especially when coupled with other forces, as the Challenger accident demonstrates.

REFERENCES

1. Gregory, K. L., "Native-View Paradigms: Multiple Cultures and Culture Conflicts in Organizations," *Administrative Science Quarterly* 28 (1983), 359–76.
2. Riley, P., "A Structurationist Account of Political Culture," *Administrative Science Quarterly* 28 (1983), 414–37.
3. Presidential Commission on the Space Shuttle Challenger Accident, *Report of the Presidential Commission on the Space Shuttle Accident,* 5 vols, Washington: GPO, 1986.
4. Housel, T. J., and Davis, W. E., "The Reduction of Upward Communication Distortion," *Journal of Business Communication* 14 (1977), 49–65.
5. Frank, A. D., "Can Water Flow Uphill?" *Training and Development Journal* 38 (1984), 118–28.
6. Rasberry, R. W., and Lemoine, L. F., *Effective Managerial Communication,* Boston: Kent, 1986, pp. 63–97.
7. Applebaum, R. L., and Anatol, K. W. E., *Effective Oral Communication for Business and the Professions,* Chicago: SRA, 1982, pp. 259–262.
8. Steele, F., *The Open Organization: The Impact of Secrecy and Disclosure on People and Organizations,* Reading, MA: Addison-Wesley, 1975.
9. Zaldivar, R. A., "Rocket Chief Was Warned and Failed to Act," *Detroit Free Press,* May 23, 1986.
10. Vance, C. C., "How to Encourage Upward Communication from Your Employees," *Association Management* 28 (1976), 56–59.
11. Driver, R. W., "Opening the Channels of Upward Communication," *Supervisory Management* 25 (1980), 24–29.
12. Macleod, J. S., "How to Unmuzzle Employees," *Employment Relations Today* 11 (1984), 49–54.
13. Thomas, P., "Plugging the Communication Channel—How Managers Stop Upward Communication," *Supervisory Management* 30 (1985), 7–10.

Darrell Huff

How to Lie with Statistics

Darrell Huff, a freelance writer, expanded this article into a book with the same title (Norton, 1954).

"The average Yaleman, Class of '24," *Time* magazine reported last year after reading something in the New York *Sun,* a newspaper published in those days, "makes $25,111 a year."

Well, good for him!

But, come to think of it, what does this improbably precise and salubrious figure mean? Is it, as it appears to be, evidence that if you send your boy to Yale you won't have to work in your old age and neither will he? Is this average a mean or is it a median? What kind of sample is it based on? You could lump one Texas oilman with two hundred hungry freelance writers and report *their* average income as $25,000-odd a year. The arithmetic is impeccable, the figure is convincingly precise, and the amount of meaning there is in it you could put in your eye.

In just such ways is the secret language of statistics, so appealing in a fact-minded culture, being used to sensationalize, inflate, confuse, and oversimplify. Statistical terms are necessary in reporting the mass data of social and economic trends, business conditions, "opinion" polls, this year's census. But without writers who use the words with honesty and understanding and readers who know what they mean, the result can only be semantic nonsense.

In popular writing on scientific research, the abused statistic is almost crowding out the picture of the white-jacketed hero laboring overtime without time-and-a-half in an ill-lit laboratory. Like the "little dash of powder, little pot of paint," statistics are making many an important fact "look like what she ain't." Here are some of the ways it is done.

The sample with the built-in bias. Our Yale men—or Yalemen, as they say in the Time-Life building—belong to this flourishing group. The exaggerated estimate of their income is not based on all members of the class nor on a random or representative sample of them. At least two interesting categories of 1924-model Yale men have been excluded.

First there are those whose present addresses are unknown to their classmates. Wouldn't you bet that these lost sheep are earning less than the boys from prominent families and the others who can be handily reached from a Wall Street office?

There are those who chucked the questionnaire into the nearest wastebasket. Maybe they didn't answer because they were not making enough money to brag about. Like the fellow who found a note clipped to his first pay check suggesting that he consider the amount of his salary confidential: "Don't worry," he told the boss. "I'm just as ashamed of it as you are."

Omitted from our sample then are just the two groups most likely to depress the average. The $25,111 figure is beginning to account for itself. It may indeed be a true figure for those of the Class of '24 whose addresses are known and who are willing to stand up and tell how much they earn. But even that requires a possibly dangerous assumption that the gentlemen are telling the truth.

To be dependable to any useful degree at all, a sampling study must use a representative sample (which can lead to trouble too) or a truly random one. If *all* the Class of '24 is included, that's all right. If every tenth name on a complete list is used, that is all right too, and so is drawing an adequate number of names out of a hat. The test is this: Does every name in the group have an equal chance to be in the sample?

You'll recall that ignoring this requirement was what produced the *Literary Digest's* famed fiasco.* When names for polling were taken only from telephone books and subscription lists, people who did not have telephones or *Literary Digest* subscriptions had no chance to be in the sample. They possibly did not mind this underprivilege a bit, but their absence was in the end very hard on the magazine that relied on the figures.

This leads to a moral: You can prove about anything you want to by letting your sample bias itself. As a consumer of statistical data—a reader, for example, of a news magazine—remember that no statistical conclusion can rise above the quality of the sample it is based upon. In the absence of information about the procedures behind it, you are not warranted in giving any credence at all to the result.

The truncated, or gee-whiz, graph. If you want to show some statistical information quickly and clearly, draw a picture of it. Graphic presentation is the thing today. If you don't mind misleading the hasty looker, or if you quite clearly *want* to deceive him, you can save some space by chopping the bottom off many kinds of graph.

---

*Editor's note: The *Literary Digest* predicted that Alfred Landon would defeat Franklin Roosevelt in the 1936 presidential election. Landon carried only two states.

Suppose you are showing the upward trend of national income month by month for a year. The total rise, as in one recent year, is 7 percent. It looks like this:

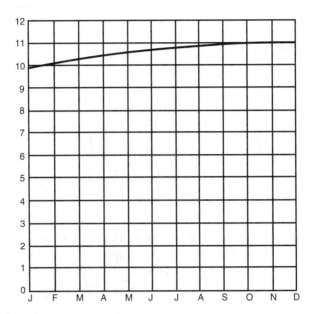

That is clear enough. Anybody can see that the trend is slightly upward. You are showing a 7 percent increase, and that is exactly what it looks like.

But it lacks schmaltz. So you chop off the bottom, this way:

The figures are the same. It is the same graph and nothing has been falsified—except the impression that it gives. Anyone looking at it can just feel prosperity throbbing in the arteries of the country. It is a subtler equivalent of editing "National income rose 7 percent" into ". . . climbed a whopping 7 percent."

It is vastly more effective, however, because of that illusion of objectivity.

The souped-up graph. Sometimes truncating is not enough. The trifling rise in something or other still looks almost as insignficant as it is. You can make that 7 percent look livelier than 100 percent ordinarily does. Simply change the proportion between the ordinate and the abscissa. There's no rule against it, and it does give your graph a prettier shape.

But it exaggerates, to say the least, something awful:

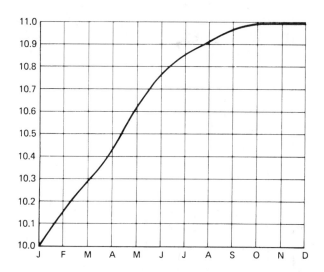

The well-chosen average. I live near a country neighborhood for which I can report an average income of $15,000. I could also report it as $3,500.

If I should want to sell real estate hereabouts to people having a high snobbery content, the first figure would be handy. The second figure, however, is the one to use in an argument against raising taxes, or the local bus fare.

Both are legitimate averages, legally arrived at. Yet it is obvious that at least one of them must be as misleading as an out-and-out lie. The $15,000-figure is a mean, the arithmetic average of the incomes of all the families in the community. The smaller figure is a median; it might be called the income of the average family in the group. It indicates that half the families have less than $3,500 a year and half have more.

Here is where some of the confusion about averages comes from. Many human characteristics have the grace to fall into what is called the "normal" distribution. If you draw a picture of it, you get a curve that is shaped like a bell. Mean and median fall at about the same point, so it doesn't make very much difference which you use.

But some things refuse to follow this neat curve. Income is one of them. Incomes for most large areas will range from under $1,000 a year to upward of $50,000. Almost everybody will be under $10,000, way over on the lefthand side of that curve.

One of the things that made the income figure for the "average Yaleman" meaningless is that we are not told whether it is a mean or a median. It is not that one type of average is invariably better than the other; it depends upon what you are talking about. But neither gives you any real information—and either may be highly misleading—unless you know which of those two kinds of average it is.

In the country neighborhood I mentioned, almost everyone has less than the average—the mean, that is—of $10,500. These people are all small farmers, except for a trio of millionaire week-enders who bring up the mean enormously. You can be pretty sure that when an income average is given in the form of a mean nearly everybody has less than that.

The insignificant difference or the elusive error. Your two children Peter and Linda (we might as well give them modish names while we're about it) take intelligence tests. Peter's IQ, you learn, is 98 and Linda's is 101. Aha! Linda is your brighter child.

Is she? An intelligence test is, or purports to be, a sampling of intellect. An IQ, like other products of sampling, is a figure with a statistical error, which expresses the precison or reliability of the figure. The size of this probable error can be calculated. For their test the makers of the much-used Revised Stanford-Binet have found it to be about 3 percent. So Peter's indicated IQ of 98 really means only that there is an even chance that it falls between 95 and 101. There is an equal probability that it falls somewhere else—below 95 or above 101. Similarly, Linda's has no better than a fifty-fifty chance of being within the fairly sizeable range of 98 to 104.

You can work out some comparisons from that. One is that there is rather better than one chance in four that Peter, with his lower IQ rating, is really at least three points smarter than Linda. A statistician doesn't like to consider a difference significant unless you can hand him odds a lot longer than that.

Ignoring the error in a sampling study leads to all kinds of silly conclusions. There are magazine editors to whom readership surveys are gospel; with a 40 percent readership reported for one article and a 35 percent for another, they demand more like the first. I've seen even smaller differences given tremendous weight, because statistics are a mystery and numbers are impressive. The same thing goes for market surveys and so-called public opinion polls. The rule is that you cannot make a valid comparison between two such figures unless you know the deviations. And unless the difference between the figures is many times greater than the probable error of each, you have only a guess that the one appearing greater really is.

Otherwise you are like the man choosing a camp site from a report of mean temperature alone. One place in California with a mean annual temperature of 61 is San Nicolas Island on the south coast, where it always stays in the comfortable range between 47 and 87. Another with a mean of 61 is in the inland desert, where the thermometer hops around from 15 to 104. The deviation from the mean marks the difference, and you can freeze or roast if you ignore it.

The one-dimensional picture. Suppose you have just two or three figures to compare—say the average weekly wage of carpenters in the United States and another country. The sums might be $60 and $30. An ordinary bar chart makes the difference graphic.

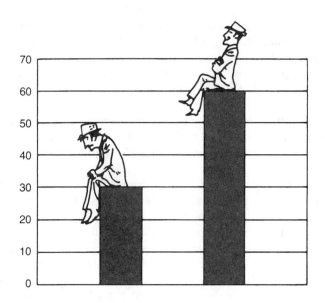

That is an honest picture. It looks good for American carpenters, but perhaps it does not have quite the oomph you are after. Can't you make that difference appear overwhelming and at the same time give it what I am afraid is known as eye-appeal? Of course you can. Following tradition, you represent these sums by pictures of money bags. If the $30 bag is one inch high, you draw the $60 bag two inches high. That's in proportion, isn't it?

The catch is, of course, that the American's money bag, being twice as tall as that of the $30 man, covers an area on your page four times as great. And since your two-dimensional picture represents an object that would in fact have three dimensions, the money bags actually would differ much more than that. The volumes of any two similar solids vary as the cubes of their heights. If the unfortunate foreigner's bag holds $30 worth of dimes, the American's would hold not $60 but a neat $240.

You didn't say that, though, did you? And you can't be blamed, you're only doing it the way practically everybody else does.

The ever-impressive decimal. For a spurious air of precision that will lend all kinds of weight to the must disreputable statistics, consider the decimal.

Ask a hundred citizens how many hours they slept last night. Come out with a total of, say, 781.3. Your data are far from precise to begin with. Most people will miss their guess by fifteen minutes or more and some will recall five sleepless minutes as half a night of tossing insomnia.

But go ahead, do your arithmetic, announce that people sleep an average of 7.813 hours a night. You will sound as if you knew precisely what you are talking about. If you were foolish enough to say 7.8 (or "almost" 8) hours it would sound like what it was—an approximation.

The semiattached figure. If you can't prove what you want to prove, demonstrate something else and pretend that they are the same thing. In the daze that follows the collision of statistics with the human mind, hardly anybody will notice the difference. The semiattached figure is a durable device guaranteed to stand you in good stead. It always has.

If you can't prove that your nostrum cures colds, publish a sworn laboratory report that the stuff killed 31,108 germs in a test tube in eleven seconds. There may be no connection at all between assorted germs in a test tube and the whatever-it-is that produces colds, but people aren't going to reason that sharply, especially while sniffling.

Maybe that one is too obvious and people are beginning to catch on. Here is a trickier version.

Let us say that in a period when race prejudice is growing it is to your advantage to "prove" otherwise. You will not find it a difficult assignment.

Ask that usual cross section of the population if they think . . . [Blacks] have as good a chance as white people to get jobs. Ask again a few months later. As Princeton's Office of Public Opinion Research has found out, people who are most unsympathetic to . . . [Blacks] are the ones most likely to answer yes to this question.

As prejudice increases in a country, the percentage of affirmative answers you will get to this question will become larger. What looks on the face of it like growing opportunity for . . . [Blacks] actually is mounting prejudice and nothing else. You have achieved something rather remarkable: the worse things get, the better your survey makes them look.

The unwarranted assumption, or *post hoc* rides again. The interrelation of cause and effect, so often obscure anyway, can be most neatly hidden in statistical data.

Somebody once went to a good deal of trouble to find out if cigarette smokers make lower college grades than non-smokers. They did. This naturally pleased many people, and they made much of it.

The unwarranted assumption, of course, was that smoking had produced dull minds. It seemed vaguely reasonable on the face of it, so it was quite widely accepted. But it really proved nothing of the sort, any more than it proved that poor grades drive students to the solace of tobacco. Maybe the relationship worked in one direction, maybe in the other. And maybe all this is only an indication that the sociable sort of fellow who is likely to take his books less than seriously is also likely to sit around and smoke many cigarettes.

Permitting statistical treatment to befog casual relationships is little better than superstition. It is like the conviction among the people of Hebrides that body lice produce good health. Observation over the centuries had taught them that people in good health had lice and sick people often did not. *Ergo,* lice made a man healthy. Everybody should have them.

Scantier evidence, treated statistically at the expense of common sense, has made many a medical fortune and many a medical article in magazines, including professional ones. More sophisticated observers finally got things straightened out in the Hebrides. As it turned out, almost everybody in those circles had lice most of the time. But when a man took a fever (quite possibly carried to him by those same lice) and his body became hot, the lice left.

Here you have cause and effect not only reversed, but intermingled.

There you have a primer in some ways to use statistics to deceive. A well-wrapped statistic is better than Hitler's "big lie": it misleads, yet it can't be pinned onto you.

Is this little list altogether too much like a manual for swindlers? Perhaps I can justify it in the manner of the retired burglar whose published reminiscences amounted to a graduate course in how to pick a lock and muffle a footfall: The crooks already know these tricks. Honest men must learn them in self-defense.

Dan Jones

Determining the Ethics of Style

Dan Jones is on the faculty of the University of Central Florida.

Doublespeak is not the product of carelessness or sloppy thinking. Indeed, most doublespeak is the product of clear thinking and is carefully designed and constructed to appear to communicate when in fact it doesn't. It is language designed not to lead but mislead. It is language designed to distort reality and corrupt thought.[1]

—*William Lutz*

Sometimes we want to be unclear. We don't know what we're talking about, and we don't want anyone else to know that. Or we do know what we're talking about, and we don't want anyone else to know what we know. On those occasions, we write unclearly deliberately, and if what we write gets the job done, then we say the writing is "good." But "good" has two meanings. Assassins can be "good" at their jobs but not be "good" people. In the same way, writing can be "good" if it gets the job done, but if the job is ethically questionable, then the writing may be bad just because it is so good.[2]

—*Joseph Williams*

What kind of behavior is "prose behavior"? Prose is usually described in a moral vocabulary—"sincere," "open" or "devious," and "hypocritical"—but is this vocabulary justified? Why, for that matter, has it been so moralistic? Why do so many people feel that bad prose threatens the foundations of civilization? And why, in fact, do we think "bad" the right word to use for it?[3]

—*Richard Lanham*

WHAT IS ETHICS?

Simply defined, "Ethics is the study of right and wrong conduct."[4] More broadly, ethics is "the discipline dealing with what is good and bad and with moral duty and obligation."[5] Ethics also means "a set of moral principles or values" or "a theory or system of moral values."[6] More broadly still, ethics may be defined as a guiding philosophy.

Vincent Ruggiero observes that "the focus of ethics is moral situations—that is, *those situations in which there is a choice of behavior involving human values* (those qualities that are regarded as good and desirable)."[7]

An ethical person, then, is someone who is governed by right conduct. But what is right conduct? What is moral? What is good? You'll have to go elsewhere for detailed answers to these questions. A detailed discussion of ethics is beyond the scope of this textbook. After all, ethics is a major branch of philosophy with over a 2500 year history in the West. And this textbook is about how to make your technical prose style more effective, not how to live more ethically. Still, many writing choices are ethical choices, so a chapter outlining these ethical concerns is necessary. . . .

ETHICS AND TECHNICAL PROSE

Just as people must make many ethical decisions throughout their lives, you must make many ethical decisions concerning what you write throughout your career. For example, are you doing your best to document a product honestly and accurately? Are you knowingly omitting any essential information? If you are unclear or imprecise, and if your poor instructions cause injury to someone, are you morally responsible? If you are in a company's marketing department, and you are asked to exaggerate a product's features, are you guilty of lying? If you know you are promising more than you can deliver in a proposal written in response to a request for a proposal (RFP), are you unethical or just keenly competitive? If you fail to point out numerous known bugs in a program in your software manual, are you a good software documentation writer or are you unethical? If you exaggerate your qualifications on your resume or in your cover letter to gain an advantage over the competition, are you unethical? These are just some of many possible scenarios faced by writers of technical prose everywhere.

Why isn't it possible just to make a list of ethical language choices that everyone could agree on and everyone could abide by? Unfortunately, it's not that easy. Suppose that you wrote the clearest instructions you could write, but someone neglected to provide some essential information and a customer is injured following your instructions. Are you unethical in this instance? Suppose because of unreasonable deadlines, you have to cut corners and you just don't have time to document some important features of a software program. Are you a bad person? Suppose you have been told by your boss in marketing that you're expected to hype the product just to keep up with the competition. Are you unethical for trying to keep your job and doing what you are told to do? As these possible scenarios and many others show, it's not always easy to determine which writing decisions are right or wrong, moral or immoral, ethical or unethical.

ETHICS AND THE PROFESSIONS

Of course, ethical concerns are not new to those who must write technical prose. Most professions and professional organizations have published ethical guidelines. The Computer Ethics Institute, for example, published the following guidelines:

The Ten Commandments of Computer Ethics

1. Thou shalt not use a computer to harm other people.
2. Thou shalt not interfere with other people's computer work.
3. Thou shalt not snoop around in other people's computer files.
4. Thou shalt not use a computer to steal.
5. Thou shalt not use a computer to bear false witness.
6. Thou shalt not copy or use proprietary software for which you have not paid.
7. Thou shalt not use other people's computer resources without authorization or proper compensation.
8. Thou shalt not appropriate other people's intellectual output.
9. Thou shalt think about the social consequences of the program you are writing or the system you are designing.
10. Thou shalt always use a computer in ways that insure consideration and respect for your fellow humans.[8]

Search for information on *ethics* using any search engine on the World Wide Web, and you'll see all kinds of databases offering other codes of ethics and all kinds of information on ethics. You'll see medical ethics, business ethics, computer ethics, military ethics, media ethics, journalism ethics, and so on. You'll see links to many professional societies and their published codes of ethics. You'll see course syllabi, papers, online journals, and much more.

It seems as though almost everyone is concerned about ethics in one way or another. Developing ethical guidelines for technical communicators parallels the challenges of doing so for computer professionals and engineers. All three groups have obligations to society, to their employers, to their clients, and to co-professionals and even professional organizations.

In *Computer Ethics*, Tom Forester and Perry Morrison list many ethical questions faced by computer professionals:

*   Is copying software really a form of stealing? What sort of intellectual property rights should software developers have?
*   Are so-called "victimless" crimes (against, e.g., banks) more acceptable than crimes with human victims? Should computer professionals be sued for lax computer security?
*   Is hacking merely a bit of harmless fun or is it a crime equivalent to burglary, forgery and/or theft? Or are hackers to be seen as guardians of our civil liberties?

- Should the creation of viruses be considered deliberate sabotage and be punished accordingly?
- Does information on individuals stored in a computer constitute an intolerable invasion of privacy? How much protection are individuals entitled to?
- Who is responsible for computer malfunctions or errors in computer programs? Should computer companies be made to provide a warranty on software?
- Is "artificial intelligence" a realistic and a proper goal for computer science? Should we trust our lives to allegedly artificially intelligent "expert" systems?
- Should we allow the workplace to be computerized if it de-skills the workforce and/or increases depersonalization, fatigue and boredom?
- Is it OK for computer professionals to make false claims about the capabilities of computers when selling systems or representing computers to the general public? Is it ethical for computer companies to "lock-in" customers to their products?
- Should, indeed, computer professionals be bound by a Code of Conduct and if so, what should it include?[9]

As you would expect, the Society for Technical Communication also has ethical guidelines:

STC Ethical Guidelines for Technical Communicators

Introduction. As technical communicators, we observe the following ethical guidelines in our professional activities. Their purpose is to help us maintain ethical practices.

Legality. We observe the laws and regulations governing our professional activities in the workplace. We meet the terms and obligations of contracts that we undertake. We ensure that all terms of our contractual agreements are consistent with STC Ethical Guidelines.

Honesty. We seek to promote the public good in our activities. To the best of our ability, we provide truthful and accurate communications. We dedicate ourselves to conciseness, clarity, and creativity, striving to address the needs of those who use our products. We alert our clients and employers when we believe material is ambiguous. Before using another person's work, we obtain permission. In cases where individuals are credited, we attribute authorship only to those who have made an original, substantive contribution. We do not perform work outside our job scope during hours compensated by clients or employers, except with their permission; nor do we use their facilities, equipment or supplies without their approval. When we advertize our services, we do so truthfully.

Confidentiality. Respecting the confidentiality of our clients, employers, and professional organizations, we release business-sensitive information only with their consent or when legally required. We acquire releases from clients and employers before including their business-sensitive information in our portfolios or before using such material for a different client or employer or for demo purposes.

Quality. With the goal of producing high-quality work, we negotiate realistic, candid agreement on the schedule, budget, and deliverables with clients and employers in the initial project planning stage. When working on the project, we fulfill our negotiated roles in a timely and responsible manner and meet the stated expectations.

Fairness. We respect cultural variety and other aspects of diversity in our clients, employers, development teams, and audiences. We serve the business interest of our clients and employers, as long as such loyalty does not require us to violate the public good. We avoid conflicts of interest in the fulfillment of our responsibilities and activities. If we are aware of a conflict of interest, we disclose it to those concerned and obtain their approval before proceeding.

Professionalism. We seek candid evaluations of our professional performance from clients and employers. We also provide candid evaluations of communication products and services. We advance the technical communication profession through our integrity, standards, and performance.[10]

Codes of conduct are valuable because they establish ideals and help define the character of a profession. These codes help to establish an atmosphere of professionalism, and they help to encourage members of a profession to act ethically even in the most difficult of circumstances. However, guidelines on paper may have little meaning if people do not value the guidelines. And, of course, no matter how thorough your written guidelines are, there will always be a situation not covered by the guidelines. One practical approach is to have a strategy for determining whether an action is ethical or unethical. The next section offers such a strategy or framework.

CRITERIA FOR JUDGING ETHICAL ACTIONS

The purpose here is to outline briefly one clear approach (of many possible approaches) in order for you to apply some criteria to language issues.

One of the most difficult aspects of ethics is how to evaluate moral issues, how to determine what is and is not ethical. Is there a strategy we can use to help us evaluate actions? Of course, there are many. (In fact, the number is probably countless.) Some of us are guided entirely by our religious beliefs. Others believe that many ethical questions cannot be completely answered by religious beliefs or by one institutional religion and its doctrine. The distinction between a religion and a system of ethics is quite complex, but, it should suffice to say, Protestants, Catholics, Buddhists, Hindus, Muslims, and Jews (to mention only most of the major religious followers) face daily ethical choices just as do sophists, stoics, existentialists, agnostics, and atheists.

In *Thinking Critically about Ethical Issues,* Vincent Ruggiero comments: "To develop a sound and meaningful basis for discussing moral issues, we must find a standard for judging the morality of actions, a standard that is acceptable to men and women of various moral perspectives. In other words, we must find a standard that reflects the principles that most ethical systems have in common."[11]

Ruggiero points to one important principle that underlies almost all ethical systems: the principle of *respect for persons.*[12] This principle has three separate requirements:

> First, that each and every person should be regarded as worthy of sympathetic consideration, and should be so treated; secondly, that no person should be regarded by another as a mere possession, or used as a mere instrument, or treated as a mere obstacle, to another's satisfaction; and thirdly, that persons are not and ought never to be treated in any undertaking as mere expendables.[13]

This principle of respect for persons is either explicit or implicit in most ethical systems.

Ruggiero suggests that three basic criteria emerge from this principle of respect for persons—*obligations, ideals,* and *consequences.* In one way or another, these three criteria are used in most ethical systems to determine what actions are ethical or unethical.[14]

Concerning obligations, "Every significant human action occurs, directly or indirectly, in a context of relationships with others. And relationships usually imply *obligations;* that is, restrictions on our behavior, demands to do something or avoid doing it."[15] One kind of obligation is a formal agreement. Other kinds include those of friendship, citizenship, and business. Obligations, of course, have to be valued: "When we say obligations bind morally, we mean they exist to be honored. To honor them is right; to dishonor them is wrong."[16]

The criterion of ideals is also essential: "In the general sense *ideals* are notions of excellence, goals that bring greater harmony in one's self and between self and others. In ethics, ideals are also specific concepts that assist us in applying the principle of respect for persons in our moral judgments. Some common ideals that figure prominently in ethical reasoning are fairness, tolerance, compassion, loyalty, forgiveness, amity, and peace."[17]

Finally, there is the criterion of consequences:

> *Consequences* are the beneficial or harmful effects that result from the action and affect the people involved, including, of course, the person performing the action. Some consequences are physical; others are emotional. Some occur immediately; others occur only with the passing of time. Some are intended by the person performing the act; others are unintended. Finally, some consequences may be obvious, and others may be very subtle and hidden by appearances. The ethicist is concerned with all significant consequences occurring in a moral context.[18]

As you can imagine, applying these three criteria to any action to determine whether or not the action is ethical is not easy. Obligations may conflict with each other. Ideals may compete with each other and many conflict with obligations, and many actions may produce many consequences, some good and some harmful. To complicate matters further, Ruggiero reminds us that we are also often confronted by moral dilemmas. A moral dilemma "is defined as any situation or predicament that arises from the impossibility of honoring all the moral values that deserve honoring. A moral dilemma exists whenever the conflicting obligations, ideals, and consequences are so very nearly equal in their importance that we feel we cannot choose among them, even though we must."[19]

Ruggiero advises us to use the following approach for determining whether an action (or action in a hypothetical case) is ethical or unethical.

First, he suggests that you study the details of the action. Are there any extenuating circumstances that make this action different from similar actions?

Second, you should identify how the three basic criteria may be relevant to the action: "Are there any *obligations?* What *ideals* are involved? What are the *consequences* of the case? Whom will they affect? In what way?"[20] Then decide if the emphasis is on one of the three criteria or all three almost equally.

Third, you should list all of the possible actions that are available.

Fourth, after reviewing what you now know from the first three steps, decide which action is most ethical.

Ruggiero believes this approach helps people analyze complex moral issues more systematically. You can use this approach to decide what is or is not ethical behavior and to explain your decisions to others.[21]

UNETHICAL LANGUAGE

If you lie to someone in a memo, letter, or manual, are you being unethical? What if you intentionally state something ambiguously? Pretentiously? Deceptively? Of course, the answers to these questions depend on a great deal. You need much more information before you can judge whether or not the action is ethical or unethical. We can use the criteria briefly explained in the preceding section to help us evaluate language choices.

Doublespeak

William Lutz's *Doublespeak* may be viewed as a valuable collection of unethical language used in everyday living, advertising, business communication, around the world, the military, the bureaucracy of government, and for nuclear weapons. Lutz defines doublespeak as the following:

> Doublespeak is language that pretends to communicate but really doesn't. It is language that makes the bad seem good, the negative appear positive, the unpleasant appear attractive or at least tolerable. Doublespeak is language that avoids or shifts responsibility, language that is at variance with its real or purported meaning. It is language that conceals or prevents thought; rather than extending thought, doublespeak limits it.
>
> Doublespeak is not a matter of subjects and verbs agreeing; it is a matter of words and facts agreeing. Basic to doublespeak is incongruity, the incongruity between what is said or left unsaid, and what really is. It is incongruity between the word and the referent, between seem and be, between the essential function of language—communication—and what doublespeak does—mislead, distort, deceive, inflate, circumvent, obfuscate.[22]

. . . [F]or Lutz there are four kinds of doublespeak: euphemisms, jargon, gobbledygook or bureaucratese, and inflated language.

A euphemism becomes doublespeak when it "is used to mislead or deceive."[23] He cites the example of the U.S. State Department announcing it would no longer use the word *killing* in annual reports on the condition of human rights in various countries: "Instead, it would use the phrase 'unlawful or arbitrary deprivation of life,' which the department claimed was more accurate. Its real purpose for using this phrase was simply to avoid discussing the embarrassing situation of government-sanctioned killings in countries that are supported by the United States and have been certified by the United States as respecting the human rights of their citizens."[24]

Jargon can also be doublespeak when it is "pretentious, obscure, and esoteric terminology used to give an air of profundity, authority, and prestige to speakers and their subject matter. Jargon as doublespeak often makes the simple appear complex, the ordinary profound, the obvious insightful. In this sense it is used not to express but impress."[25] Lutz offers the example of National Airlines inserting a footnote to its annual report to stockholders. Because of an airplane crash, National Airlines made an after-tax insurance benefit of $1.7 million dollars. In its footnote in the annual report National Airlines refers to "the involuntary conversion of a 727."[26] In this way National Airlines recognizes the crash of the airplane and the profit, but avoids referring to the accident or the deaths.

Gobbledygook or bureaucratese is doublespeak because this kind of language "is simply a matter of piling on words, of overwhelming the audience with words, the bigger the words and the longer the sentences the better."[27] Lutz cites many examples including the words of Jesse Moore, NASA's associate administrator, when asked during an investigation of the *Challenger* disaster in 1986 if the performance of the shuttle program was improving with each launch or remaining the same. Moore replied:

> I think our performance in terms of the liftoff performance and in terms of the orbital performance, we knew more about the envelope we were operating under, and we have been pretty accurately staying in that. And so I would say the performance has not by design drastically improved. I think we have been able to characterize the performance more as a function of our launch experience as opposed to it improving as a function of time.[28]

Inflated language is also a kind of doublespeak because it "is designed to make the ordinary seem extraordinary; to make everyday things seem impressive; to give an air of importance to people, situations, or things that would not normally be considered important; to make the simple seem complex."[29] Examples of inflated language include Chrysler stating that it "initiates a career alternative enhancement program" instead of admitting it is letting five thousand employees go. Another example is a hospital bill sent to a family for "negative patient care outcome" instead of admitting the patient died.[30]

In all of these examples and many others that Lutz provides, you can see prose that is deliberately ambiguous, pretentious, or deceptive.

According to Lutz, one of "the most chilling and terrifying" uses of doublespeak occurred in 1981 when then Secretary of State Alexander Haig testified before congressional committees about the murder of three American nuns and a Catholic lay worker in El Salvador. The four women were raped and shot at close range. And there was ample evidence that soldiers of the Salvadoran government committed the crime. Secretary Haig made the following comment before the House Foreign Affairs Committee:

> I'd like to suggest to you that some of the investigations would lead one to believe that perhaps the vehicle the nuns were riding in may have tried to run a roadblock, or may accidentally have been perceived to have been doing so, and there'd been an exchange of fire and then perhaps those who inflicted the casualties sought to cover it up. And this could have been at a very low level of both competence and motivation in the context of the issue itself. But the facts on this are not clear enough for anyone to draw a definitive conclusion.[31]

In follow-up questioning the next day, Haig was asked to clarify his comment. He finally said, "What I meant was that if one fellow starts shooting, then the next thing you know they all panic."[32] Lutz suggests that Haig was deliberately vague to imply "that the women were in some way responsible for their own fate."[33] He uses vague expressions such as "would lead one to believe" and "may accidentally have been perceived to have been doing so." He avoids the word "kill" and instead says "inflicted the casualties." In effect, Lutz argues, "the secretary of state has become an apologist for rape and murder."[34]

How can you apply the three criteria of obligation, ideals, and consequences to this action? Haig had an obligation to his employer, the State Department and U.S. government, to diffuse a volatile situation. As a representative of the government, he also had an obligation to tell the public the truth. Ideals involved here range from the public's right to know, the government's need to maintain a good relationship with other governments, to individual commitment to government service. The consequences of Haig's testimony include increased cynicism on the part of the public and increased distrust of public officials, unbelievable and ineffective testimony, and damage to the credibility of the State Department, to mention only a few. Certainly, the bad consequences outweigh any good consequences in this particular action. Therefore, Haig's testimony was clearly unethical.

As for other examples of potentially unethical communication, consider warranties and disclaimers from software developers. As Forester and Morrison note, "Given the incidence of faulty software and system failure in general, it is perhaps not surprising that software developers rarely provide their clients or purchasers with warranties of any substance. Indeed, if one compared this situation with that of almost any other form of purchased goods, then it is difficult not to be amused and perhaps even incredulous."[35] Consider the following example:

> We don't claim Interactive EasyFlow is good for anything—if you think it is, great, but it's up to you to decide. If Interactive EasyFlow doesn't work: tough. If you lose a million because Interactive EasyFlow messes up, it's you that's out of the million, not us. If you don't like this disclaimer: tough. We reserve the right to do the absolute minimum provided by law, up to and including nothing.
>
> This is basically the same disclaimer that comes with all software packages, but ours is in plain English and theirs is in legalese.
>
> We didn't really want to include a disclaimer at all, but our lawyers insisted. We tried to ignore them, but they threatened us with the shark attack at which point we relented.[36]

Is such a disclaimer fair? Ethical?

Other Strategies for Lying with Language

Of course, companies, governments, and people have other ways of using the language unethically besides deliberately choosing to be ambiguous, deceptive, or pretentious. Essential or potentially damaging information may be omitted in a document. What you are told may be true, but you may not be told the whole story. Also, information supporting your prose or verbal comments—statistics, graphics, and photographs, for example—may be distorted in numerous subtle ways. The ways of lying through language are almost countless.

THE ETHICS OF STYLE

. . . Richard Lanham's "The Official Style" . . . [is] an example of a genuine style. In the last chapter of his book *Revising Prose*, Lanham addresses the ethics of style. In this chapter Lanham is concerned with the question of why we should be concerned about revising "The Official Style" into a plain style. Lanham believes there are two kinds of answers to this question.

First, as Lanham explains, "Generally, it helps if you write better prose. It makes for a better statement of purpose when you apply for law school or a job; later on, it will help you write a better legal brief or progress report or memo."[37] Yet Lanham also realizes that writing in a plain style is not always appropriate: "Where The Official Style is mandated by rule or custom (as in some bureaucratic situations it is), plain prose may sound simple-minded or even flip. The sensible procedure here: learn both languages, the plain and The Official Style."[38] Lanham is a pragmatist when it comes to effective writing. He knows, as anyone who writes frequently knows, "*Stylistic* judgment is, last as well as first, always *political* judgment."[39] Different situations or contexts determine different styles. We must adjust the style to suit the occasion.

Second, Lanham offers a more complex answer

> The second kind of answer to "Why bother?" is simply, "Are you willing to sign your name to what you have written? To present yourself in public—whether it matters to anyone else or not—as this kind of person?" In a sense, it is a simple question: "Whatever the advantage—or disadvantage—ought I do this?" The primary moral question: If everyone else is committing murder, ought I do the same? Do you choose to encounter the world on its terms or on your own? A simple question but one we must all answer for ourselves.[40]

For Lanham, the ethics of style is a matter of personal ethics. Lanham speaks of the "primary ground for morality, the self."[41] He refers to this self as both a dynamic and static entity: "It is static when we think of ourselves as having central, fixed selves independent of our surroundings, an 'I' we can remove from society without damage, a central self inside our heads. But it becomes dynamic when we think of ourselves as actors playing social roles, a series of roles that vary with the social situation in which we find ourselves."[42] Lanham observes, "Our complex identity comes from the constant interplay of these two kinds of self. Our final 'self' is usually a mixed one, few of us being completely the same in all situations. . . . If we were completely sincere we would always say exactly what we think—and cause social chaos. If we were always acting an appropriate role, we would be certifiably insane."[43]

Lanham's static and dynamic selves correspond closely to ethos and persona. . . . Ethos is the character of the writer, the writer's integrity, competency, objectivity. Persona is the mask or assumed identity of a writer. In the process of writing, ethos and persona are in constant interplay. Whenever a writer tries to be sincere, at least part of his or her ethos is usually apparent. At the same time, the act of writing requires playing various social roles, requires assumed selves or identities or personas. But where does ethos end and persona begin in the act of writing? Where does ethos end and persona begin in the self? As we mentioned earlier, Lanham suggests our identity lies in the interplay between the static and dynamic selves. And this is the case for ethos and persona.

SUMMARY

Simply defined, ethics is the study of right and wrong conduct. Ethics concerns all kinds of behaviors, not just decisions concerning language. But when decisions concerning language are involved, both ethical and stylistic decisions are involved. Writers of technical prose are faced daily with all kinds of ethical choices, and it's not always easy to make the most ethical decision. Professional or organizational ethical guidelines may provide some help, but, usually, a more systematic approach is more helpful.

Knowing how to determine obligations, ideals, and consequences for a particular action can be particularly helpful in judging the morality of an action involving language. Also, you can use a four-step approach to judge the morality of a language decision—studying the details of an action, identifying how the three basic criteria are relevant to the action, listing all possible actions that are available, and reviewing what you know after these first three steps and deciding which action is most ethical.

Often a style may be ambiguous, deceptive, or pretentious and not be unethical. More often, however, your style may be unethical when you deliberately use ambiguous, deceptive, or pretentious language for questionable obligations or ideals and if there are damaging consequences. The four kinds of doublespeak—euphemisms, jargon, gobbledygook or bureaucratese, and inflated language—are often good examples of unethical language. Of course, there are other strategies for lying with language, by omitting essential or damaging information, for example.

The ethics of style cannot be separated from the ethics of the self. As Lanham suggests, the ethics of style is a matter of personal ethics. Who you are as a writer is a matter of your static and dynamic selves, your ethos and persona.

Ultimately, a bureaucratic style is often an unethical style because it undermines both the ethos of the self and the necessary respect for others that underlies most ethical systems. It depersonalizes the self and society.

NOTES

1. William Lutz, *Doublespeak* (New York: Harper & Row, 1989) 18–19.
2. Joseph Williams, *Style*, 4th ed. (New York: HarperCollins, 1994) 134.
3. Richard Lanham, *Revising Prose*, 3rd ed. (New York: Macmillan, 1992) 96–97.

4. Vincent Ryan Ruggiero, *Thinking Critically about Ethical Issues,* 3rd ed. (Mountain View, CA: Mayfield, 1992) 4.

5. *Merriam-Webster's Tenth Collegiate Dictionary and Thesaurus.* Electronic Edition. CD-ROM, 1995.

6. *Merriam-Webster's Tenth Collegiate Dictionary and Thesaurus.* Electronic Edition. CD-ROM, 1995.

7. Ruggiero, p. 5.

8. Computer Ethics Institute Home Page, http://www.cpsr.org:80/dox/cei.html

9. Tom Forester and Perry Morrison, *Computer Ethics: Cautionary Tales and Ethical Dilemmas in Computing* (Cambridge, MA: MIT P, 1992) 4–5.

10. *STC Membership Directory 1995–96* 42:3A (September 1995) xi.

11. Ruggiero, p. 53.

12. Ruggiero, p. 54.

13. Ruggiero quoting Errol E. Harris, p. 54.

14. Ruggiero, p. 54.

15. Ruggiero, p. 55.

16. Ruggiero, p. 55.

17. Ruggiero, pp. 55–56.

18. Ruggiero, p. 56.

19. Ruggiero, p. 91.

20. Ruggiero, p. 57.

21. Ruggiero, p. 58.

22. Lutz, pp. 1–2.

23. Lutz, p. 3.

24. Lutz, p. 3.

25. Lutz, p. 4.

26. Lutz, p. 4.

27. Lutz, p. 5.

28. Lutz, pp. 5–6.

29. Lutz, p. 6.

30. Lutz, p. 6.

31. Lutz, pp. 17–18.

32. Lutz, p. 18.

33. Lutz, p. 18.

34. Lutz, p. 18.

35. Forester and Morrison, p. 76.

36. Cited in Forester and Morrison, p. 77.

37. Lanham, p. 96.

38. Lanham, p. 96.

39. Lanham, p. 96.

40. Lanham, p. 96.

41. Lanham, p. 97.

42. Lanham, p. 97.

43. Lanham, p. 97.

Carolyn D. Rude

Legal and Ethical Issues in Editing

Carolyn D. Rude is a member of the faculty at Texas Tech University.

Laws and codes of ethics aim to protect the good of a society as well as to protect individual rights. In technical publication, intellectual property laws govern copyright, trade secrets, and trademarks. Laws also require organizations to accept responsibility for product safety. Editors work with other members of product development teams and legal experts to verify adherence to these laws and the representation of them in the text through warnings and notices of copyright, trademarks, or patents.

Codes of ethics are less formally encoded than laws, as are the sanctions for violating them, but, like laws, they aim to protect individuals and groups from harm and to provide opportunities by creating a work environment in which individuals can achieve. They may be encoded as statements of professional responsibility, such as the "Ethical Guidelines" by the Society for Technical Communication, or derive from cultural values of right and wrong, such as the values that individuals should not harm other persons and that they have responsibility for maintaining the quality of the environment.

Legal and ethical issues pertain both to individuals and to organizations. Editors can be most effective as individuals if corporate policies establish commitment to legal and ethical behavior and if corporate procedures allow for review of products and documents by a variety of knowledgeable people, not just the editor or even just the legal department.

This . . . [selection] reviews legal and ethical issues that pertain to technical publication and suggests corporate policies to encourage ethical behavior.

LEGAL ISSUES IN EDITING

Editors share responsibility with writers, researchers, and managers for protecting documents legally and for ensuring that they do not violate intellectual property, product safety, and libel laws. Editors verify that permissions to reprint portions of other people's publications are obtained, and they review instructions and warnings for potentially dangerous products. Using the symbols ©, ®, and ™ indicates copyrights, registered patents, or trademarks.

The editor can fulfill legal responsibilities more readily and thoroughly if the organization's policy manual and style manual include policies and guidelines about intellectual property, product safety, and misrepresentation.

Intellectual Property: Copyright, Trademarks, Patents, Trade Secrets

The law of many countries, the United States among them, recognizes that one can own intellectual property just as one can own land or a business. These protections are intended as incentives to develop ideas or products that may improve the quality of life for the community. Intellectual property includes original works of fiction or nonfiction, artwork and photographs, recordings, computer programs, and any other expression that is fixed in some form—printed, recorded, or posted on the World Wide Web. These expressions are protected by copyright law. Other intellectual property includes work protected as trademarks, patents, and trade secrets. The owner of intellectual property has the legal right to determine if and where the work may be reproduced or used.

Copyright

The United States Copyright Act of 1976 protects authors of "original works of authorship," whether or not the works are published. The Copyright Act gives the owner of a copyright the right to reproduce and distribute the work and to prepare derivative works based on the copyrighted work. No one else has these rights, and reproducing work copyrighted by someone else violates the copyright law. As editor, you verify permission to use copyrighted material and take steps to protect documents published by your company.

Ownership

Copyrights belong to the author who created the work unless the author wrote the work to meet responsibilities of employment. In the case of "work for hire," the employer owns the copyright. Most technical writers and editors work for hire, and the manuals and reports they write are the intellectual property of their employers. Some publishers require authors to surrender the copyright to them. Works by the U.S. government are not eligible for copyright protection; they are in the "public domain" and can be used by other people without permission. Collections with contributions by multiple authors are generally protected by a single copyright, but the sections may also be copyrighted individually.

The copyright extends 70 years beyond the copyright owner's death. Works written for hire are protected 95 years beyond publication. When the copyright expires, works are in the public domain and may be reproduced and distributed by others unless a copyright is renewed.

Copyright Notice, Registration, and Deposit

Copyright is automatic in the United States as soon as a work exists in fixed form, and protection does not require a notice or registration. However, registration with the Copyright Office gives maximum legal protection. Registration requires sending an application form (Form TX for most technical documents), a fee ($30 in the year 2001), and two copies of the document to the Register of Copyrights in the Copyright Office.

For the best protection, a published work should include a notice of copyright. In a book, the notice usually appears on the verso page of the spread that contains the title page and is sometimes called the copyright page. The notice includes the symbol ©, abbreviation "Copr.," or the word "Copyright"; the year of publication; and the owner's name. Here is an example: © 2001 Longman.

You can get more detailed information on U.S. copyrights from the Copyright Office, either in print or on the World Wide Web (see Further Reading). Particularly useful publications are Circular 1, "Copyright Basics," and Circular 3, "Copyright Notice." You can download application forms from the website.

Copyrightable work published in the United States is subject to mandatory deposit for use in the Library of Congress. Some publications are exempt from this requirement. Circular 7d gives more information on what must be deposited and what is exempt.

International Copyright Protection

Copyright in one country does not automatically extend to another. Use of works in a particular country depends on the laws of that country. Countries that have signed one or both of two multilateral treaties, the Universal Copyright Convention (UCC) and the Berne Convention, offer some protection to work published in member countries. Circular 38a lists countries that maintain copyright relations with the United States. Some countries offer little or no copyright protection for work published in other countries.

Permissions and "Fair Use"

Because work is protected by copyright, permission must be obtained to reproduce sections of someone else's work. Usually the writer requests permission from the copyright holder, but editors verify that permissions have been acquired before the document goes to print. The permission ought to exist in writing, whether in a letter or on a form. The correspondence will establish exactly what will be reprinted and where. The copyright owner may charge a fee for use of the material. If permission is denied, the material cannot be used. Permission must be acquired for each use. Permission to use copyrighted material in one publication does not give one the right to use it elsewhere.

A request for permission to reprint copyrighted material should include the following information:

- Title, author, and edition of the materials to be reprinted.
- Exact material to be used: include page numbers and/or a photocopy.
- How it will be used: nature of the document in which it will be reprinted, author, intended audience, publisher, where the material will appear (for example, quoted in a chapter, cited in a footnote).

Fair use allows some copying for educational or other noncommercial purposes. For example, you may photocopy an article in a journal to study for your research paper. But your professor cannot copy whole sections of a book for the class so that the class won't have to buy the book because doing so would deprive the book's publisher of sales of its intellectual property. In a work setting, it is best to be cautious about copying, especially if you will distribute the work widely and profit from this distribution.

Copyright prohibits duplication of software for multiple users unless an organization has purchased a site license. The terms of use are often listed on the software package.

Copyright and Online Publication

The internet and World Wide Web have provided the public with wonderful access to information. An ethic of openness, sharing, and access has encouraged individuals and organizations to offer documents, statistics, and other information to be freely used. This ethic and the ease of copying material distributed through the internet or web makes some people think that it is all right to do so, but material on the web and even email messages are protected by the same copyright laws that protect print, especially when the information has commercial value. Because publishing on the web makes material available without purchase, you are not depriving an author of sales if you copy material from the web for your personal use and knowledge. Such use is probably fair use. But if you distribute something you found on the web to make money or to people who would otherwise read it on the web and see the advertisements that accompany it, or if you copy the words and use them as your own without citation, then you are violating copyright law. . . .

Cyberspace law is a new area that is still developing. For the present it is prudent to assume that the laws for print publication apply to online text. You need to get permission to use material from the web in other contexts. You may even need permission to link from your website to another site, especially if you wish to link to pages deep in the site, bypassing the home page where the advertisements may appear.

Trademarks, Patents, and Trade Secrets

Trademarks are brand names, phrases, graphics, or logos that identify products. The rainbow-colored apple is a trademark of Apple Corporation, as is the name Macintosh. If the marks are registered with the United States Patent and Trademark Office (PTO), no one else can use those particular marks to represent their own products. Editors look for proper representation of trademarks that may be referred to in the text. The symbol ® next to the mark means that the mark is formally registered and certified with the PTO. The symbol ™ indicates a mark registered on a state basis only or one that has not been officially placed on the Principal Register in the PTO. Trademarks are capitalized in print. Dictionaries identify words that are registered trademarks. Product literature is a good way to find out about whether to use the ® or ™. A typical procedure is to use the symbol on first use of each trademark or to list all trademarks used in a publication in the front matter. Constant repetition of the symbol in the text may become distracting, and the law does not require it.

Patents protect inventions in the way that copyright protects expressions. Their significance to editors is that the text records patent registration with the symbol ® just as it does for trademarks.

Law also protects trade secrets, such as the specifications for a new product or a customer list. Employees owe a "duty of trust" to current and former employers. It is illegal for a company to hire you to find out what a competitor is planning, and it is illegal for you to give trade secrets of a former employer to a new employer or of your current employer to anyone else. Documents produced under government contract projects may be classified and require secrecy. For either private or government work, you may have to use special protections to keep information in your computer files secure. The duty of trust represents one exception to the First Amendment protection of free speech.

Product Safety and Liability

According to U.S. law, companies and individuals must assume responsibility for safety of the products as they are used or even misused by consumers. The products themselves must be designed to be as safe as possible, but because design itself cannot ensure safety, instructions for use and warnings must be complete and clear. The instructions and warnings must cover use, anticipated misuse, storage, and disposal. Manufacturers cannot avoid responsibility with disclaimers (statements that the manufacturer is not responsible for misuse or accidents).

Instructions, Safety Labels, and the Duty to Warn

The first strategy of documenting safe use of a product is to write clear and complete instructions. Clarifying procedures where safety is an issue should be part of document planning. Editors may wish to request a summary of hazards so that they don't have to rely on the instructions alone to identify them. In reviewing a draft, editors rely on the principles of organization, style, visual design, and illustrations . . . , as well as an attitude of vigilance and care. Ambiguity must be clarified. For example, if a procedure calls for "adequate ventilation," an editor may query whether "adequate" means "fresh" air or if circulating air will suffice. Taking the perspective of the reader, the editor may note some gaps that a writer misses.

If there are hazards of using products, manufacturers and suppliers must warn of the risks unless the product is common and its hazards well known. Instructions do not constitute warnings, nor can a warning substitute for instructions. A warning calls attention to a particular procedure verbally, visually, or both. Some standards for identifying different categories of risks and for symbols and colors to identify them have been developed by the American National Standards Institute (ANSI), The Occupational Safety and Health Administration (OSHA), and the International Organization for Standardization (ISO). For example, the word "danger" and the color red are used only when serious injury or death may result. "Warning" (orange) and "caution" (yellow) warn of less serious risks.

Safety labels should be attached to products where users will see them before and as they use the product. This principle may seem obvious but is not always followed. For example, a warning to thaw a turkey slowly in the refrigerator to avoid food poisoning was placed inside the frozen cavity of the turkey where it could not be found until the turkey was already thawed!

The Editor's Legal Responsibility

Like everyone else who is involved in product distribution, writers and editors have some responsibility in the eyes of the law for safe use of a product. They can be named in product liability lawsuits, though the usual practice is to name the company. A sense of professional responsibility as much as respect for the law encourages editors to be careful in the legal edit.

Libel, Fraud, and Misrepresentation

Libel is a defamatory statement without basis in fact that shames or lowers the public reputation of an identifiable person. People who can prove libel may win damages from a publisher. The possibility of being sued for libel worries editors of fiction and periodicals more than it does technical editors because people are more likely to be discussed, referred to, or otherwise cited in works of fiction and in periodicals than in technical documents. Nevertheless, all editors should read alertly for facts to verify the accuracy of any negative statements that may be made about individuals.

Fraud and misrepresentation deceive the public. Misrepresentation may occur in labeling products or making claims about them or in claims about credentials of individuals or about data.

ETHICAL ISSUES IN EDITING

The requirements for ethical conduct may restrain companies and communicators more than laws. For example, one long-distance phone company trademarked the name, "I Don't Care." When consumers were asked to choose their long-distance company and responded, "I don't care," they were enrolled in the company with that name—whose rates were significantly higher than the rates of other companies. The company complied with the law in choosing its name, but consumers would probably not find the practice ethical because it takes advantage of people. Likewise the contractor who read the specification for "stainless steel nuts and bolts" and used a cheaper, less durable metal for the bolts acted within the law (arguing that "stainless steel" modified only "nuts") but did not act ethically, considering that the bolts caused the structure to fail.

Editors and others in technical communication have responsibilities that extend beyond meeting the requirements of the law. These responsibilities derive from some values that seem universal, no matter what their philosophical or religious bases, such as the responsibility not to harm others. This principle enables people to live together in groups and with some trust in others. It may reflect the wish for survival ("I won't hurt you if you won't hurt me") as well as more generous and lofty principles of mutual responsibility. Other bases for ethical choice derive from cultural values and values that are defined by the particular profession. These values anticipate the consequences of actions and help people make judgments in particular cases.

Professional Codes of Conduct

Technical communicators, like physicians, lawyers, professors, and clergy, consider themselves professionals rather than staff or laborers. Such classification brings privileges of status and material benefits—and responsibilities as well. Professionals accept responsibility for their work rather than deferring to a superior. It is unprofessional in any setting to "pass the buck," or to blame someone else for a problem for which the professional person bears some responsibility. The editor's ethical responsibilities increase along with other responsibilities—a comprehensive editor is more responsible than a copyeditor. Yet all editors have a part in ensuring the quality and integrity of documents. . . .

One point of conflict and confusion in the ethics of technical communication relates to the professional value of "objectivity." Objectivity seems to represent truth uncontaminated by the bias of opinions. The goal of objectivity is that the person (the "subject") should get out of the way and let "facts speak for themselves" (the facts are presumed to be objects separate from the subject). Should the communicator remain neutral, taking no position but rather translating technical information accurately for the public? Furthermore, urged to adapt to an audience, should the communicator defer to an audience's wishes? These slippery words—*objectivity, neutral, translation, adaptation*—appear in various discussions that attempt to define the values and ethics of technical communication.

One approach to answering these questions is to consider whether facts can ever speak for themselves and whether anyone can ever be neutral. . . . [R]eaders interpret based on their prior experiences and knowledge, and different readers will find different meanings in the same texts. The object can never be entirely separated from the subject, or the person who observes and interprets the object. Furthermore, the acts of arranging information in sentences or on a page influence interpretation. As you saw in the chapters on style, for example, the independent clause is more emphatic than a dependent clause, and a writer chooses to emphasize a point by placing it in an independent clause. Presumably the writer will aim to reflect the content in this choice of structure, but still the writer is making a choice. Writing is never entirely neutral, even when the subject is materials and procedures. The claim of neutrality may be avoidance of commitment, an excuse not to get involved.

The nature of the expertise that enables people to call themselves "professionals" also influences answers to these questions. Expertise generally requires specialized education and perhaps experience. As a consumer of services by professionals, such as your physician, you expect information about options even if you claim the right to make the ultimate choice. You respect your own limitations enough to listen to the options and welcome them.

Likewise, your client or manager may depend on your expertise. It would be unethical to withhold it because in so doing you might jeopardize their ability to make the best decision. For example, your client or manager may request the use of a particular type of software that you know is less efficient than another or may not know the environmental consequences of petroleum-based inks for printing compared to soy inks. If you remain silent, they may never know the options. Or you may resist making recommendations in a feasibility study because the reader has higher status in the organization than you do. In all those cases, your professional expertise carries with it a responsibility to inform a client or reader, even when the reader may not want to hear your information. Your judgment represents the collected wisdom and knowledge of the profession represented by your expertise, not something arbitrary and personal.

Ethics in technical communication involves more than reporting accurately what a subject matter expert says. It requires more than passively telling the truth. It also requires a proactive position in offering information and helping to guide the decisions of people who lack the specialized knowledge of the communicator and therefore do not see all the consequences of choices.

Misrepresentation of Content or Risks

Because of the serious, informative purpose of technical communication, readers expect it to be truthful. They may approach advertising with enough skepticism to question claims that are made, but because readers expect technical communication to be accurate, writers and editors have a particular responsibility for accuracy. Misrepresentations deceive readers and may cause them to make mistakes in procedures, interpretations, or decisions.

Intentional misrepresentations, such as faulty data in research or false claims about ingredients or features, are illegal, but misrepresentation is also an ethical issue because there is no certainty that instances of misrepresentation will be identified and penalized by the law.

Misrepresentation may be unintentional, as in a graph in which the segments are not mathematically defined or perspective that inadvertently emphasizes one piece of information. Editors have an ethical responsibility to understand the structure and style of visuals as well as words so that they can correct unintentional misrepresentation.

Misrepresentation includes omission as well as inaccuracy. Omissions are especially significant when there is danger of harm. Risks of purchasing or using a product should be documented. The editor may have to ask the writer directly what the risks are and not just wait to be told.

Environmental Ethics

Environmental ethics, like all other ethics, are based on the principle of responsibility for the social good as balanced against short-term prosperity. As consumers of paper, electricity, and inks, technical publications groups use products that harm the environment and, by extension, human health. The use of these products is ethical according to cultural values, but misuse is not because the danger to health and safety increases disproportionately to the benefits of use. Corporate policies can encourage product choices that are environmentally friendly. Policies as well as individual practice can minimize waste as from excessive photocopying and printing.

The paper and publishing industries spill a higher percentage of industrial toxins into the air and water than any other industry except for the chemicals and metals industries. Many of those toxins are dioxins used to bleach paper. Dioxins cause cancers and birth defects. Recycling paper and using recycled paper cuts down on waste because the manufacture of recycled paper is a less toxic process than manufacture of new paper, and it also consumes fewer trees. Uncoated papers are more readily recycled than coated papers, and white paper is more readily recycled than colored. White, uncoated paper is thus preferable except in special circumstances. Processes that minimize multiple draft copies should be encouraged.

Other choices of materials and processes can also minimize environmental and health consequences. Inks made from soybean oil are made from a replenishable agricultural product rather than from petroleum, and UV-curable inks eliminate solvents in manufacturing. Conservative use of electricity and plastic is environmentally as well as economically responsible. Electricity used to run computers, printers, and photocopiers may be generated by fossil fuels, which contribute to global warming. Plastic wrappers for manuals and other packaging will probably become landfill and may be superfluous. Thus, choices about papers, inks, and other consumables affect not just whether readers get the information they need but also the general health and welfare. As with all ethical issues, responsible corporations and individuals look beyond immediate needs and convenience to long-term and broad consequences of choices.

Bases for Ethical Decisions

Ethical decisions can be based on principles (such as honesty), law, professional standards, and assessment of consequences.

In a survey of technical communicators, Sam Dragga identified eight types of explanations for decisions about the ethics of a communication decision (see Further Reading). The most common reason was *consequences:* what happens to readers as a result of that choice. The extent to which a choice is unethical is proportional to the degree of deception or injury to the reader.

Other explanations are document *specifications* and *common practice.* Specifications refer to document descriptions, such as number of pages. Common practice establishes some expectations. If it is common practice in an employee evaluation to emphasize achievements and de-emphasize limitations, then the potential deception of doing so seems less grave than misrepresentation of data in a research report. Readers may not be deceived by emphasis if practice leads them to expect it.

Technical communicators generally rejected the explanation that the burden for accurate interpretation rests on readers rather than writers.

If an important basis for ethical choice is consequences, then editors and writers must consider consequences broadly—long-term as well as immediate consequences; to secondary as well as primary readers; to the environment as well as to the corporation; and so forth.

ESTABLISHING POLICIES FOR LEGAL AND ETHICAL CONDUCT

An editor or writer working for a company is only one person among many who are responsible for products and their documentation. Although any professional person should accept personal responsibility for the safety and welfare of others, ethical behavior in a corporate setting is also a social and cultural issue. To increase chances of the organization and its employees following high legal and ethical standards, and to minimize the potential for conflict of an individual employee against the organization as a whole, the company should have policies defining what constitutes ethical communication, procedures to include training of staff in ethical issues, and document review for adherence to the policies.

A first step for an employee is to review existing policy statements and the house style manual. If potentially useful sections are missing, or if some require revision, a proposal to research and revise the manuals could be developed. Such revisions should be the responsibility of a group of people representing different components of the organization, including technical communicators, engineers, managers, and lawyers. Developing or revising these policies as a group helps to create the sense of group responsibility for adherence to high standards of legal and ethical behavior.

A style and publications procedures manual might cover these policies and procedures:

- identification of trademarks,
- registering copyrighted materials,
- requesting permission to use material copyrighted by others and forms or letter templates for doing so,
- visual and verbal identification of warnings and hazards,
- accurate representation of verbal and visual information,
- ethical use of resources and procedures to minimize waste,
- review of ethical and legal issues in each document. The review should be the responsibility of various employees, including but not limited to the editor.

The employee policy manual should include commitments to

- legal and ethical conduct by the organization and its employees with examples specifically related to publications, including commitment to accurate representation of a product's value and limitations and instructions and warnings that accommodate the needs of users;
- organizational as well as individual responsibility for legal and ethical conduct.

Such statements affirm corporate responsibility and encourage individuals to act ethically by creating an environment in which ethical actions are expected and respected. They provide guidance for making ethical decisions on a case-by-case basis. They may prevent problems from occurring and support editors if problems do occur.

SUMMARY

Law and ethics balance the rights of individuals in a society with the needs of the society as a whole. They aim to encourage achievement while protecting individuals or groups from harm. Different countries interpret these rights differently depending on their values. Intellectual property, product liability, and misrepresentation laws apply to technical communication. Because technical communication aims to influence the people who use documents, it is never ethically neutral. Part of the definition of what makes communication "good" is the contextual one of ethics: whether a document accomplishes its purpose of giving information or directions without harming or unduly jeopardizing people directly or indirectly. A good publications department conducts itself according to high legal and ethical standards not just because the law requires it to but also because professional responsibility and cultural values require concern for the welfare of others. Comprehensive editing requires a procedure for review of legal and ethical standards before, during, and at the conclusion of document development.

FURTHER READING

"Copyright Basics" (Circular 1) and "Copyright Notice" (Circular 3). Copyright Office, Publications Section, LM–455, Library of Congress, Washington, DC 20559. Also available at www.loc.gov/copyright/.

Dombrowski, Paul (1999). *Ethics in technical communication*. New York: Allyn & Bacon.

Dragga, Sam (1996). Is it ethical? A survey of opinion on principles and practices of document design. *Technical Communication, 43.3,* 255–265.

Johannesen, Richard L. (1996). *Ethics in human communication* (4th ed.). Prospect Heights, IL: Waveland.

Lessig, Larry, Post, David, & Volokh, Eugene (1996). Cyberspace law for non-lawyers. Alan Lewine (Ed.). www.ssrn.com/cyberlaw/.

Acknowledgments

GUIDE TO NONSEXIST LANGUAGE. By the University of Wisconsin-Extension Equal Opportunities Program Office and Department of Agricultural Journalism. Reprinted with the kind permission of the Board of Regents of the University of Wisconsin System and the University of Wisconsin-Extension.

Part Three

MAKING YOUR CORRESPONDENCE GET RESULTS. By David V. Lewis. From *Secrets of Successful Writing, Speaking, and Listening* by David V. Lewis (AMACOM a division of the American Management Association, New York, 1982). Reprinted with the kind permission of the author.

LETTERS THAT SELL (penned anonymously by the late Robert Stewart). Copyright © 1974 The Royal Bank of Canada and The RBC Financial Group. Reprinted with the permission of the Corporate Communications Division of The RBC Financial Group from *The Royal Bank* (now *The RBC*) *Monthly Letter* May 1974. The entire run of *The RBC Letter* is available on-line at http://www.rbc.com/community/letter/index.html

"I HAVE SOME BAD NEWS FOR YOU." By Allan A. Glatthorn. From *Writing for Success* by Allan A. Glatthorn (Scott, Foresman and Company, 1985). Reprinted with the kind permission of the author.

HOW TO WRITE BETTER MEMOS. By Harold K. Mintz. Copyright © 1970 by Chemical Week Associates, 110 Williams Street, New York, NY 10038-3901. Reprinted by special permission from *Chemical Engineering* January 26, 1970, pp. 136–139.

HOW TO USE BOTTOM-LINE WRITING IN CORPORATE COMMUNICATIONS. By John S. Fielden and Ronald E. Dulek. Copyright © 1984 by the Trustees at Indiana University, Kelley School of Business. Reprinted with permission from *Business Horizons* July–August 1984, pp. 24–30.

ELECTRONIC MAIL. By Heidi Shultz. Copyright © 2000 by Allyn & Bacon. Reprinted by permission of Pearson Education, Inc. Excerpt from *The Elements of Electronic Communication* by Heidi Schultz, pp. 19–62.

Part Four

AUDIENCE ANALYSIS: THE PROBLEM AND A SOLUTION. By J. C. Mathes and Dwight W. Stevenson. Copyright © 1991 by Allyn & Bacon. Reprinted by permission of Pearson Education, Inc. from *Designing Technical Reports*, 2nd ed. by J. C. Mathes and Dwight W. Stevenson, pp. 27–48.

WHAT TO REPORT. By Richard W. Dodge. Reprinted from *Westinghouse Engineer*, 22 (July–September, 1962), by permission of the Westinghouse Electric Corporation.

THE WRITING OF ABSTRACTS. By Christian K. Arnold. Copyright © 2003 by IEEE, The Institute of Electrical and Electronics Engineering. Reprinted with permission from *IRE Transactions on Engineering Writing & Speech* December 1961, pp. 80–82.

TEN REPORT WRITING PITFALLS: HOW TO AVOID THEM. By Vincent Vinci. Copyright © 1975 by Chemical Week Associates, 110 Williams Street, New York, NY 10038-3901. Reprinted by special permission from *Chemical Engineering* December 22, 1975, pp. 45–48.

CREATING TABLES AND ILLUSTRATIONS. By Walter E. Oliu, Charles T. Brusaw, and Gerald J. Alred. Copyright © 2001 by Bedford/St. Martin's. Reprinted from *Writing That Works*, 7th ed. by Walter E. Oliu, Charles T. Brusaw, and Gerald J. Alred. Used with permission of Bedford/St. Martin's.

Part Five

Part Six

Index